Lecture Notes in Mathematics　　2089

Editors:
J.-M. Morel, Cachan
B. Teissier, Paris

W0037650

For further volumes:
http://www.springer.com/series/304

Christoph Kawan

Invariance Entropy for Deterministic Control Systems

An Introduction

 Springer

Christoph Kawan
Institute of Mathematics
University of Augsburg
Augsburg, Germany

ISBN 978-3-319-01287-2   ISBN 978-3-319-01288-9 (eBook)
DOI 10.1007/978-3-319-01288-9
Springer Cham Heidelberg New York Dordrecht London

Lecture Notes in Mathematics ISSN print edition: 0075-8434
ISSN electronic edition: 1617-9692

Library of Congress Control Number: 2013948764

Mathematics Subject Classification (2010): 94A17, 93C10, 93C15

Printed on acid-free paper

Springer is part of Springer Science+Business Media (www.springer.com)

Para Amar e Helena,
  os outros dois cabeças do monstro

Foreword

The topic of mathematical control theory is the exploration of the possibilities and the limitations of changes in dynamical systems due to inputs. Hence connections to the theory of dynamical systems are immanent. In this monograph Christoph Kawan adds a new angle to this connection by bringing to bear concepts and techniques from the global theory of topological and differentiable dynamical systems upon the problem to determine minimal data rates, a very timely subject in control. Here the combination with arguments from nonlinear control seems particularly noteworthy.

This field has developed over the last few years and the present text shows that it has reached a certain maturity. At the same time, I hope that the many open problems will lead to further fruitful investigations.

Augsburg, Germany   Fritz Colonius
June 2013

Preface

This book is supposed to serve as an introduction to the theory of *invariance entropy* which is related to the control task of making a subset in the state space of a control system invariant. Inspired by the seminal work of Nair et al. [85] about notions of entropy measuring the complexity of certain control tasks, and the Bowen–Dinaburg characterizations of topological entropy in metric spaces, Fritz Colonius created the concept of invariance entropy in 2007. At that time, I started to write my Ph.D. thesis under his supervision at the Mathematical Institute of the University of Augsburg and had the pleasure and great opportunity to work on this new topic in the field of information-based control. The text at hand presents the theory obtained in five fruitful years of research in Augsburg and during two research stays in Campinas (Brazil) in August 2010 and in the period from September to November 2011. There I had the chance to work with Luiz San Martin who showed great interest in our research and contributed several important ideas. In this text, the theory as presented so far in the articles [23, 63–65] and in the thesis [62] is also put on a new level of generality. We work with a fairly general definition of control systems which is basically the one that can be found in Sontag's book [102]. Despite the fact that this definition treats discrete- and continuous-time systems simultaneously, the emphasis in this text clearly lies on continuous-time systems given by differential equations. However, where it is no great deal to prove a result also in discrete time and/or in a purely topological setting, we do not hesitate to do so.

The central motivation behind the theory presented in this book comes from the need to deal with communication constraints in digitally networked control systems. Here the assumption of classical control theory that information can be transmitted within control loops instantaneously, lossless, and with arbitrary precision is no longer satisfied. Realistic mathematical models of many important real-world communication and control networks have to take into account general data-rate constraints in the communication channels, time delays, partial loss of information, and variable network topologies. This raises the question about the smallest possible information rate above which a given control task can be solved. Though networked control systems can have a complicated topology, consisting of multiple sensors, controllers, and actuators, a first step towards understanding the problem of minimal data rates is to analyze the simplest possible network topology, consisting of one controller and one dynamical system connected by a digital channel with a certain rate in bits per unit time. The problem to determine such minimal data rates has been considered for more than 20 years. Early landmarks are the papers by Delchamps [33] who considered quantized information for stabilization and proposed to use statistical methods from ergodic theory and by Wong and Brockett [113] who discussed stabilization of linear systems via coding. From the wealth of literature on this topic there should also be mentioned Tatikonda and Mitter [107], Delvenne [34], Fagnani and Zampieri [41], Liberzon and Hespanha [76], Matveev and Savkin [79], De Persis [36], Savkin [96], and Xie [114]. In these works, mainly linear systems (both deterministic and stochastic) have been considered, and despite different formulations and assumptions, the results therein show that the minimal data rate for stabilization only depends on the unstable open-loop eigenvalues of the system and therefore is independent of the parameters of the coding and control scheme. Nonlinear systems have been considered in [76], where the authors show that global asymptotic stabilization at an equilibrium can be accomplished by using sampled encoded measurements of the state, with a data rate larger than the product of the right-hand side Lipschitz constant and the dimension of the state space. Furthermore, nonlinear systems in feedforward form have been treated in [36], where a hybrid controller is constructed which achieves stabilization at data rates arbitrarily close to zero, in spite of arbitrarily large communication delays. Different control problems for nonlinear systems are treated in [96], namely observability and robustness. Here a systematic approach in terms of a quantity similar to topological entropy of classical dynamical systems leads to a description of the minimal data rate. The research monograph [79] by Matveev and Savkin provides various results concerning state estimation and control of linear and nonlinear systems over channels of limited capacity, including several data rate theorems. In particular, the minimal data rate for observability is related to a notion of topological entropy of the control system. There is much more literature in this field and I apologize to many authors in advance for not mentioning their contributions. A comprehensive and detailed survey with an excellent overview of the literature up to the year 2007 can be found in Nair et al. [86].

The first systematic approach to the problem of minimal data rates for set-invariance and stabilization of (deterministic, nonlinear) control systems was presented in the outstanding paper [85] by Nair, Evans, Mareels, and Moran, which introduced the notion of *topological feedback entropy*. This quantity, which is defined in terms of the open-loop control system, is a measure for the smallest data rate a communication channel connecting a coder and a controller is allowed to have if the system is supposed to solve the control task of rendering a compact subset of the state space invariant. Furthermore, a local version of feedback entropy at an equilibrium is defined which measures the smallest possible data rate for local uniform asymptotic stabilization, and its value is determined by the unstable eigenvalues of the linearization at the corresponding equilibrium.

The definition of topological feedback entropy is similar to the open-cover definition of topological entropy for classical dynamical systems by Adler et al. [1].

The difference, however, is that for topological feedback entropy only such open covers of the given compact set K are considered which can be made invariant in the sense that to each member of the cover a control sequence can be assigned which allows to steer from every state in this open set into the interior of K. Then the entropy of that cover is defined analogously as in the open-cover definition of topological entropy, but the topological feedback entropy of K is defined as the infimum (instead of the supremum) over all such invariant open covers. Looking at this definition, one expects that topological feedback entropy has some properties that are similar to the properties of topological entropy, but that, on the other hand, the similarity is not going too far.

The richness and maturity of the entropy theory in topological and smooth dynamics is based in first line on the variety of alternative definitions which are available next to the open-cover definition. There are the definitions of entropy in terms of separated and spanning sets introduced by Dinaburg [37] and independently by Bowen [10]. Another alternative definition due to Bowen [12] resembles Hausdorff dimension. Arguably the most powerful characterization is given by the variational principle which asserts that the topological entropy is the supremum over the metric entropies with respect to all invariant probability measures of the given system. For topological feedback entropy it was not clear if there was any alternative approach until the concept of invariance entropy, defined as follows, was introduced. For a compact and controlled invariant set Q of a continuous-time control system, one counts for every positive time τ the number of open-loop control functions which are necessary to stay in Q up to time τ from any initial state. Then the exponential growth rate of these minimal numbers as τ tends to infinity defines the entropy. The intuition behind this definition is that a controller which receives a certain amount of information about the state, say n bits, can generate at most 2^n different control functions to steer the system on a finite time interval, and hence the minimal number of control functions needed to accomplish the control task on this time interval is a measure for the necessary amount of information.

The definition of invariance entropy is close in spirit to the Bowen–Dinaburg definition of topological entropy via spanning sets, and because of its conceptual simplicity it allows to draw plenty of more or less obvious consequences immediately. As it turns out, for each one of the properties of topological entropy which are usually considered as elementary the invariance entropy has an analogous property. For linear control systems the analogy goes even far enough that one can use Bowen's formula for the topological entropy of a linear map to give an analogous formula for the invariance entropy.

By its definition invariance entropy measures how fast the number of open-loop control functions grows which are needed to stay in Q for longer and longer times. But next to this obvious meaning it indeed turns out to coincide with topological feedback entropy after the appropriate adaptations to the setting in which the latter is defined, and in this sense invariance entropy is really an alternative way of defining topological feedback entropy.

Before I start to give a description of the book's contents, I provide an overview of the mathematical tools used therein. These mainly come from the classical theory of dynamical systems, including differential-geometric methods and concepts from ergodic and dimension theory, as well as from geometric control theory. In particular, the applied techniques and results have their origins in the following sources:

- the work on entropy in dynamical systems by Adler et al. [1], Bowen [10,11], Ito [60], Kolyada and Snoha [70], and many others;
- the work in dimension theory of dynamical systems by Douady and Oesterlé [38], Temam [108], Boichenko, Leonov, and Reitmann [8,9], Franz [44], Gelfert [49,50], and Noack [87];
- the work of Nair et al. [85] on topological feedback entropy;
- the control-theoretic work of Colonius and Kliemann (and coauthors) [21,25,26], in particular the theory of control and chain control sets for systems given by differential equations;
- the work of Sontag [100, 101] and Coron [30] on controllability and regularity for control systems given by differential equations;
- the work of Bowen [13], Bowen and Ruelle [14], Young [115], and Liu [78] in ergodic theory of hyperbolic dynamical systems.

The contents of the book are briefly sketched as follows:

The first chapter serves as the introduction of basic control-theoretic notions. As mentioned before, we work with a very general definition of control systems due to Sontag, but we restrict ourselves to time-invariant and complete systems. This definition is given in Sect. 1.1. After that, several particular classes of systems are defined, namely topological, linear, and smooth systems. Section 1.2 establishes the notion of smooth systems given by differential equations which constitute the most important subclass of smooth systems in this book. In Sect. 1.3, the reader is reminded of elementary control-theoretic notions such as orbits, accessibility, and controlled invariant sets. In Sect. 1.4, the control flow of a control-affine system is introduced and its regularity properties are analyzed. Furthermore, control sets and chain control sets are defined and their basic properties are studied. Finally, Sect. 1.5 treats the linearization of a smooth system given by differential equations along a trajectory and the notion of regular control functions.

In Chap. 2, the central notion of invariance entropy for topological time-invariant systems is established and discussed. Also a related notion, named outer invariance entropy, is introduced which in general is only a lower bound for the actual invariance entropy, but in some respect is better behaved. After proving a list of elementary properties in Sects. 2.1 and 2.2, as a first nontrivial example, the invariance entropy of a scalar linear system given by differential equations is computed. Here for the first time a volume growth argument is used to derive a lower bound, which in different variations appears in all of the following chapters and is one of the main ideas in the theory developed in this book. In the last two sections, the relations between invariance entropy and topological feedback entropy as well as minimal data rates are discussed. The central idea here consists in an alternative characterization of invariance entropy in terms of the entropies of the so-called invariant covers of the given controlled invariant set. This leads to the main results, which are the data rate theorem for invariance entropy and a result which relates both entropies to each other. Additionally, a proof of the data rate theorem for topological feedback entropy is given.

Chapter 3 contains the linear theory. The first main result of this chapter gives a formula for the outer invariance entropy of a linear system. As one expects, under appropriate assumptions, this quantity is given by the sum of the logarithms of the unstable eigenvalues. This corresponds with a multitude of results in the control literature which provide formulas for the minimal data rates for stabilization of linear systems. An important ingredient in the proof of this result is Bowen's formula for the topological entropy of a linear map. The second main result provides an estimate from below for the invariance entropy of an inhomogeneous bilinear system. This lower bound is expressed in terms of the minimal volume growth rate on an invariant subbundle of the control flow of the associated homogeneous system. In continuous time, one can use Selgrade's theorem to choose this subbundle such that the volume growth rate becomes maximal. In this case, the growth rate reduces to the sum of the unstable eigenvalues again if one considers the special case of a linear system.

In Chap. 4, the development of the nonlinear theory begins. In Sect. 4.1, we first prove a result for topological systems, which gives an upper bound for the entropy in terms of a Lipschitz constant and the upper capacitive dimension of the considered subset of the state space. This result is proved in pretty much the same way as the analogous result for topological entropy which has its origins in Kushnirenko [72] and Ito [60] and is nowadays considered as Folklore. The topological result is then adapted to smooth systems on Riemannian manifolds, both in continuous and in discrete time. In the continuous-time case, an appropriate Lipschitz constant can be described in terms of the maximal eigenvalues of the symmetrized covariant derivatives of the right-hand side vector fields. In Sect. 4.2, a general lower bound for a smooth system on a Riemannian manifold with invertible dynamics is given. Here again the volume growth argument is used which leads to an expression in terms of the functional determinants of the transition maps. In the case of a smooth system given by differential equations, the Liouville formula can be used to relate this expression to the divergence of the right-hand side vector fields.

In Chap. 5, the invariance entropy of sets with additional controllability properties is investigated. For simplicity, the main result of this chapter is only proved for smooth systems given by differential equations. This result gives an upper bound for the invariance entropy of a control set in terms of the sum of unstable Lyapunov exponents of a regular periodic trajectory inside the given set. The proof is basically an adaptation of the proof for a result about topological feedback entropy in Nair et al. [85]. Here for the first time classical control-theoretic methods for nonlinear systems enter the scene, and the interplay between the global controllability on the control set and the local controllability along the periodic trajectory is exploited to give the announced result. In general, we are not able to answer the question whether a control set contains regular periodic trajectories. However, for strongly accessible real-analytic systems, Sontag's theorem about universally regular controls yields the existence of plenty of such trajectories, and a more general result of Coron yields such trajectories under considerably weaker assumptions. These trajectories can be used to show that the assumptions of regularity and periodicity in the upper bound theorem can be weakened under a weak partial hyperbolicity condition. This is carried out in Sect. 5.2.

In Chap. 6, another variant of the volume argument is used to achieve tighter lower bounds for the invariance entropy. The basic idea used here again stems from the classical theory of dynamical systems, more precisely from the theory of escape rates which is closely related to the classical entropy theory and the thermodynamic formalism. Section 6.1 explains this idea in detail. Basically, we use the fact that the invariance entropy is bounded from below by a uniform escape rate from the considered set. This allows to adapt methods from the classical dynamical systems theory to describe the lower bound in terms of volume growth rates and expressions close to topological entropy. Accordingly, instead of control-theoretic assumptions as in Chap. 5, here additional dynamical assumptions have to be imposed on the system, namely, hyperbolicity conditions of weaker or stronger form. The most important ingredients used in this chapter are two volume lemmas for Bowen-balls, the classical one by Bowen and Ruelle [14], in its nonautonomous version proved by Liu [78], and another one by Franz [44] and Gelfert [49, 50].

Finally, Chap. 7 presents examples for the application of the nonlinear theory developed in the preceding three chapters to particular classes of systems. Section 7.1 treats one-dimensional control-affine systems which turn out to be the most nicely behaved class of nonlinear systems. Under appropriate regularity assumptions, here the invariance entropy of a control set can be expressed in terms of the infimum of the Lyapunov spectrum over the control set. If the given system has only one control vector field, this expression can be reformulated in terms of the drift and control vector fields and their derivatives. As an application, a model for the inverse pendulum is studied and the invariance entropy for the region of stabilizability is computed. In Sect. 7.2, we consider the class of nonlinear systems which are uniformly expanding, that is, the systems whose trajectories for a fixed control function exponentially diverge from each other at a rate which is independent of the control function. The main result for this class of systems gives an almost-formula for the invariance entropy of a control set. Section 7.3 again treats inhomogeneous bilinear systems given by differential equations and gives an improvement over the lower estimate of Chap. 3 by using the methods introduced in Chap. 6, and an almost-formula in the case of a control set. Finally, Sect. 7.4 treats projective systems, that is, control-affine systems on projective space which are induced by bilinear systems in Euclidean space. Under the assumption of local accessibility, a complete picture of the maximal regions of controllability of such systems is available. In particular, the control sets with nonempty interior (called *main control sets*) and the chain control sets can be described via the semigroup of the bilinear system and its control flow. Under a hyperbolicity assumption, we are able to provide a formula for the invariance entropy of the open control set in terms of quantities that can be computed directly from the right-hand side of the bilinear system. A thorough analysis of the chain and main control sets shows that these possess a partially hyperbolic structure. Under specific assumptions about the spectrum of the bilinear system, they have a uniformly hyperbolic structure, which allows to apply the main results of Chaps. 5

and 6. However, there are still some unsolved problems that remain if one wants to give a formula for the invariance entropy of all of these sets.

My intention was to keep the book to a large extent self-contained. However, there are some well-known results whose proofs are not given such as Krener's theorem about accessibility, Sontag's theorem about existence of universally regular control functions, Selgrade's theorem, and the existence of finest Morse decompositions. I assume that the reader is familiar with the material taught in standard courses on linear algebra, real analysis, set-theoretic topology, functional analysis, and measure theory. Some supplementary material can be found in the two appendices, mostly without proofs. Appendix A treats some more advanced linear and multilinear algebra as well as basics about differentiable manifolds and Carathéodory differential equations. In Appendix B, some topics related to dynamical systems are covered, in particular, chain recurrence, linear flows on vector bundles, topological entropy, and (sub-)additive cocycles. I hope that the reader who is familiar with the concepts treated in the appendices can skip reading them and may only have to check for the notation introduced there.

Given the subject matter, it is natural that the presented theory is rather incomplete and leaves many questions open. At the end of each chapter, one finds some questions that might be interesting for further research. My hope is that this text is of use in a further development of a systematic analysis of minimal data rate problems in control, and that both mathematicians working in control theory and in dynamical systems will find the problems in this area appealing from an application-oriented and a purely mathematical point of view.

There is a long list of people who I have to thank for their direct or indirect contributions to this piece of work. First of all, there is Fritz who initiated the whole thing, who put and still puts a great effort in pushing it forwards, and never gets tired in developing new ideas and discussing technical details. Then there are the people who showed interest in our work, helped us solving mathematical problems, or contributed new ideas: Tim Bremer, Tomasz Downarowicz, Roberta Fabbri, Ryuichi Fukuoka, Isabell Graf, Lars Grüne, Uwe Helmke, Anne-Marie Hoock, Pei-Dong Liu, Peter Nagel, Girish Nair, Claudio de Persis, Luiz San Martin, Alexandre Santana, Adriano da Silva, Marco Spadini, Ursula Weinhuber, and Fabian Wirth. For proofreading parts of the manuscript I owe thanks to Isabell Graf, Peter Quast, Alexandre Santana, and Helena Soares. Last but in no respect least, there are the people who I owe thanks not so much for their help with mathematical problems, but for their friendship, their hospitality, and their help and support with all kinds of "real problems": My parents, my friends Daniela, Helmut, Ingrid, Isabell, Thomas, and Torben, as well as my Brazilian housemates Amar, Helena, Henrique, Juliana, and Marcos (not to forget Peter, our neighbor). I also acknowledge the financial support of the following grants: DFG grants Co 124/17-1 and 17-2 within DFG Priority Program 1305 and FAPESP grant no. 11/03140-2.

Augsburg, Germany   Christoph Kawan
June 2013

Contents

Acronyms

$\emptyset$	The empty set.
A^c	The complement of a subset $A \subset X$, that is, $A^c = X \backslash A$.
$\#A$	For a finite set A, this notation stands for the number of elements in A; if A is infinite, we set $\#A := \infty$.
$A \subset B$	Set inclusion: A is a (not necessarily proper) subset of B.
$A \subsetneq B$	A is a proper subset of B.
Y^X	For sets X and Y, we denote by Y^X the set of all maps $f : X \to Y$.
$f(x) \equiv g(x)$	The maps f and g coincide for all x on their common domain.
$f(\cdot + t)$	If $f : \mathbb{T} \to X$ is a map with $\mathbb{T} \in \{\mathbb{Z}, \mathbb{R}\}$ and $t \in \mathbb{T}$, this notation is used for the map $s \mapsto f(s + t)$, $\mathbb{T} \to X$.
$f(t\cdot)$	If $f : \mathbb{T} \to X$ is a map with $\mathbb{T} \in \{\mathbb{Z}, \mathbb{R}\}$ and $t \in \mathbb{T}$, this notation is used for the map $s \mapsto f(st)$, $\mathbb{T} \to X$.
$\omega_1 \omega_2$	The concatenation of two maps $\omega_1 : [\sigma, \tau) \to U$ and $\omega_2 : [\tau, \mu) \to U$, defined by

$$\omega_1 \omega_2(t) := \begin{cases} \omega_1(t) \text{ if } t \in [\sigma, \tau), \\ \omega_2(t) \text{ if } t \in [\tau, \mu). \end{cases}$$

ω^μ	If $\omega : [\sigma, \tau) \to U$ is some map, then $\omega^\mu : [\mu + \sigma, \mu + \tau) \to U$ is defined by $\omega^\mu(t) \equiv \omega(t - \mu)$.
$f\vert_A$	The restriction of a map $f : X \to Y$ to a subset $A \subset X$.
f^n	The n-th iterate of a map $f : X \to X$, defined inductively by $f^0 := \mathrm{id}_X$ and $f^{n+1} := f \circ f^n$.
$f \times g$	The Cartesian product of two maps $f : X \to X$ and $g : Y \to Y$, that is, $(f \times g)(x, y) = (f(x), g(y))$, $f \times g : X \times Y \to X \times Y$.
id_X	The identity map on a set X, $\mathrm{id}_X(x) \equiv x$.
$\mathbb{1}_A$	The characteristic function of a set A,

$$\mathbb{1}_A(x) = \begin{cases} 1 \text{ if } x \in A, \\ 0 \text{ if } x \notin A. \end{cases}$$

$\mathbb{Z}$	The set of all integers.
$\mathbb{N}$	The set of all positive integers.
$\mathbb{Q}$	The set of all rational numbers.
$\mathbb{R}$	The set of all real numbers.
$\mathbb{T}$	This notation simultaneously stands for $\mathbb{Z}$ and $\mathbb{R}$.
$\mathbb{T}_+$	The set of all nonnegative elements of $\mathbb{T}$.
$\mathbb{R}^d$	The d-dimensional Euclidean space $\mathbb{R}^d = \mathbb{R} \times \cdots \times \mathbb{R}$ (d copies).
$B(x, \varepsilon)$	In a metric space, $B(x, \varepsilon)$ denotes the open ball of radius ε centered at x.
$N_\varepsilon(A)$	In a metric space, $N_\varepsilon(A)$ denotes the open ε-neighborhood of a set $A \subset X$, that is, the union of all balls $B(x, \varepsilon)$ with $x \in A$.
$\mathrm{dist}(x, A)$	In a metric space (X, ϱ), $\mathrm{dist}(x, A)$ denotes the distance from a point x to a nonempty set $A \subset X$, defined by $\mathrm{dist}(x, A) := \inf_{a \in A} \varrho(x, a)$.
$\mathrm{diam}\, A$	The diameter of a nonempty subset of a metric space (X, ϱ), $\mathrm{diam}\, A = \sup_{x,y \in A} \varrho(x, y)$.
$\mathrm{int}\, A$	The interior of a subset A of a topological space.
$\mathrm{cl}\, A$	The closure of a subset A of a topological space.
∂A	The boundary of a subset A of a topological space.
$\mathrm{supp}\, f$	The support of a continuous function $f : X \to \mathbb{R}$, that is, $\mathrm{supp}\, f := \mathrm{cl}\{x \in X : f(x) \neq 0\}$.
$\lfloor \cdot \rfloor$	For a real number x, we denote by $\lfloor x \rfloor$ the integer part of x, that is, the unique integer such that $x - \lfloor x \rfloor \in [0, 1)$.
$\sigma(T)$	For a linear operator $T : X \to X$, we write $\sigma(T)$ for the spectrum of T, that is, for the set of all eigenvalues.
$\lambda_{\max}(T)$	The maximal eigenvalue of a linear self-adjoint operator T on a Euclidean space X.
$\sigma_i(T)$	The i-th singular value of a linear operator $T : X \to Y$ between Euclidean spaces of the same dimension d, where $\sigma_1(T) \geq \cdots \geq \sigma_d(T)$.
$\mathscr{L}(X, Y)$	For normed vector spaces X and Y, we write $\mathscr{L}(X, Y)$ for the set of all bounded linear maps $T : X \to Y$.
$\|T\|$	If $(X, |\cdot|_X)$ and $(Y, |\cdot|_Y)$ are normed vector spaces and $T \in \mathscr{L}(X, Y)$, by $\|T\|$ we denote the operator norm of T, that is, $\|T\| = \sup_{|x|_X = 1} |Tx|_Y$.
$\ker T$	The kernel of a linear operator $T : X \to Y$, $\ker T = T^{-1}(0)$.
$\mathrm{im}\, T$	The image of a linear operator $T : X \to Y$, $\mathrm{im}\, T = T(X)$.
$\det T$	The determinant of a linear operator T between oriented Euclidean spaces of the same dimension.
$\mathrm{tr}\, T$	The trace of a linear operator T.
T^*	The adjoint operator of a linear operator $T : X \to Y$ between Euclidean spaces of the same dimension.

$\delta_{ij} = \delta_i^j = \delta_j^i$ This notation stands for the Kronecker-Delta, that is,

$$\delta_{ij} = \begin{cases} 1 \text{ if } i = j, \\ 0 \text{ if } i \neq j. \end{cases}$$

$I = I_d$ The $d \times d$ identity matrix, $I = (\delta_{ij})$.

GL(X)   If X is a vector space, GL(X) denotes the group of all linear automorphisms of X.

O(d)   This notation stands for the orthogonal group of $\mathbb{R}^d$.

$E^\perp$ The orthogonal complement of a subspace E of a Euclidean space.

$x^\perp$ The orthogonal complement of the one-dimensional subspace spanned by a nonzero element x of a Euclidean space.

$\angle(x, y)$ The angle between two nonzero vectors x, y in a Euclidean space.

$\mathrm{D}f(x)$ If $f : \mathbb{R}^n \supset D \to \mathbb{R}^m$ is a map defined on an open set D, which is differentiable at x, we write $\mathrm{D}f(x)$ for the Jacobi-matrix of f at x.

$\mathbf{0}$ A boldface zero stands for a constant function $\omega : I \to X$ from an interval I to a vector space X, which is identically 0, that is, $\mathbf{0}(t) = 0 \in X$ for all $t \in I$.

$T_p M$ The tangent space of a $\mathscr{C}^k$-manifold at $p \in M$.

TM The tangent bundle of a $\mathscr{C}^k$-manifold.

π_{TM} The base point projection from TM to M, which sends a tangent vector $v \in T_p M$ to its base point p.

$\mathscr{C}^r(M, N)$ If M and N are $\mathscr{C}^r$-manifolds, then $\mathscr{C}^r(M, N)$ stands for the set of all $\mathscr{C}^r$-maps $f : M \to N$.

$\mathscr{X}^r(M)$ If M is a $\mathscr{C}^{r+1}$-manifold, $\mathscr{X}^r(M)$ stands for the space of all $\mathscr{C}^r$-vector fields on M.

S^d The unit sphere of dimension d, that is, $\mathrm{S}^d = \{x \in \mathbb{R}^{d+1} : |x| = 1\}$.

$\mathrm{S}(X)$ The unit sphere in a Euclidean space $(X, \langle \cdot, \cdot \rangle)$, $\mathrm{S}(X) = \{x \in X : \langle x, x \rangle = 1\}$.

$\mathbb{P}^d$ The d-dimensional real projective space which is defined as the quotient space of $\mathbb{R}^{d+1} \setminus \{0\}$ by the equivalence relation $x \sim y$ iff $y = \alpha x$ for some nonzero $\alpha \in \mathbb{R}$.

$\mathscr{L}(\gamma)$ The length of a piecewise $\mathscr{C}^1$ or locally absolutely continuous curve $\gamma : I \to M$, where $I \subset \mathbb{R}$ is an interval and M a $\mathscr{C}^k$-manifold.

$\exp_p$ If (M, g) is a Riemannian $\mathscr{C}^3$-manifold, $\exp_p$ denotes the Riemannian exponential map at $p \in M$.

$\nabla f(p)$ The covariant derivative of a $\mathscr{C}^1$-vector field f on a Riemannian manifold (M, g) at $p \in M$.

$S\nabla f(p)$ The symmetrized covariant derivative of a $\mathscr{C}^1$-vector field f at p, that is, $S\nabla f(p) = (1/2)(\nabla f(p) + \nabla f(p)^*)$.

$\mathrm{D}X/\mathrm{d}t$ The covariant derivative of a vector field X along a curve.

vol	The Riemannian volume on a Riemannian manifold.
$L^p(I, \mathbb{R}^m)$	The Banach space of all L^p-functions from an interval $I \subset \mathbb{R}$ to $\mathbb{R}^m$.
$\text{ess sup}_x f(x)$	The essential supremum of a measurable real-valued function $f : X \to \mathbb{R}$, $\text{ess sup}_{x \in X} f(x) = \inf_{N: \mu(N)=0} \sup_{x \in X \setminus N} f(x)$.
$\|f\|_{[0,\tau]}$	This notation stands for the L^∞-norm of a function $f \in L^\infty([0, \tau], \mathbb{R}^m)$.
$n(\varepsilon, K)$	The minimal number of ε-balls necessary to cover a totally bounded subset K of a metric space.
$\overline{\dim}_C(X)$	The upper capacitive dimension of a metric space X.
$h_{\text{top},\varrho}(K, f)$	The topological entropy of a uniformly continuous map $f : X \to X$ on a compact subset K of a metric space (X, ϱ).
$h_{\text{top},\varrho}(f)$	The topological entropy of $f : X \to X$ on (X, ϱ).
$\mathcal{V}_1 \oplus \mathcal{V}_2$	The Whitney sum of two subbundles $\mathcal{V}_1$ and $\mathcal{V}_2$ of some vector bundle $\mathcal{W}$.
$\text{rk}\,\mathcal{V}$	The rank of a vector bundle $\mathcal{V}$.

Chapter 1
Basic Properties of Control Systems

This introductory chapter provides the necessary background on control systems that we need for the development of the entropy theory. In Sect. 1.1, we give the definition of a control system which is basically the one from Sontag's book [102]. We also introduce particular classes of (time-invariant) systems, namely topological, linear, and smooth systems. Section 1.2 introduces the subclass of smooth systems which our main focus is on, the class of smooth systems given by differential equations. The third section serves for the introduction of several useful control-theoretic notions. In Sect. 1.4, we prove elementary properties of the control flow associated with a control-affine system. Moreover, we establish the notions of control and chain control sets, and we give the proofs for some elementary properties of these objects. Most of this material stems from the book of Colonius and Kliemann [25]. Finally, in Sect. 1.5, the linearization of a smooth system given by differential equations along a fixed trajectory and related notions are studied.

1.1 Basic Definitions

We start with the definition of control systems (throughout the book simply called *systems*), which is basically the one that can be found in Sontag's book, see [102, Definition 2.1.2]. However, we restrict ourselves to complete systems (that is, systems whose trajectories are defined for all times), which reduces the number of necessary axioms. Since Sontag's definition treats discrete- and continuous-time systems simultaneously, and for continuous-time systems completeness means completeness with respect to a certain class of control functions (usually, essentially bounded functions), we have to add this class to the data defining the system. This again complicates matters a little bit. Before we give the definition, we need to introduce some notation.

A set $\mathbb{T} \in \{\mathbb{Z}, \mathbb{R}\}$ is called a *time set*. The set $\mathbb{T}_+$ is defined as $\mathbb{T}_+ := \{t \in \mathbb{T} : t \geq 0\}$. When the time set is understood from the context, all intervals are assumed to be restricted to $\mathbb{T}$. For example,

C. Kawan, *Invariance Entropy for Deterministic Control Systems*, Lecture Notes in Mathematics 2089, DOI 10.1007/978-3-319-01288-9_1,
© Springer International Publishing Switzerland 2013

$$[\sigma, \tau) = \{t \in \mathbb{T} \ : \ \sigma \leq t < \tau\}.$$

Let U be any nonempty set and $\mathscr{U}$ a subset of $U^{\mathbb{T}}$ (the set of all mappings from $\mathbb{T}$ to U). Then for all $\sigma, \tau \in \mathbb{T}$ with $\sigma \leq \tau$ we define

$$\mathscr{U}[\sigma, \tau) := \{\omega|_{[\sigma,\tau)} \ : \ \omega \in \mathscr{U}\}.$$

For $\sigma = \tau$, we can think of this set as containing only one element $\omega_0 : \emptyset \to U$, the *empty function*, which we denote by $\diamond$.

If $\sigma, \tau, \mu \in \mathbb{T}$ with $\sigma \leq \tau \leq \mu$, $\omega_1 \in U^{[\sigma,\tau)}$, and $\omega_2 \in U^{[\tau,\mu)}$, the *concatenation* $\omega \in U^{[\sigma,\mu)}$ of ω_1 and ω_2, also denoted by $\omega_1\omega_2$, is defined as

$$\omega(t) = (\omega_1\omega_2)(t) := \begin{cases} \omega_1(t) & \text{if } t \in [\sigma, \tau), \\ \omega_2(t) & \text{if } t \in [\tau, \mu). \end{cases}$$

Now we can give the definition of a system.

Definition 1.1. A *system* $\Sigma = (\mathbb{T}, X, U, \mathscr{U}, \phi)$ consists of:

- A set $\mathbb{T} \in \{\mathbb{Z}, \mathbb{R}\}$ called the *time set*;
- A nonempty set X called the *state space*;
- A nonempty set U called the *control-value space*;
- A nonempty set $\mathscr{U} \subset U^{\mathbb{T}}$ called the *set of admissible control functions*;
- A map $\phi : \mathscr{D}_\phi \to X$, called the *transition map*, which is defined on

$$\mathscr{D}_\phi := \{(\tau, \sigma, x, \omega) \ : \ \sigma, \tau \in \mathbb{T}, \ \sigma \leq \tau, \ x \in X, \ \omega \in \mathscr{U}[\sigma, \tau)\},$$

such that the following properties hold:

(S1) The set $\mathscr{U}$ is shift-invariant, that is, if $\omega \in \mathscr{U}$ and $\mu \in \mathbb{T}$, then also the shifted function $\omega(\mu + \cdot), t \mapsto \omega(\mu + t)$, is an element of $\mathscr{U}$;

(S2) If σ, τ, μ are any three elements of $\mathbb{T}$ so that $\sigma \leq \tau \leq \mu$, if $\omega_1 \in \mathscr{U}[\sigma, \tau)$ and $\omega_2 \in \mathscr{U}[\tau, \mu)$, and if $x \in X$ so that

$$\phi(\tau, \sigma, x, \omega_1) = x_1 \quad \text{and} \quad \phi(\mu, \tau, x_1, \omega_2) = x_2,$$

then $\omega = \omega_1\omega_2$ is an element of $\mathscr{U}[\sigma, \mu)$ and satisfies

$$\phi(\mu, \sigma, x, \omega) = x_2;$$

(S3) For each $\sigma \in \mathbb{T}$ and each $x \in X$, the empty function $\diamond \in \mathscr{U}[\sigma, \sigma)$ satisfies $\phi(\sigma, \sigma, x, \diamond) = x$.

The elements of $\mathbb{T}$ are called *times*, the elements of X *states*. In the case that $\mathbb{T} = \mathbb{Z}$, the *admissible control functions*, that is, the elements of $\mathscr{U}$, are also called *control sequences*. For $x \in X$, $\omega \in \mathscr{U}$, and $\sigma \in \mathbb{T}$, we call the map $t \mapsto \phi(t, \sigma, x, \omega|_{[\sigma,t)})$, $[\sigma, \infty) \to X$, a *trajectory*.

We give a short explanation of this definition: The time set $\mathbb{T}$ contains all the times at which the system can be evaluated. If $\mathbb{T} = \mathbb{Z}$, we call Σ a *discrete-time system*, if $\mathbb{T} = \mathbb{R}$, a *continuous-time system*. The elements of the state space X represent all the possible configurations of the system. If Σ is the model for a mechanical system, these might consist of (generalized) coordinates and velocities. In this case, X is a subset, usually a submanifold, of some Euclidean space $\mathbb{R}^n$. The set U of control values is the codomain of the admissible control functions, the elements of $\mathscr{U}$. A control function can be seen as a "generalized force" which is used to influence the behavior of the system in a certain way. The transition map ϕ describes the evolution of the system in time under the influence of the applied control. More precisely, starting from an initial state x at time σ using the control function ω, at time τ the state of the system is $\phi(\tau, \sigma, x, \omega)$.

Axiom (S1) becomes important for time-invariant systems which are introduced below. For these systems, the axiom allows to define a simplified transition map with only three inputs and the nice property that together with the shift flow on $\mathscr{U}$ it constitutes a skew-product system on the extended state space $\mathscr{U} \times X$, usually called the *control flow* associated with the system (cf. Sect. 1.4). In Sontag's definition of a system, Axiom (S1) is missing, since he does not include an explicitly given set $\mathscr{U}$ of admissible control functions into the data defining a system but rather into the domain of the transition map. Sontag calls Axiom (S2) the *semigroup axiom* and Axiom (S3) the *identity axiom*. These axioms reflect the well-known properties of systems given by differential equations $\dot{x} = f(t, x, \omega(t))$ or difference equations $x_{k+1} = f(k, x_k, \omega_k)$, following in the case of Axiom (S2) from uniqueness of solutions, and in the case of Axiom (S3) from their definition. The first two axioms of Sontag's definition (the *nontriviality axiom* and the *restriction axiom*) are missing in ours, since they are only needed for systems which are not complete, meaning that trajectories may be defined only on bounded time intervals.

Remark 1.1. Also in Hinrichsen and Pritchard [58, Definition 2.1.1] a very general definition of systems is given, which uses a slightly different terminology. Except for some minor differences (for instance, the time set in their definition is just an arbitrary nonempty subset of $\mathbb{R}$), the only essential difference is that Hinrichsen and Pritchard include an output space and an output map into the data defining a system. Since systems with outputs are not treated in this book, there is no need for us to do this.

Throughout the book, we only consider time-invariant systems whose state space is a metrizable topological space (or even a differentiable manifold) and whose transition map is continuous (or even differentiable) with respect to the time and the state variables. The following definition of time-invariant systems is taken from Sontag [102, Definition 2.1.9].

Definition 1.2. A system $\Sigma = (\mathbb{T}, X, U, \mathscr{U}, \phi)$ is *time-invariant* if for each $\omega \in \mathscr{U}[\sigma, \tau)$, each $x \in X$, and each $\mu \in \mathbb{T}$, the translation

$$\omega^\mu \in \mathscr{U}[\sigma + \mu, \tau + \mu), \quad \omega^\mu(t) := \omega(t - \mu),$$

satisfies

$$\phi(\tau, \sigma, x, \omega) = \phi(\tau + \mu, \sigma + \mu, x, \omega^{\mu}). \tag{1.1}$$

Given a time-invariant system, we do not need to take care about the initial time of a trajectory, that is, we can drop the second argument in the transition map ϕ of the system.

Definition 1.3. Given a time-invariant system Σ as above, we define

$$\varphi(t, x, \omega) := \phi\left(t, 0, x, \omega|_{[0,t)}\right), \quad \varphi : \mathbb{T}_+ \times X \times \mathscr{U} \to X.$$

We call φ the *reduced transition map* or just the *transition map* (if it is clear that the system is time-invariant) of the time-invariant system. We denote such a system also by $(\mathbb{T}, X, U, \mathscr{U}, \varphi)$. The system is called a *topological system* if the state space X is a metrizable topological space and for each $\omega \in \mathscr{U}$ the map

$$\varphi_{\omega} : \mathbb{T}_+ \times X \to X, \quad \varphi_{\omega}(t, x) := \varphi(t, x, \omega),$$

is continuous, where we consider the standard topology on $\mathbb{T}_+$.

For an interval $I \subset \mathbb{T}$, we often use the handy notation

$$\varphi(I, x, \omega) = \{\varphi(t, x, \omega) \,:\, t \in I\}.$$

Moreover, we use the notations

$$\varphi_{t,\omega}(x) = \varphi_{\omega}(t, x) = \varphi_t(x, \omega) = \varphi(t, x, \omega),$$

depending on which arguments we regard as fixed in the corresponding context. If ω is a constant control function with value $u \in U$, we often simply write $\varphi(t, x, u)$ instead of $\varphi(t, x, \omega)$, that is, we do not distinguish between the control value u and the constant control function $\omega(t) \equiv u$ in our notation.

Remark 1.2. For a topological system, we do not require that the state space is endowed with a metric, but only that it is metrizable. The reason for this is that a lot of constructions are independent of a specific metric. However, often it is useful to have a metric. In those cases, naturally we assume that any considered metric is compatible with the given topology.

Remark 1.3. In the special case where the control-value space U of a time-invariant system has only one element u, the set $\mathscr{U}$ of admissible control functions necessarily also consists of only one element, the constant control function $\omega(t) \equiv u$. In this case, there is no possibility to influence the behavior of the system. Following Sontag, we call such systems *classical dynamical systems*. The transition map in this case can be regarded as a map $\Phi : \mathbb{T}_+ \times X \to X$ which satisfies the semigroup properties $\Phi(0, x) = x$ and $\Phi(t + s, x) = \Phi(t, \Phi(s, x))$ for all $t, s \in \mathbb{T}_+$ and $x \in X$.

In order to describe the properties of the reduced transition map φ, we introduce the *shift flow* on the set $\mathscr{U}$ of admissible control functions, that is, the map[1]

$$\Theta : \mathbb{T} \times \mathscr{U} \to \mathscr{U}, \quad (t, \omega) \mapsto \Theta_t \omega := \omega(t + \cdot).$$

By Axiom (S1), Θ is well-defined. It is easy to verify that Θ is a dynamical system, that is, it satisfies $\Theta_0 \omega = \omega$ and $\Theta_{t+s} \omega = \Theta_t \Theta_s \omega$ for all $t, s \in \mathbb{T}$ and $\omega \in \mathscr{U}$.

Proposition 1.1. *For a time-invariant system Σ with reduced transition map φ the following assertions hold:*

(i) *For all $(x, \omega) \in X \times \mathscr{U}$ it holds that $\varphi(0, x, \omega) = x$.*
(ii) *The map φ is a cocycle over the dynamical system Θ, that is, for all $t, s \in \mathbb{T}_+$, $x \in X$, and $\omega \in \mathscr{U}$ it holds that*

$$\varphi(t + s, x, \omega) = \varphi(s, \varphi(t, x, \omega), \Theta_t \omega).$$

(iii) *For each $t \in \mathbb{T}_+$, $\varphi(t, x, \omega)$ does not depend on the values of ω outside of $[0, t)$. That is, if $\omega_1, \omega_2 \in \mathscr{U}$ satisfy $\omega_1(s) = \omega_2(s)$ for all $s \in [0, t)$, then $\varphi(t, x, \omega_1) = \varphi(t, x, \omega_2)$.*

Proof. Assertions (i) and (iii) are immediate from the definition of φ and the properties of ϕ. To show the cocycle property, fix $t, s \in \mathbb{T}_+$, $x \in X$, and $\omega \in \mathscr{U}$. Then, by definition of φ,

$$\varphi(s, \varphi(t, x, \omega), \Theta_t \omega) = \phi\left(s, 0, \phi(t, 0, x, \omega|_{[0,t)}), (\Theta_t \omega)|_{[0,s)}\right).$$

Using (1.1), this gives

$$\varphi(s, \varphi(t, x, \omega), \Theta_t \omega) = \phi\left(t + s, t, \phi(t, 0, x, \omega|_{[0,t)}), \left(\Theta_t \omega|_{[0,s)}\right)^t\right).$$

Using that $\left((\Theta_t \omega)|_{[0,s)}\right)^t (\tau) = \Theta_t \omega(\tau - t) = \omega(\tau)$, we obtain

$$\varphi(s, \varphi(t, x, \omega), \Theta_t \omega) = \phi\left(t + s, t, \phi(t, 0, x, \omega|_{[0,t)}), \omega|_{[t,t+s)}\right).$$

By Axiom (S2), this reduces to

$$\varphi(s, \varphi(t, x, \omega), \Theta_t \omega) = \phi\left(t + s, 0, x, \omega|_{[0,t)} \omega|_{[t,t+s)}\right)$$
$$= \phi\left(t + s, 0, x, \omega|_{[0,t+s)}\right) = \varphi(t + s, x, \omega),$$

which concludes the proof. □

---

[1] Usually, by a *flow* one understands a continuous-time dynamical system. Essentially, we also use the terminology in this way, but we make an exception here.

From the properties (i) and (ii) of the reduced transition map, described in the preceding proposition, it follows that the map

$$\mathbb{T}_+ \times (\mathscr{U} \times X) \to \mathscr{U} \times X, \quad (t, (\omega, x)) \mapsto (\Theta_t \omega, \varphi(t, x, \omega)),$$

is a skew-product semiflow on $\mathscr{U} \times X$. For the class of control-affine systems, to be introduced later, this semiflow has some special features which allow to analyze the control-theoretic properties of the system by studying the dynamical properties of the semiflow.

For a time-invariant system, the properties of φ listed in Proposition 1.1 characterize φ as a reduced transition map, which is shown next.

Proposition 1.2. *Let $(\mathbb{T}, X, U, \mathscr{U}, \varphi)$ be a quintuple such that $\mathbb{T}$, X, U, and $\mathscr{U}$ are as in Definition 1.1, and φ is a map from $\mathbb{T}_+ \times X \times \mathscr{U}$ to X which has the properties (i)–(iii) in Proposition 1.1. Then there exists a map ϕ which is the (full) transition map of a time-invariant system $(\mathbb{T}, X, U, \mathscr{U}, \phi)$ such that φ is the associated reduced transition map.*

Proof. There is only one possibility how ϕ can be defined. For $(\tau, \sigma, x, \omega)$ with $\tau, \sigma \in \mathbb{T}$, $\tau \geq \sigma$, $x \in X$, and $\omega \in \mathscr{U}[\sigma, \tau)$, we choose $\overline{\omega} \in \mathscr{U}$ such that ω is the restriction of $\overline{\omega}$ to $[\sigma, \tau)$. Then we define

$$\phi(\tau, \sigma, x, \omega) := \varphi(\tau - \sigma, x, \Theta_\sigma \overline{\omega}).$$

Since by assumption $\varphi(\tau - \sigma, x, \Theta_\sigma \overline{\omega})$ does not depend on the values of $\overline{\omega}$ outside of $[\sigma, \tau)$, the map ϕ is well-defined. To show that Axiom (S2) holds, let $\sigma \leq \tau \leq \mu$ be elements of $\mathbb{T}$ and $\omega_1 \in \mathscr{U}[\sigma, \tau)$, $\omega_2 \in \mathscr{U}[\tau, \mu)$. Due to Axiom (S2) of Definition 1.1 we can find $\overline{\omega} \in \mathscr{U}$ such that $\overline{\omega}|_{[\sigma,\tau)} = \omega_1$ and $\overline{\omega}|_{[\tau,\mu)} = \omega_2$, or equivalently, $\overline{\omega}|_{[\sigma,\mu)} = \omega_1 \omega_2$. Using property (ii) in Proposition 1.1, we obtain

$$\phi(\mu, \tau, \phi(\tau, \sigma, x, \omega_1), \omega_2) = \varphi(\mu - \tau, \varphi(\tau - \sigma, x, \Theta_\sigma \overline{\omega}), \Theta_\tau \overline{\omega})$$
$$= \varphi(\mu - \tau, \varphi(\tau - \sigma, x, \Theta_\sigma \overline{\omega}), \Theta_{\tau - \sigma} \Theta_\sigma \overline{\omega})$$
$$= \varphi(\mu - \sigma, x, \Theta_\sigma \overline{\omega})$$
$$= \phi(\mu, \sigma, x, \overline{\omega}|_{[\sigma,\mu)})$$
$$= \phi(\mu, \sigma, x, \omega_1 \omega_2).$$

This proves that Axiom (S2) is satisfied. The validity of Axiom (S3) easily follows from property (i) in Proposition 1.1. Hence, $\Sigma := (\mathbb{T}, X, U, \mathscr{U}, \phi)$ is a system in the sense of Definition 1.1. Time-invariance is obvious. $\square$

Next we define the important class of linear time-invariant systems.

Definition 1.4. A time-invariant system $\Sigma = (\mathbb{T}, X, U, \mathscr{U}, \varphi)$ is called *linear* (over the field $\mathbb{K} \in \{\mathbb{R}, \mathbb{C}\}$) if there are integers $d, m \geq 1$ such that the following properties hold:

- The state space X is a d-dimensional vector space over $\mathbb{K}$;
- The control-value space U is an m-dimensional vector space over $\mathbb{K}$;
- There exists a linear subspace $\mathscr{V}$ of $U^{\mathbb{T}}$ with $\mathscr{U} \subset \mathscr{V}$ and a map $\tilde{\varphi} : \mathbb{T}_+ \times X \times \mathscr{V} \to X$ such that $(\mathbb{T}, X, U, \mathscr{V}, \tilde{\varphi})$ is a topological time-invariant system, for fixed $t \in \mathbb{T}_+$, the map $\tilde{\varphi}(t, \cdot, \cdot) : X \times \mathscr{V} \to X$ is $\mathbb{K}$-linear, and $\varphi(t, x, \omega) = \tilde{\varphi}(t, x, \omega)$ whenever the left-hand side is defined.

Often we denote the extended transition map $\tilde{\varphi}$ also by φ.

This definition of linear systems seems a little bit complicated when compared to the usual definitions. The reason for this complication is that we also want to regard systems as linear which come from differential equations of the form $\dot{x} = Ax + Bu$ or difference equations $x_{k+1} = Ax_k + Bu_k$, where the set of admissible controls is not a linear space, for instance, because the values of these functions are restricted to a bounded subset of some vector space U.

The property of linear systems that the extended system with transition map $\tilde{\varphi}$ is a *topological* system, has an important consequence for continuous-time systems, described in the next proposition.

Proposition 1.3. *Let* $\Sigma = (\mathbb{R}, X, U, \mathscr{U}, \varphi)$ *be a continuous-time linear system. Then there exists a linear operator* $A \in \mathscr{L}(X, X)$ *such that*

$$\varphi(t, x, \omega) = e^{At}x + \varphi(t, 0, \omega)$$

holds for all $(t, x, \omega) \in \mathbb{R}_+ \times X \times \mathscr{U}$.

Proof. This follows from Proposition B.3. Indeed, consider the extended system $(\mathbb{R}, X, U, \mathscr{V}, \tilde{\varphi})$. Let $\mathbf{0} \in \mathscr{V}$ denote the constant control function $\mathbf{0}(t) \equiv 0 \in U$. Then, by linearity of $\tilde{\varphi}(t, \cdot, \cdot)$, we have

$$\varphi(t, x, \omega) = \tilde{\varphi}(t, x, \omega) = \tilde{\varphi}(t, x, \mathbf{0}) + \tilde{\varphi}(t, 0, \omega).$$

The family of maps $T(t) := \tilde{\varphi}_{t, \mathbf{0}} : X \to X$, $t \in \mathbb{R}_+$, satisfies the assumptions of Proposition B.3 and therefore $T(t) = e^{At}$ for a unique $A \in \mathscr{L}(X, X)$. $\square$

Finally, we introduce the class of smooth systems.

Definition 1.5. A topological time-invariant system $\Sigma = (\mathbb{T}, M, U, \mathscr{U}, \varphi)$ is called *smooth* if it satisfies the following properties:

- The state space M is a connected $\mathscr{C}^1$-manifold;
- For all $t \in \mathbb{T}_+$ and $\omega \in \mathscr{U}$, the map $\varphi_{t, \omega} : M \to M$ is of class $\mathscr{C}^1$.

Example 1.1. Let M be a connected $\mathscr{C}^1$-manifold, Ω some nonempty set, and $F : M \times \Omega \to M$, $(x, \omega) \mapsto F_\omega(x)$, a map which is of class $\mathscr{C}^1$ in its first argument. Then for each $\omega \in \Omega^{\mathbb{Z}}$, $\omega = (\omega_k)_{k \in \mathbb{Z}}$, and for all $k \in \mathbb{Z}_+$, $x \in M$, we can define

$$\varphi(k, x, \omega) := \begin{cases} x & \text{for } k = 0, \\ F_{\omega_{k-1}} \circ \cdots \circ F_{\omega_1} \circ F_{\omega_0}(x) & \text{for } k \geq 1. \end{cases}$$

It is easy to verify that $\Sigma := (\mathbb{Z}, M, \Omega, \Omega^{\mathbb{Z}}, \varphi)$ is a smooth system. This system is called the *system given by the difference equations*

$$x_{k+1} = F(x_k, \omega_k), \quad \omega_k \in \Omega.$$

In the following section, we introduce the class of smooth systems given by differential equations on which we mainly concentrate in the development of the entropy theory.

1.2   Systems Given by Differential Equations

In this section, we describe a special class of smooth systems, namely systems generated by ordinary differential equations of the form

$$\dot{x}(t) = F(x(t), \omega(t)), \quad \omega \in \mathscr{U}.$$

Here, $F : M \times \mathbb{R}^m \to TM$ is a continuous map, where M denotes a d-dimensional $\mathscr{C}^k$-manifold and TM its tangent bundle, such that for fixed $u \in \mathbb{R}^m$ the map $F_u := F(\cdot, u) : M \to TM$ is a vector field on M (that is, $F_u(x)$ is an element of the tangent space $T_x M$ for each x). The following lemma allows to apply the theory of Carathéodory differential equations to such families of equations (cf. Sect. A.3).

Lemma 1.1. *Let M be a connected $\mathscr{C}^k$-manifold, $k \geq 2$, and $F : M \times \mathbb{R}^m \to TM$ a continuous map which satisfies $F(x, u) \in T_x M$ for all $(x, u) \in M \times \mathbb{R}^m$. Moreover, assume that F is of class $\mathscr{C}^r$ in its first argument ($r \in \{1, \ldots, k-1\}$), that is, for each local representation $F = F^i(x, u)\partial_i \phi(x)$, the coordinate functions $y \mapsto F^i(\phi^{-1}(y), u)$ have partial derivatives up to order r which are continuous in (y, u). Then for every function $\omega \in L^\infty(\mathbb{R}, \mathbb{R}^m)$ the map*

$$f_\omega : \mathbb{R} \times M \to TM, \quad f_\omega(t, x) := F(x, \omega(t)),$$

is a locally integrally $\mathscr{C}^r$-vector field.

Proof. From $F(x, u) \in T_x M$ for all (x, u) it follows that $f_\omega(t, x) \in T_x M$ for all (t, x). Now let $\alpha : M \to T^* M$ be a continuous one-form. In order to show that f_ω is a Carathéodory vector field, we have to prove that $\alpha \cdot f_\omega : \mathbb{R} \times M \to \mathbb{R}$ is a Carathéodory function, that is, continuous in x and measurable in t. For fixed $t \in \mathbb{R}$, it is clear that the map $x \mapsto \alpha(x)f_\omega(t, x) = \alpha(x)F(x, \omega(t))$ is continuous, since both α and F are continuous. Since ω is measurable and α, F are continuous, it also follows that $t \mapsto \alpha(x)F(x, \omega(t))$ is measurable for each fixed x. Now assume that

F is of class $\mathscr{C}^r$ in the first argument and α is of class $\mathscr{C}^r$. Then f_ω is of class $\mathscr{C}^r$ in the second argument. Hence, to prove that f_ω is locally integrally of class $\mathscr{C}^r$, it remains to show that

$$X_1 \cdots X_r(\alpha \cdot f_\omega(t, \cdot))$$

is locally integrally bounded for all $\mathscr{C}^r$-vector fields $X_1, \ldots, X_r$ and all $t \in \mathbb{R}$. Since every compact set $K \subset M$ can be covered by finitely many chart domains, it suffices to verify integral boundedness on every sufficiently small domain of a chart (ϕ, U). Let us choose the chart (ϕ, U) such that $\phi(U)$ is the ball $B(0, 1)$ of radius 1 around $0 \in \mathbb{R}^d$. If $g : U \to \mathbb{R}$ is a $\mathscr{C}^r$-function and $X_k = X_k^{i_k} \partial_{i_k} \phi$ for every $k \in \{1, \ldots, r\}$, an elementary computation shows that $X_1 \cdots X_r(g)$ is a sum of terms which are products of partial derivatives of both the components $X_k^{i_k}$ of the vector fields and the function g from order 0 up to order r. For $g = \alpha \cdot f_\omega(t, \cdot)$, these partial derivatives have the form

$$\frac{\partial^k (\alpha \cdot f_\omega(t, \cdot))}{\partial \phi^{i_1} \cdots \partial \phi^{i_k}}(x), \quad x \in U, \ 1 \le k \le r.$$

From the assumptions on α and F it follows that the partial derivatives up to order r of the function

$$y \mapsto \alpha(\phi^{-1}(y))F(\phi^{-1}(y), u), \quad \phi(U) \to \mathbb{R},$$

exist and are continuous in (y, u). For every $\omega \in L^\infty(\mathbb{R}, \mathbb{R}^m)$ and every continuous function $h : \phi(U) \times \mathbb{R}^m \to \mathbb{R}$, the associated function $h' : \phi(U) \times \mathbb{R} \to \mathbb{R}$, $(y, t) \mapsto h(y, \omega(t))$, has the property that $|h'(y, t)| \le \psi(t)$ for all $y \in \mathrm{cl}\, B(0, 1/2)$ and $t \in \mathbb{R}$ with a locally integrable function ψ. This is shown as follows: Since ω is essentially bounded, there exists a compact set $\Omega \subset \mathbb{R}^m$ such that $\omega(t) \in \Omega$ for almost all $t \in \mathbb{R}$. Then the constant function $\psi(t) :\equiv \max_{(y,u) \in \mathrm{cl}\, B(0,1/2) \times \Omega} h(y, u)$ has the desired property. Now, taking for h the partial derivatives of $\alpha \cdot f_\omega(t, \cdot)$, we obtain the result. $\qquad\square$

Proposition 1.4. *In addition to the assumptions of the preceding lemma, let $\Omega \subset \mathbb{R}^m$ be a nonempty set and[2]*

$$\mathscr{U} := \{\omega \in L^\infty(\mathbb{R}, \mathbb{R}^m) \ : \ \omega(t) \in \Omega \text{ a.e.}\}.$$

Further suppose that for every $(\omega, x_0) \in \mathscr{U} \times M$ the solution of the initial value problem $\dot{x}(t) = F(x(t), \omega(t))$, $x(0) = x_0$, exists for all times $t \in \mathbb{R}$. Then a smooth continuous-time time-invariant system is given by $\Sigma = (\mathbb{R}, M, \mathbb{R}^m, \mathscr{U}, \varphi)$, where for each $x \in M$ and $\omega \in \mathscr{U}$ the trajectory $t \mapsto \varphi(t, x, \omega)$ is defined as

---

[2]Throughout the book, the abbreviation "a.e." stands for "*(Lebesgue) almost everywhere*".

the unique solution of $\dot{x}(t) = F(x(t), \omega(t))$ *with* $\varphi(0, x, \omega) = x$. *Furthermore, the maps* $\varphi_{t,\omega} : M \to M$ *are* $\mathscr{C}^r$-*diffeomorphisms.*

Proof. We have to prove the following six properties of Σ:

(i) The set $\mathscr{U}$ is shift-invariant;
(ii) If $\sigma \leq \tau \leq \mu$, $\omega_1 \in \mathscr{U}[\sigma, \tau)$, and $\omega_2 \in \mathscr{U}[\tau, \mu)$, then $\omega_1 \omega_2 \in \mathscr{U}[\sigma, \mu)$;
(iii) The map φ satisfies the cocycle property;
(iv) $\omega_1(t) = \omega_2(t)$ a.e. on $[0, \tau)$ implies $\varphi(\tau, x, \omega_1) = \varphi(\tau, x, \omega_2)$;
(v) For every (t, ω) the map $\varphi_{t,\omega}$ is a $\mathscr{C}^r$-diffeomorphism;
(vi) The map $(t, x) \mapsto \varphi_\omega(t, x)$ is continuous for every $\omega \in \mathscr{U}$.

Properties (i) and (ii) are obvious by definition of $\mathscr{U}$. The cocycle property of φ follows from uniqueness of solutions guaranteed by the flow box theorem A.1. In fact, for fixed $(x, \omega) \in M \times \mathscr{U}$ and $s \in \mathbb{R}_+$, consider the curves $\xi(t) := \varphi(t + s, x, \omega)$ and $\eta(t) := \varphi(t, \varphi(s, x, \omega), \Theta_s \omega)$. Then $\xi(0) = \varphi(s, x, \omega) = \eta(0)$, and

$$\dot{\xi}(t) = F(\varphi(t + s, x, \omega), \omega(t + s)) = F(\xi(t), \Theta_s \omega(t)),$$
$$\dot{\eta}(t) = F(\varphi(t, \varphi(s, x, \omega), \Theta_s \omega), \Theta_s \omega(t)) = F(\eta(t), \Theta_s \omega(t))$$

for almost all t. This implies $\xi(t) = \eta(t)$ for all $t \in \mathbb{R}_+$. Property (iv) also follows from the definition of $\varphi(t, x, \omega)$ as the solution of the corresponding differential equation, and property (v) again follows from the flow box theorem. To prove (vi), consider a Riemannian metric g on M with induced distance function ϱ. Choose $\omega \in \mathscr{U}$, a point $x \in M$, and a time $t \in \mathbb{R}_+$. Consider sequences $x_n \to x$ and $t_n \to t$. Then the triangle inequality gives

$$\varrho\left(\varphi_\omega(t_n, x_n), \varphi_\omega(t, x)\right) \leq \varrho\left(\varphi_\omega(t_n, x_n), \varphi_\omega(t, x_n)\right) + \varrho\left(\varphi_\omega(t, x_n), \varphi_\omega(t, x)\right).$$

The second term converges to zero, since $\varphi_{t,\omega}$ is continuous. To show that the first one becomes small, let us assume without loss of generality that $t_n < t$ for all n. Then

$$\varrho\left(\varphi_\omega(t_n, x_n), \varphi_\omega(t, x_n)\right) \leq \mathscr{L}\left(\varphi_\omega(\cdot, x_n)|_{[t_n, t]}\right)$$
$$= \int_{t_n}^t |F(\varphi_\omega(s, x_n), \omega(s))| \, ds$$
$$\leq |t - t_n| \operatorname*{ess\,sup}_{s \in [t_n, t]} |F(\varphi_\omega(s, x_n), \omega(s))|.$$

To conclude the proof, it suffices to show that for small $\delta, \beta > 0$ the image of $[t - \beta, t + \beta] \times \operatorname{cl} B(x, \delta)$ under φ_ω is relatively compact. We have $\varphi_\omega([t - \beta, t + \beta] \times \operatorname{cl} B(x, \delta)) = \varphi_{\Theta_{t-\beta}\omega}([0, 2\beta] \times \varphi_{t-\beta,\omega}(\operatorname{cl} B(x, \delta)))$ and by continuity of φ in x we may assume that $\varphi_{t-\beta,\omega}(\operatorname{cl} B(x, \delta)) \subset \operatorname{cl} B(\varphi_{t-\beta,\omega}(x), \varepsilon)$ for a small $\varepsilon > 0$. Hence, it remains to prove that sets of the form $\varphi_\mu([0, \gamma] \times \operatorname{cl} B(z, \rho))$ with $\mu \in \mathscr{U}$ and small $\gamma, \rho > 0$ are relatively compact. To this end, choose a cut-off function

$\theta : M \rightarrow [0, 1]$ of class $\mathscr{C}^1$ with compact support and $\theta(x) \equiv 1$ on cl $B(z, 2\rho)$. Consider the corresponding differential equation $\dot{x}(t) = \theta(x(t)) F(x(t), \mu(t))$ and denote its solutions by $\psi(t, x)$. Using that μ is essentially bounded (say $|\mu(t)| \leq \zeta$), for all $t \in [0, \gamma]$ and $w \in$ cl $B(z, \rho)$ we find

$$\varrho(\psi(t, w), w) \leq \mathscr{L}\left(\psi(\cdot, w)|_{[0,t]}\right) \leq \gamma \sup_{(a,u) \in \text{supp } \theta \times \text{cl } B(0, \zeta)} |(\theta F)(a, u)|.$$

Choosing γ small enough, we obtain

$$\varrho(\psi(t, w), z) \leq \varrho(\psi(t, w), w) + \varrho(w, z) \leq \rho + \rho = 2\rho,$$

which implies $\psi(t, w) = \varphi_\mu(t, w)$ for all $t \in [0, \gamma]$. Since cl $B(x, 2\rho) \subset$ supp θ is compact, this concludes the proof.   □

Systems which arise by differential equations as in the preceding proposition are called *smooth systems given by differential equations*. Their precise definition is as follows.

Definition 1.6. A *smooth system given by differential equations* is a time-invariant system of the form $\Sigma = (\mathbb{R}, M, \mathbb{R}^m, \mathscr{U}, \varphi)$ which satisfies the following properties:

(i) The state space M is a connected $\mathscr{C}^2$-manifold;
(ii) The set $\mathscr{U}$ of admissible control functions has the form

$$\mathscr{U} = \{\omega \in L^\infty(\mathbb{R}, \mathbb{R}^m) \ : \ \omega(t) \in \Omega \text{ a.e.}\}$$

for some nonempty set $\Omega \subset \mathbb{R}^m$;
(iii) There exists a continuous map $F : M \times \mathbb{R}^m \rightarrow TM$ which is of class $\mathscr{C}^1$ in its first argument (that is, the partial derivative exists and is continuous in (x, u)), such that for each $u \in \mathbb{R}^m$, $F_u := F(\cdot, u) : M \rightarrow TM$ is a $\mathscr{C}^1$-vector field and for every $(x_0, \omega) \in M \times \mathscr{U}$ the trajectory $t \mapsto \varphi(t, x_0, \omega)$ is the unique solution of the initial value problem

$$\dot{x}(t) = F(x(t), \omega(t)), \quad x(0) = x_0.$$

The map F is called the *right-hand side*, and the set Ω the *control range* of the system.

Remark 1.4.

- The assumption that M be connected guarantees that we can endow M with a Riemannian metric such that the induced Riemannian distance ϱ gives M the structure of a metric space. This is used in particular in Chaps. 4–6. Sometimes we have to impose stronger differentiability assumptions on the state space M and the right-hand side F. For instance, to have well-defined geodesics and

Riemannian exponential maps of class $\mathscr{C}^1$, we have to assume that M is a $\mathscr{C}^3$-manifold and the Riemannian metric is of class $\mathscr{C}^2$ (cf. Sect. A.2 in Appendix A). If we want to linearize the system along a trajectory, we additionally have to require that F is continuously differentiable with respect to the second argument. In this case, $F \in \mathscr{C}^1(M \times \mathbb{R}^m, TM)$ (cf. Sect. 1.5).

- The assumption that all solutions are defined for all times is in general restrictive. However, the majority of the results in this book on systems given by differential equations deals with the behavior of a system inside of a compact set or in its vicinity. In these cases, the assumption of completeness goes without loss of generality.
- Instead of essentially bounded control functions only, we could also allow locally integrable ones. However, this would result in technical difficulties at some points. Moreover, mostly we assume that the control range is bounded and in this case every control function is necessarily an element of $L^\infty(\mathbb{R}, \mathbb{R}^m)$.
- It is clear that smooth systems given by differential equations satisfy Definition 1.5, that is, they constitute a particular class of smooth systems.

The transition map φ of a smooth system given by differential equations extends to a map on $\mathbb{R} \times M \times \mathscr{U}$, that is, it is also defined for negative times. As follows easily from the definition of $\varphi(\cdot, x, \omega)$ as the solution of the corresponding differential equation, this extended map also satisfies the cocycle property:

$$\varphi(t + s, x, \omega) = \varphi(t, \varphi(s, x, \omega), \Theta_s \omega) \quad \text{for all } t, s \in \mathbb{R}, \ x \in M, \ \omega \in \mathscr{U}.$$

This will be used several times without further saying.

A particularly important class of smooth systems given by differential equations consists of the linear ones. For these systems, an explicit expression for the transition map is available due to the variation-of-constants formula (see Proposition A.2). The easy proof of the following proposition is left to the reader.

Proposition 1.5. *Let (A, B) be a pair of matrices such that $A \in \mathbb{R}^{d \times d}$ and $B \in \mathbb{R}^{d \times m}$ for some $d, m \in \mathbb{N}$. Moreover, let $\Omega \subset \mathbb{R}^m$ be a nonempty set and $\mathscr{U} := \{\omega \in L^\infty(\mathbb{R}, \mathbb{R}^m) : \omega(t) \in \Omega \text{ a.e.}\}$. Then the solutions of the differential equations*

$$\dot{x}(t) = Ax(t) + B\omega(t), \quad \omega \in \mathscr{U},$$

define a smooth system $\Sigma = (\mathbb{R}, \mathbb{R}^d, \mathbb{R}^m, \mathscr{U}, \varphi)$ (given by differential equations), which is also a linear system. The transition map has the form

$$\varphi(t, x, \omega) = e^{At} x + \int_0^t e^{A(t-s)} B\omega(s) ds$$

for all $(t, x, \omega) \in \mathbb{R} \times \mathbb{R}^d \times \mathscr{U}$.

Finally, we present an approximation result which follows, for instance, from Grasse and Sussmann [52, Theorem 2.24].

Proposition 1.6. *Let* $\Sigma = (\mathbb{R}, M, \mathbb{R}^m, \mathcal{U}, \varphi)$ *be a smooth system given by differential equations and assume that* ϱ *is a metric on* M. *Let* $(x_0, \omega_0) \in M \times \mathcal{U}$ *and* $\tau > 0$. *Then for every* $\varepsilon > 0$ *there is* $\delta > 0$ *and a piecewise constant control function* $\omega \in \mathcal{U}$ *such that* $\varrho(x, x_0) < \delta$ *implies*

$$\varrho(\varphi(t, x, \omega), \varphi(t, x_0, \omega_0)) < \varepsilon \quad \text{for all } t \in [0, \tau].$$

1.3 Some Control-Theoretic Notions

This section serves for the introduction of several useful notions for the qualitative analysis of systems.

Given a topological time-invariant system $\Sigma = (\mathbb{T}, X, U, \mathcal{U}, \varphi)$ and a state $x \in X$, we introduce the sets

$$\mathcal{O}_{\leq \tau}^+(x) := \{ y \in X \mid \exists \omega \in \mathcal{U}, \, t \in [0, \tau] : \, y = \varphi(t, x, \omega) \}, \quad \tau > 0,$$

$$\mathcal{O}^+(x) := \bigcup_{\tau \in \mathbb{T} \cap (0, \infty)} \mathcal{O}_{\leq \tau}^+(x).$$

The set $\mathcal{O}_{\leq \tau}^+(x)$ is called the *set of points reachable from* x *up to time* τ; the set $\mathcal{O}^+(x)$ is called the *positive orbit of* x. Moreover, we define

$$\mathcal{O}_{\leq \tau}^-(x) := \{ y \in X \mid \exists \omega \in \mathcal{U}, \, t \in [0, \tau] : \, x = \varphi(t, y, \omega) \}, \quad \tau > 0,$$

$$\mathcal{O}^-(x) := \bigcup_{\tau \in \mathbb{T} \cap (0, \infty)} \mathcal{O}_{\leq \tau}^-(x).$$

The set $\mathcal{O}_{\leq \tau}^-(x)$ is called the *set of points controllable to* x *within time* τ and $\mathcal{O}^-(x)$ is called the *negative orbit of* x. Moreover, for every $\tau > 0$ we define the *set of points reachable from* x *at time* τ by

$$\mathcal{O}_{\tau}(x) := \{ y \in X \mid \exists \omega \in \mathcal{U} : \, y = \varphi(\tau, x, \omega) \}.$$

For questions of controllability of a system, the notion of *local accessibility* is essential.

Definition 1.7. The system Σ is called *locally accessible from* $x \in X$ if the sets int $\mathcal{O}_{\leq \tau}^+(x)$ and int $\mathcal{O}_{\leq \tau}^-(x)$ are nonempty for every $\tau > 0$. It is called *locally accessible* if it is locally accessible from every point $x \in X$.

For smooth systems given by differential equations whose right-hand side vector fields are of class $\mathscr{C}^\infty$, the well-known theorem of Krener gives a Lie-algebraic criterion for local accessibility (see Colonius and Kliemann [25, Theorem A.4.4] for this version).

Proposition 1.7. *Consider a smooth system given by differential equations with right-hand side F and control range Ω, and assume that F_u is a $\mathscr{C}^\infty$-vector field for every $u \in \mathbb{R}^m$. Define*

$$\mathscr{F} := \{F_u \mid u \in \Omega\} \subset \mathscr{X}^\infty(M).$$

Let $\mathscr{L}(\mathscr{F}) \subset \mathscr{X}^\infty(M)$ be the smallest Lie algebra containing the set $\mathscr{F}$ and $\Delta_{\mathscr{L}(\mathscr{F})}(x) := \{f(x) : f \in \mathscr{L}(\mathscr{F})\}$ for all $x \in M$. Then, if $\Delta_{\mathscr{L}(\mathscr{F})}(x) = T_x M$ for all $x \in M$, the system is locally accessible.

Remark 1.5. If the system is real-analytic, that is, if M is a real-analytic manifold and the vector fields F_u are real-analytic, Krener's criterion for local accessibility is not only sufficient but also necessary. This criterion is also known as the *Lie algebra rank condition.*

In the theory developed in this book, subsets of the state space which can be made invariant by choosing appropriate control functions are of particular interest. If such a set contains only one element, this element is called an equilibrium.

Definition 1.8. Let $\Sigma = (\mathbb{T}, X, U, \mathscr{U}, \varphi)$ be a time-invariant system. A pair $(u, x) \in U \times X$ is called an *equilibrium pair* if the constant control function $\omega_u(t) :\equiv u$ is admissible and $\varphi(t, x, \omega_u) = x$ for all $t \in \mathbb{T}_+$.

For a smooth system given by differential equations, it is obvious that (u, x) is an equilibrium pair if and only if $u \in \Omega$ and $F(x, u) = 0$.

Definition 1.9. Consider a time-invariant system $\Sigma = (\mathbb{T}, X, U, \mathscr{U}, \varphi)$. A set $Y \subset X$ is called *controlled invariant* if for every $x \in Y$ there exists $\omega \in \mathscr{U}$ such that $\varphi(\mathbb{T}_+, x, \omega) \subset Y$.

An important class of nonlinear smooth systems given by differential equations is the class of control-affine systems because of the special features of these systems described below.

Definition 1.10. A smooth system $\Sigma = (\mathbb{R}, M, \mathbb{R}^m, \mathscr{U}, \varphi)$ given by differential equations is called *control-affine* if the right-hand side F has the form

$$F(x, u) = f_0(x) + \sum_{i=1}^{m} u_i f_i(x)$$

with vector fields $f_0, f_1, \ldots, f_m \in \mathscr{X}^1(M)$, where $u_1, \ldots, u_m$ are the components of u in the standard basis of $\mathbb{R}^m$, and if the control range Ω is a compact and convex set. The vector field f_0 is called the *drift vector field* and $f_1, \ldots, f_m$ the *control vector fields* of the system, respectively.

Remark 1.6. Note that the right-hand side of a control-affine system automatically satisfies $F \in \mathscr{C}^k(M \times \mathbb{R}^m, TM)$ if $f_0, \ldots, f_m \in \mathscr{X}^k(M)$, $k \in \mathbb{N} \cup \{\infty\} \cup \{\omega\}$.

Proposition 1.8. *For a control-affine system with vector fields* $f_0, f_1, \ldots, f_m$ *of class* $\mathscr{C}^\infty$ *and a control range* Ω *with* $0 \in \text{int}\,\Omega$ *the Lie algebra rank condition at a point* $x \in M$ *is satisfied if and only if the Lie algebra* $\mathscr{L} = \mathscr{L}(f_0, f_1, \ldots, f_m)$ *generated by* $f_0, f_1, \ldots, f_m$ *has full rank at* x, *that is,* $\Delta_{\mathscr{L}}(x) = T_x M$.

Proof. The set $\mathscr{F} = \{f_0 + \sum_{i=1}^m u_i f_i : u \in \Omega\}$ contains f_0, since $0 \in \Omega$. The smallest Lie algebra $\mathscr{L}(\mathscr{F})$ containing $\mathscr{F}$ is a vector space and hence contains also the differences of elements in $\mathscr{F}$. Since Ω has nonempty interior, it thus contains vector fields of the form $c f_i$ for small $c \neq 0$ and $i = 1, \ldots, m$. Therefore, $\mathscr{L}(\mathscr{F})$ contains $f_0, f_1, \ldots, f_m$, which implies $\mathscr{L}(f_0, f_1, \ldots, f_m) \subset \mathscr{L}(\mathscr{F})$. On the other hand, the Lie algebra $\mathscr{L}(f_0, f_1, \ldots, f_m)$ contains all the linear combinations of the vector fields $f_0, f_1, \ldots, f_m$, so $\mathscr{F} \subset \mathscr{L}(f_0, f_1, \ldots, f_m)$. Since $\mathscr{L}(\mathscr{F})$ is the smallest Lie algebra containing $\mathscr{F}$, it follows that $\mathscr{L}(f_0, f_1, \ldots, f_m) = \mathscr{L}(\mathscr{F})$, which proves the assertion. $\square$

For control-affine systems, the set $\mathscr{U}$ of admissible control functions becomes a compact metrizable space with the weak*-topology of $L^\infty(\mathbb{R}, \mathbb{R}^m) = L^1(\mathbb{R}, \mathbb{R}^m)^*$. Moreover, the associated *control flow*

$$\Phi : \mathbb{R} \times (\mathscr{U} \times M) \to \mathscr{U} \times M, \quad \Phi(t, (\omega, x)) = (\Theta_t \omega, \varphi(t, x, \omega)),$$

is continuous and we have the following result. (Both is proved in Sect. 1.4.)

Proposition 1.9. *The shift flow* $\Theta : \mathbb{R} \times \mathscr{U} \to \mathscr{U}$ *associated with a control-affine system is chain transitive.*

In particular, the chain transitivity of Θ is used when we apply Selgrade's theorem to the control flow of a bilinear system in Sects. 3.2, 7.3, and 7.4.

Proposition 1.10. *Let* $\Sigma = (\mathbb{R}, M, \mathbb{R}^m, \mathscr{U}, \varphi)$ *be a control-affine system and* $Q \subset M$ *a compact controlled invariant set. Then the forward lift of* Q *to* $\mathscr{U} \times M$, *which is defined by*

$$\mathscr{Q} := \{(\omega, x) \in \mathscr{U} \times M : \varphi(\mathbb{R}_+, x, \omega) \subset Q\},$$

is a compact forward-invariant set for the control flow, that is, $\Phi_t(\mathscr{Q}) \subset \mathscr{Q}$ *for all* $t \geq 0$.

Proof. Take a sequence (ω_n, x_n) in $\mathscr{Q}$ with $(\omega_n, x_n) \to (\omega, x) \in \mathscr{U} \times Q$ and let $t \geq 0$. Then $\varphi(t, x_n, \omega_n) \in Q$ for all n, which implies $\varphi(t, x, \omega) \in Q$ by compactness of Q and continuity of φ. This shows that $\mathscr{Q}$ is closed in $\mathscr{U} \times Q$ and hence compact. Forward-invariance under Φ is a trivial consequence of the cocycle property of the transition map φ. $\square$

With a similar argument, one proves the following proposition.

Proposition 1.11. *If* A *is a controlled invariant set of a control-affine system, then also* cl A *is controlled invariant.*

Finally, we introduce some notions associated with linear systems given by differential equations

$$\dot{x}(t) = Ax(t) + B\omega(t), \quad \omega \in \mathscr{U},$$

where $A \in \mathbb{R}^{d \times d}$ and $B \in \mathbb{R}^{d \times m}$. It is well-known that in the case $\mathscr{U} = L^{\infty}(\mathbb{R}, \mathbb{R}^m)$ such a system is *completely controllable*, that is, for any two states $x, y \in \mathbb{R}^d$ there exist $\omega \in \mathscr{U}$ and $\tau \geq 0$ with $\varphi(\tau, x, \omega) = y$, if and only if the pair of matrices (A, B) is controllable in the sense of the following definition.

Definition 1.11. A pair (A, B) of matrices $A \in \mathbb{R}^{d \times d}$ and $B \in \mathbb{R}^{d \times m}$ is said to be *controllable* if the so-called *controllability matrix*

$$\left[B \,|\, AB \,|\, A^2 B \,|\, \cdots \,|\, A^{d-1} B \right] \in \mathbb{R}^{d \times md}$$

has full rank.

This criterion for controllability is known as the *Kalman rank condition*. If the control range Ω is not the whole of $\mathbb{R}^m$ but a compact and convex subset with $0 \in \operatorname{int} \Omega$, the Kalman rank condition guarantees the existence of a unique control set $D \subset \mathbb{R}^d$ with nonempty interior, which is given by $D = \mathscr{O}^-(0) \cap \operatorname{cl} \mathscr{O}^+(0)$. In this case, complete controllability holds on $\operatorname{int} D$. Control sets in general are maximal sets of controlled invariance and approximate controllability. They are introduced and discussed in the next section.

1.4   Control Flow and Control Sets

In this section, we introduce the control flow of a control-affine system and the important notions of control and chain control sets.

The Control Flow of a Control-Affine System

To every time-invariant system $\Sigma = (\mathbb{T}, X, U, \mathscr{U}, \varphi)$ we can associate a skew-product semiflow on $\mathscr{U} \times X$, called the *control semiflow* of Σ, given by

$$\Phi : \mathbb{T}_+ \times (\mathscr{U} \times X) \to \mathscr{U} \times X, \quad \Phi_t(\omega, x) := (\Theta_t \omega, \varphi(t, x, \omega)),$$

where $\Theta : \mathbb{T} \times \mathscr{U} \to \mathscr{U}$ denotes the shift flow introduced in Sect. 1.1. We leave it to the reader to verify the semigroup properties $\Phi_0 = \operatorname{id}$ and $\Phi_t \circ \Phi_s = \Phi_{t+s}$. For smooth systems given by differential equations, where φ is defined on $\mathbb{R} \times M \times \mathscr{U}$, the control semiflow can be extended to a flow, then called the *control flow*. With respect to the L^{∞}-topology on $\mathscr{U}$, this flow is not continuous (in fact, not even measurable), since not every trajectory of the shift flow is continuous. However, for the class of control-affine systems, one can choose a topology on $\mathcal{U}$ which implies not only continuity of Φ but also compactness of $\mathcal{U}$. In the following, we will describe this construction.

Consider a control-affine system $\Sigma = (\mathbb{R}, M, \mathbb{R}^m, \mathcal{U}, \varphi)$ given by differential equations

$$\dot{x}(t) = f_0(x(t)) + \sum_{i=1}^{m} \omega_i(t) f_i(x(t)), \quad \omega \in \mathcal{U},$$

with vector fields $f_0, f_1, \ldots, f_m \in \mathscr{X}^1(M)$ on a connected $\mathscr{C}^3$-manifold M. Recall that the control range Ω of such a system is assumed to be compact and convex. The aim of this subsection is to prove that the associated control flow $\Phi : \mathbb{R} \times (\mathcal{U} \times M) \to \mathcal{U} \times M$ is continuous when the set $\mathcal{U}$ of admissible control functions is endowed with the weak*-topology of $L^\infty(\mathbb{R}, \mathbb{R}^m) = L^1(\mathbb{R}, \mathbb{R}^m)^*$ which turns it into a compact metrizable space. Before we start proving this, we first recall some facts from functional analysis.

Let X be a normed vector space over $\mathbb{R}$ and X^* its dual space, that is, the vector space of all continuous linear functionals on X. With the norm $|x^*| := \sup_{|x|=1} |x^*x|$, X^* becomes a Banach space. The *weak*-topology* on X^* is defined as the smallest topology such that for each $x \in X$ the linear functional $x^* \mapsto x^*x$ on X^* is continuous with respect to that topology. The following proposition summarizes some properties of the weak*-topology.

Proposition 1.12. *Let X be a normed vector space.*

(i) *(Banach–Alaoglu): The unit ball $B_{X^*} := \{x^* \in X^* : |x^*| \leq 1\}$ is compact with respect to the weak*-topology. Consequently, subsets of X^* that are bounded and weakly* closed are weakly* compact (cf. Megginson [80, Theorem 2.6.18 and Corollary 2.6.19]).*

(ii) *The relative weak*-topology on B_{X^*} is induced by a metric if and only if X is separable (cf. Megginson [80, Theorem 2.6.23]).*

(iii) *The family of all sets of the form $\{y^* \in X^* : |(x^* - y^*)x| < 1\}$ with $x^* \in X^*$ and $x \in X$ is a subbasis of the weak*-topology. If x_0^* is an element of a weakly* open set U, then there are finitely many elements $x_1, \ldots, x_n \in X$ such that $\bigcap_{i=1}^{n}\{y^* \in X^* : |(x_0^* - y^*)x_i| < 1\} \subset U$ (cf. Megginson [80, Proposition 2.4.12]).*

(iv) *If X is a separable Banach space, then a convex subset of X^* is weakly* closed if and only if it is weakly* sequentially closed (cf. Megginson [80, Corollary 2.7.13]).*

We further need to know one fact about the weak topology. If X is a normed vector space, the *weak topology* of X is the smallest topology for the space such that every element of the dual space X^* is continuous with respect to that topology. The following proposition can be found in Megginson [80, Theorem 2.5.16].

Proposition 1.13. *The closure and the weak closure of a convex subset of a normed space are the same. In particular, a convex subset of a normed space is closed if and only if it is weakly closed.*

Finally, we use the fact that $L^1(\mathbb{R}, \mathbb{R}^m)$ is an infinite-dimensional separable Banach space whose dual space $L^1(\mathbb{R}, \mathbb{R}^m)^*$ is isometrically isomorphic to $L^\infty(\mathbb{R}, \mathbb{R}^m)$ via

$$\omega \mapsto \left(T_\omega : \mu \mapsto \int_{\mathbb{R}} \langle \mu(t), \omega(t) \rangle \, dt \right), \quad L^\infty(\mathbb{R}, \mathbb{R}^m) \to L^1(\mathbb{R}, \mathbb{R}^m)^*.$$

On $L^1(\mathbb{R}, \mathbb{R}^m)^*$ we consider the weak*-topology which is the smallest topology such that for every $\mu \in L^1(\mathbb{R}, \mathbb{R}^m)$ the linear functional $S_\mu : f \mapsto f\mu$, $S_\mu : L^1(\mathbb{R}, \mathbb{R}^m)^* \to \mathbb{R}$, is continuous. With the identification of $L^\infty(\mathbb{R}, \mathbb{R}^m)$ and $L^1(\mathbb{R}, \mathbb{R}^m)^*$, this means that for each $\mu \in L^1(\mathbb{R}, \mathbb{R}^m)$ the functional

$$L^\infty(\mathbb{R}, \mathbb{R}^m) \to \mathbb{R}, \quad \omega \mapsto \int_{\mathbb{R}} \langle \mu(t), \omega(t) \rangle \, dt,$$

is continuous. For convergence of sequences with respect to the weak*-topology we use the common notation $\omega_n \overset{*}{\rightharpoonup} \omega$.

Proposition 1.14. *The set $\mathscr{U}$ endowed with the relative weak*-topology is a compact path-connected metrizable space. A metric on $\mathscr{U}$ compatible with the weak*-topology is given by*

$$\varrho(\omega_1, \omega_2) = \sum_{k=1}^{\infty} \frac{1}{2^k} \frac{\left| \int_{\mathbb{R}} \langle \omega_1(t) - \omega_2(t), x_k(t) \rangle \, dt \right|}{1 + \left| \int_{\mathbb{R}} \langle \omega_1(t) - \omega_2(t), x_k(t) \rangle \, dt \right|}, \tag{1.2}$$

where $\{x_k\}$ is an arbitrary countable dense subset of $L^1(\mathbb{R}, \mathbb{R}^m)$.

Proof. The dual space of $L^1(\mathbb{R}, \mathbb{R}^m)$ is $L^\infty(\mathbb{R}, \mathbb{R}^m)$ and $L^1(\mathbb{R}, \mathbb{R}^m)$ is separable. Hence, to show that $\mathscr{U}$ is compact and metrizable, by Proposition 1.12 (i), it suffices to prove that $\mathscr{U}$ is bounded and weakly* closed. Boundedness is obvious. Using Proposition 1.12 (iv), it remains to show that every sequence ω_n in $\mathscr{U}$ with weak*-limit $\omega_\infty \in L^\infty(\mathbb{R}, \mathbb{R}^m)$ satisfies $\omega_\infty \in \mathscr{U}$. For every compact interval $I \subset \mathbb{R}$, the sequence $\omega_n|_I$ of restrictions is a sequence in the Hilbert space $L^2(I, \mathbb{R}^m)$, which is weakly convergent, that is, $\int_I \langle \omega_n(t), \alpha(t) \rangle \, dt$ converges for every $\alpha \in L^2(I, \mathbb{R}^m)$. This follows from

$$\int_I \langle \omega_n(t), \alpha(t) \rangle \, dt = \int_{\mathbb{R}} \langle \omega_n(t), \mathbb{1}_I(t) \alpha(t) \rangle \, dt \to \int_{\mathbb{R}} \langle \omega_\infty(t), \mathbb{1}_I(t) \alpha(t) \rangle \, dt,$$

since $\mathbb{1}_I \cdot \alpha : \mathbb{R} \to \mathbb{R}^m$ is an L^1-function, which is a consequence of Hölder's inequality (cf. Dunford and Schwartz [40, Lemma III.3.2])

$$\int_{\mathbb{R}} |\mathbb{1}_I(t)\alpha(t)|\,\mathrm{d}t = \int_{\mathbb{R}} \mathbb{1}_I(t)|\alpha(t)|\mathrm{d}t \le \|\mathbb{1}_I\|_{L^2} \cdot \|\alpha\|_{L^2}$$

$$= \lambda(I)^{1/2}\|\alpha\|_{L^2} < \infty.$$

We claim that the set $\mathscr{U}(I) = \{\omega|_I : \omega \in \mathscr{U}\}$ is weakly closed in $L^2(I, \mathbb{R}^m)$ implying that $\omega_\infty(t) \in \Omega$ for almost all $t \in I$, and consequently $\omega_\infty \in \mathscr{U}$. Indeed, $\mathscr{U}(I)$ is a convex and norm-closed set. This follows from convexity of Ω and the well-known fact that an L^2-convergent sequence has a subsequence which converges pointwise almost everywhere. Hence, Proposition 1.13 implies that $\mathscr{U}(I)$ is weakly closed. To prove that $\mathscr{U}$ is path-connected, it suffices to show that for $\omega_1, \omega_2 \in \mathscr{U}$ the straight line $c(s) := (1-s)\omega_1 + s\omega_2$, $c : [0, 1] \to \mathscr{U}$, which is well-defined because Ω is convex, is a continuous path. To this end, let $s_n \to s \in [0, 1]$ be a sequence and let $x \in L^1(\mathbb{R}, \mathbb{R}^m)$. Then one easily sees that

$$\left| \int_{\mathbb{R}} \langle c(s_n) - c(s), x\rangle \mathrm{d}t \right| = |s - s_n| \left| \int_{\mathbb{R}} \langle \omega_1 - \omega_2, x\rangle \mathrm{d}t \right| \to 0.$$

The proof that a metric on $\mathscr{U}$ is given by (1.2) can be found, for instance, in Dunford and Schwartz [40, Theorem V.5.1]. $\qquad\square$

Remark 1.7. Note that $\mathscr{U}$ has infinite topological dimension if Ω consists of more than one point.

Proposition 1.15. *The shift flow* $\Theta : \mathbb{R} \times \mathscr{U} \to \mathscr{U}$ *is continuous.*

Proof. Consider sequences $t_n \to t$ in $\mathbb{R}$ and $\omega_n \overset{\star}{\rightharpoonup} \omega$ in $\mathscr{U}$. Then for all $x \in L^1(\mathbb{R}, \mathbb{R}^m)$ we have

$$\left| \int_{\mathbb{R}} \langle \omega_n(t_n + \tau), x(\tau)\rangle \,\mathrm{d}\tau - \int_{\mathbb{R}} \langle \omega(t + \tau), x(\tau)\rangle \,\mathrm{d}\tau \right|$$

$$\le \left| \int_{\mathbb{R}} \langle \omega_n(t_n + \tau) - \omega_n(t + \tau), x(\tau)\rangle \,\mathrm{d}\tau \right|$$

$$+ \left| \int_{\mathbb{R}} \langle \omega_n(t + \tau) - \omega(t + \tau), x(\tau)\rangle \,\mathrm{d}\tau \right|$$

$$= \left| \int_{\mathbb{R}} \langle \omega_n(\tau), x(\tau - t_n)\rangle \,\mathrm{d}\tau - \int_{\mathbb{R}} \langle \omega_n(\tau), x(\tau - t)\rangle \,\mathrm{d}\tau \right|$$

$$+ \left| \int_{\mathbb{R}} \langle \omega_n(\tau) - \omega(\tau), x(\tau - t)\rangle \,\mathrm{d}\tau \right|.$$

The second summand in the last term converges to zero because $\omega_n \overset{\star}{\rightharpoonup} \omega$; the first one can be estimated by

$$\leq \max_{u \in \Omega} |u| \int_{\mathbb{R}} |x(\tau - t_n) - x(\tau - t)| \, d\tau.$$

It is a well-known fact that the integral converges to zero as $t_n \to t$ (cf. Dunford and Schwartz [40, Chap. IV, Theorem 8.20]). $\square$

At this point, we are able to give the proof of chain transitivity of Θ.

Proof (of Proposition 1.9). It suffices to show that the periodic functions are dense in $\mathcal{U}$, since these are the periodic points of Θ and every periodic point is chain recurrent. Then, since the chain recurrent set is closed, it is equal to $\mathcal{U}$. Since $\mathcal{U}$ is connected, the result follows from Proposition B.1. Now take $\omega_0 \in \mathcal{U}$ and let W be a neighborhood of ω_0. Then, by Proposition 1.12 (iii), there are $x_1, \ldots, x_k \in L^1(\mathbb{R}, \mathbb{R}^m)$ such that

$$\left\{ \omega \in L^\infty(\mathbb{R}, \mathbb{R}^m) \; : \; \begin{array}{c} |\int_{\mathbb{R}} \langle \omega_0(t) - \omega(t), x_j(t) \rangle dt| < 1 \text{ for } j = 1, \ldots, k \\ \text{and } \omega(t) \in \Omega \text{ a.e.} \end{array} \right\} \subset W.$$

Since $x_j \in L^1(\mathbb{R}, \mathbb{R}^m)$, there is $T > 0$ such that for $j = 1, \ldots, k$

$$\int_{\mathbb{R} \setminus [-T, T]} |x_j(t)| dt < \frac{1}{\operatorname{diam} \Omega}.$$

(We may assume that Ω consists of more than one point, so $\operatorname{diam} \Omega > 0$.) Define a periodic control function by setting $\omega_p(t) := \omega_0(t)$ for all $t \in [-T, T]$ and extending ω_p $2T$-periodically to $\mathbb{R}$. Then $\omega_p \in \mathcal{U}$ and

$$\left| \int_{\mathbb{R}} \langle \omega_0(t) - \omega_p(t), x_j(t) \rangle dt \right| = \left| \int_{\mathbb{R} \setminus [-T, T]} \langle \omega_0(t) - \omega_p(t), x_j(t) \rangle dt \right|$$

$$\leq \operatorname{diam} \Omega \int_{\mathbb{R} \setminus [-T, T]} |x_j(t)| dt < 1.$$

Hence, $\omega_p \in W$ as desired. $\square$

Before we return to the continuity proof of Φ, we provide some additional information about the dynamical properties of the shift flow. Since we do not use these properties, we omit their proofs.

Proposition 1.16. *The shift flow* $\Theta : \mathbb{R} \times \mathcal{U} \to \mathcal{U}$ *has the following properties:*

(i) Θ *is topologically mixing, that is, for any two nonempty open sets* $A, B \subset \mathcal{U}$ *there exists* $\tau > 0$ *with* $\Theta_{-\tau}(A) \cap B \neq \emptyset$;

(ii) Θ *is topologically transitive, that is, there is a point in* $\mathcal{U}$ *whose* ω-*limit set is equal to* $\mathcal{U}$;

(iii) If Ω contains more than one point, then Θ has sensitive dependence on initial conditions, that is, there is $\delta > 0$ such that for every $\omega \in \mathcal{U}$ and every neighborhood N of ω there are $\mu \in N$ and $\tau > 0$ with $\varrho(\Theta_\tau \omega, \Theta_\tau \mu) > \delta$, where ϱ denotes a metric on $\mathcal{U}$;

(iv) If Ω contains more than one point, then Θ has infinite topological entropy.

It remains to show continuity of the cocycle φ. However, we will not only show continuity of φ but also of its derivatives of arbitrarily high order with respect to the state variable. To this end, we first prove a more general result. Consider a family of parameter-dependent differential equations of the form

$$\dot{x}(t) = g_0(t, x(t), p) + \sum_{i=1}^{m} \omega_i(t) g_i(t, x(t), p), \quad \omega \in \mathcal{U}, \ p \in P, \qquad (1.3)$$

where $\mathcal{U}$ is the set of admissible control functions of the control-affine system, P is a locally compact metric space, and $g_i : \mathbb{R} \times V \times P \to V, i = 0, 1, \dots, m$, are continuous mappings, where $(V, |\cdot|)$ is a normed vector space of finite dimension over $\mathbb{R}$. Furthermore, we assume that $g_i(t, \cdot, p) : V \to V$ is continuously differentiable for each $(t, p) \in \mathbb{R} \times P$ and $i = 0, 1, \dots, m$ with a derivative depending continuously on (t, x). Then it follows from the flow box theorem A.1 that for every $(x, \omega, p) \in V \times \mathcal{U} \times P$ there exists a unique solution $\phi(t; x, \omega, p)$ with $\phi(0; x, \omega, p) = x$, and we assume that this solution is defined on $\mathbb{R}$.

Proposition 1.17. *The mapping*

$$\phi : \mathbb{R} \times V \times \mathcal{U} \times P \to V, \quad (t, x, \omega, p) \mapsto \phi(t; x, \omega, p),$$

is continuous, where on $\mathcal{U}$ we consider the weak*-topology.

Proof. The proof is subdivided into three steps.

Step 1. Fix $\tau > 0$ and consider sequences $p^n \to p^*$ in P, $\omega^n \overset{*}{\to} \omega^*$ in $\mathcal{U}$, and $x^n \to x^*$ in V. Consider the corresponding solutions $\xi^n(t) := \phi(t; x^n, \omega^n, p^n)$ on $[0, \tau]$. In the first two steps of the proof, let us assume that there exists a compact set $K \subset V$ with

$$\xi^n(t) \in K \quad \text{for all } n \in \mathbb{N} \text{ and } t \in [0, \tau].$$

On $\mathscr{C}^0([0, \tau], V)$ we consider the compact-open topology, that is, the topology of uniform convergence. We show that the set $\{\xi^n\}_{n \in \mathbb{N}}$ is relatively compact in $\mathscr{C}^0([0, \tau], V)$ by applying the theorem of Arzelà–Ascoli.[3] To this end, let $t_1, t_2 \in [0, \tau]$ with $t_1 < t_2$ (without loss of generality). Then

---

[3]Recall: A subset $\mathscr{F}$ of the function space $\mathscr{C}^0(X, Y)$ of continuous maps from a compact topological space X into a metric space Y is relatively compact if and only if $\mathscr{F}$ is equicontinuous and each of the sets $\{f(x)\}_{f \in \mathscr{F}}, x \in X$, has a compact closure.

$$|\xi^n(t_2) - \xi^n(t_1)| = \left| \int_{t_1}^{t_2} \left[g_0\left(s, \xi^n(s), p^n\right) + \sum_{i=1}^{m} \omega_i^n(s) g_i\left(s, \xi^n(s), p^n\right) \right] ds \right|$$

$$\leq \int_{t_1}^{t_2} \left(|g_0\left(s, \xi^n(s), p^n\right)| + \sum_{i=1}^{m} |\omega_i^n(s)| \, |g_i\left(s, \xi^n(s), p^n\right)| \right) ds.$$

Using the assumption that P is locally compact, we find a compact neighborhood W of p^* which contains all p^n. Since the set $X := [0, \tau] \times W \times K \times \Omega$ is compact, we find

$$|\xi^n(t_2) - \xi^n(t_1)| \leq (t_2 - t_1) \max_{(t,p,x,u) \in X} \left[|g_0(t, x, p)| + \sum_{i=1}^{m} |u_i| |g_i(t, x, p)| \right].$$

This proves that the set $\{\xi^n\}_{n \in \mathbb{N}}$ is equicontinuous. From the assumption that $\xi^n(t) \in K$ it follows that for each $t \in [0, \tau]$ the set $\{\xi^n(t)\}_{n \in \mathbb{N}}$ is relatively compact. Hence, the Arzelà–Ascoli theorem can be applied and there exists a convergent subsequence $\xi^{k_n} \to \psi \in \mathscr{C}^0([0, \tau], V)$.

Step 2. We claim that $\psi(t) = \phi(t; x^*, \omega^*, p^*)$ for all $t \in [0, \tau]$. To show this, fix $t \in [0, \tau]$ and note that

$$\xi^{k_n}(t) = x^{k_n} + \int_0^t \left[g_0\left(s, \xi^{k_n}(s), p^{k_n}\right) + \sum_{i=1}^{m} \omega_i^{k_n}(s) g_i\left(s, \xi^{k_n}(s), p^{k_n}\right) \right] ds$$

$$= x^{k_n} + \int_0^t g_0\left(s, \xi^{k_n}(s), p^{k_n}\right) ds + \int_0^t \sum_{i=1}^{m} \omega_i^{k_n}(s) g_i\left(s, \psi(s), p^*\right) ds$$

$$+ \int_0^t \sum_{i=1}^{m} \omega_i^{k_n}(s) \left[g_i\left(s, \xi^{k_n}(s), p^{k_n}\right) - g_i\left(s, \psi(s), p^*\right) \right] ds.$$

Since g_0 is uniformly continuous on the compact set $[0, t] \times K \times W$ and $p^{k_n} \to p^*, \xi^{k_n} \to \psi$, it follows that

$$\int_0^t g_0\left(s, \xi^{k_n}(s), p^{k_n}\right) ds \to \int_0^t g_0\left(s, \psi(s), p^*\right) ds.$$

The last term can be estimated by

$$\left| \int_0^t \sum_{i=1}^{m} \omega_i^{k_n}(s) \left[g_i\left(s, \xi^{k_n}(s), p^{k_n}\right) - g_i\left(s, \psi(s), p^*\right) \right] ds \right|$$

$$\leq \sum_{i=1}^{m} \max_{u \in \Omega} |u_i| \int_0^t \left| g_i\left(s, \xi^{k_n}(s), p^{k_n}\right) - g_i\left(s, \psi(s), p^*\right) \right| ds.$$

With the same argument as above we find that the integrals converge to zero. Finally, we can write

$$\int_0^t \sum_{i=1}^m \omega_i^{k_n}(s) g_i \left(s, \psi(s), p^* \right) ds = \int_0^\tau \sum_{i=1}^m \omega_i^{k_n}(s) \mathbb{1}_{[0,t]}(s) g_i \left(s, \psi(s), p^* \right) ds.$$

Now weak*-convergence $\omega^{k_n} \overset{*}{\rightharpoonup} \omega^*$ implies that this term converges to

$$\int_0^\tau \sum_{i=1}^m \omega_i^*(s) \mathbb{1}_{[0,t]}(s) g_i \left(s, \psi(s), p^* \right) ds = \int_0^t \sum_{i=1}^m \omega_i^*(s) g_i \left(s, \psi(s), p^* \right) ds.$$

Hence, taking all the limits, we obtain

$$\psi(t) = x^* + \int_0^t \left[g_0 \left(s, \psi(s), p^* \right) + \sum_{i=1}^m \omega_i^*(s) g_i \left(s, \psi(s), p^* \right) \right] ds.$$

Since this holds for every $t \in [0, \tau]$, the assertion that $\psi(t) \equiv \phi(t; x^*, \omega^*, p^*)$ is proved. We claim that not only the subsequence ξ^{k_n} but also the sequence ξ^n converges to ψ in $\mathscr{C}^0([0, \tau], V)$. Assume to the contrary that this is wrong. Then there are $\delta > 0$ and a subsequence ξ^{l_n} with $\max_{t \in [0,\tau]} |\xi^{l_n}(t) - \psi(t)| \geq \delta$ for all $n \in \mathbb{N}$. The sequence ξ^{l_n} has a convergent subsequence, converging to some $\tilde{\psi} \in \mathscr{C}^0([0, \tau], V)$ (using the arguments of Step 1). With the same arguments as before we see that $\tilde{\psi} = \phi(\cdot; x^*, \omega^*, p^*)$, a contradiction. This concludes the second step.

Step 3. Now we are in position to give the actual continuity proof. Consider sequences $t^n \to t^*$, $x^n \to x^*$, $p^n \to p^*$, and $\omega^n \overset{*}{\rightharpoonup} \omega^*$. We show that $\phi(t^n; x^n, \omega^n, p^n) \to \phi(t^*; x^*, \omega^*, p^*)$, which proves continuity of ϕ at $(t^*, x^*, p^*, \omega^*)$. Without loss of generality, we may assume that $t^* > 0$. Furthermore, we may assume that there is $\alpha > 0$ with $t^n \in [0, t^* + \alpha]$ for all $n \in \mathbb{N}$. Let K be a compact neighborhood of the compact set $\phi([0, t^* + \alpha]; x^*, \omega^*, p^*)$. Then take a cut-off function $\theta : V \to [0, 1]$ of class $\mathscr{C}^1$ with $\theta(x) \equiv 1$ on K and $\theta(x) \equiv 0$ on the complement of another compact set $\tilde{K} \supset K$. Consider the family of differential equations given by

$$\dot{x}(t) = \theta(x(t)) g_0(t, x(t), p) + \sum_{i=1}^m \omega_i(t) \theta(x(t)) g_i(t, x(t), p), \quad \omega \in \mathscr{U}, \ p \in P.$$

The solutions $\tilde{\phi}(\cdot; x, \omega, p)$ of these equations are unique and defined on $\mathbb{R}$. It can easily be seen that if $\tilde{\phi}([0, \tau]; x, \omega, p) \subset K$ for some $\tau > 0$, then $\tilde{\phi}(t; x, \omega, p) = \phi(t; x, \omega, p)$ for all $t \in [0, \tau]$. In particular, $\tilde{\phi}(t; x^*, \omega^*, p^*) = \phi(t; x^*, \omega^*, p^*)$ for all $t \in [0, t^* + \alpha]$. Moreover, all such solutions stay in the compact set $\tilde{K}$ for all $t \in \mathbb{R}$. By what we have shown in Step 2, the sequence of maps $\tilde{\phi}(\cdot; x^n, \omega^n, p^n)$ converges to $\tilde{\phi}(\cdot; p^*, \omega^*, x^*)$ uniformly on $[0, t^* + \alpha]$. We have

$$\left|\tilde{\phi}(t^n; x^n, \omega^n, p^n) - \tilde{\phi}(t^*; x^*, \omega^*, p^*)\right|$$

$$\leq \left|\tilde{\phi}(t^n; x^n, \omega^n, p^n) - \tilde{\phi}(t^n; x^*, \omega^*, p^*)\right|$$

$$+ \left|\tilde{\phi}(t^n; x^*, \omega^*, p^*) - \tilde{\phi}(t^*; x^*, \omega^*, p^*)\right|$$

$$\leq \sup_{t \in [0, t^* + \alpha]} \left|\tilde{\phi}(t; x^n, \omega^n, p^n) - \tilde{\phi}(t; x^*, \omega^*, p^*)\right|$$

$$+ \left|\tilde{\phi}(t^n; x^*, \omega^*, p^*) - \tilde{\phi}(t^*; x^*, \omega^*, p^*)\right|,$$

and hence, we see that $\tilde{\phi}(t^n; x^n, \omega^n, p^n) \to \tilde{\phi}(t^*; x^*, \omega^*, p^*)$. Moveover, for sufficiently large n we have $\tilde{\phi}([0, t^* + \alpha]; x^n, \omega^n, p^n) \subset K$, and thus $\tilde{\phi}(t^n; x^n, \omega^n, p^n) = \phi(t^n; x^n, \omega^n, p^n)$. This completes the proof. □

Now let us apply Proposition 1.17 to the control-affine system Σ. We may assume that the state space M is the Euclidean space $\mathbb{R}^d$. Having shown continuity of φ in this case, the general case follows by the standard procedure of writing the system in local coordinates and patching together appropriate chains of charts along solutions, using the cocycle property. The family of differential equations which defines the control-affine system is obviously of the form (1.3). Here $g_i(t, x, p) = f_i(x)$, $V = \mathbb{R}^d$, and $P = \emptyset$. Hence, $\varphi : \mathbb{R} \times \mathbb{R}^d \times \mathscr{U} \to \mathbb{R}^d$ is continuous. Assuming higher regularity of the drift and control vector fields $f_0, f_1, \ldots, f_m$, we can show continuity of the derivatives $D^k \varphi_{t,\omega}$ inductively.

Before we prove our main result, let us recall some basics concerning higher derivatives of maps between multi-dimensional spaces (a good reference is Lang [73]). If $f : U \to \mathbb{R}^m$ is a continuously differentiable map defined on an open subset U of $\mathbb{R}^n$, the derivative $Df(x)$ at a point $x \in U$ is an element of $\mathscr{L}(\mathbb{R}^n, \mathbb{R}^m)$ and hence, the derivative of f can be regarded as a map $Df : U \to \mathscr{L}(\mathbb{R}^n, \mathbb{R}^m)$. Knowing that $\mathscr{L}(\mathbb{R}^n, \mathbb{R}^m) \cong \mathbb{R}^{m \times n} \cong \mathbb{R}^{mn}$, and assuming that f is twice differentiable, the second derivative analogously is a map $D^2 f : U \to \mathscr{L}(\mathbb{R}^n, \mathscr{L}(\mathbb{R}^n, \mathbb{R}^m))$. The latter space is canonically isomorphic to the space $\mathscr{L}^2(\mathbb{R}^n, \mathbb{R}^m)$ of bilinear mappings from $\mathbb{R}^n \times \mathbb{R}^n$ into $\mathbb{R}^m$ via $\alpha \mapsto ((v, w) \mapsto \alpha(v)w)$, $\mathscr{L}(\mathbb{R}^n, \mathscr{L}(\mathbb{R}^n, \mathbb{R}^m)) \to \mathscr{L}^2(\mathbb{R}^n, \mathbb{R}^m)$. Analogously, the k-th derivative of f (if it exists) can be regarded as a map $D^k f : U \to \mathscr{L}^k(\mathbb{R}^n, \mathbb{R}^m)$, where $\mathscr{L}^k(\mathbb{R}^n, \mathbb{R}^m)$ is the notation for the space of k-linear maps $\mathbb{R}^{nk} = \mathbb{R}^n \times \cdots \times \mathbb{R}^n \to \mathbb{R}^m$.

We know that the first derivative of the map φ with respect to the state variable satisfies the variational equation (cf. Proposition A.3). Assuming higher regularity of the right-hand side, also the higher derivatives of $\varphi_{t,\omega}$ exist and satisfy certain differential equations, described for instance in Siegmund [98, Satz 1.16] which essentially reads as follows.

Proposition 1.18. *Let $I \subset \mathbb{R}$ be a nonempty interval, $k \in \mathbb{N}$, and $f : I \times \mathbb{R}^d \to \mathbb{R}^d$ a map such that for all $(t, x) \in I \times \mathbb{R}^d$ the k-th derivative $D^k f_t$ exists (where $f_t = f(t, \cdot)$), and both f and $D^j f_t$ for $j = 1, \ldots, k$ are Carathéodory functions, that is, continuous in x and measurable in t. Moreover, assume that there are locally integrable function $l_j : I \to \mathbb{R}_+$, $j = 0, 1, \ldots, k$, such that for all $x \in \mathbb{R}^d$*

$$|f(t, 0)| \leq l_0(t),$$

$$\|\mathrm{D}^j f_t(x)\| \leq l_j(t), \quad j = 1, \ldots, k.$$

Then the Carathéodory differential equation

$$\dot{x}(t) = f(t, x(t))$$

has for each $(\tau_0, \xi_0) \in I \times \mathbb{R}^d$ a unique solution $\lambda(\cdot; \tau_0, \xi_0) : I \to \mathbb{R}^d$ satisfying the initial condition $x(\tau_0) = \xi_0$. The map $\lambda : I \times I \times \mathbb{R}^d \to \mathbb{R}^d$ is k times continuously differentiable with respect to $\xi \in \mathbb{R}^d$ and the following statements hold:

(a) *The derivative $\mathrm{D}_\xi \lambda(t; \tau_0, \xi_0)$, regarded as a function of t, is the unique solution of the initial value problem*

$$\dot{x}(t) - \mathrm{D}_x f(t, \lambda(t; \iota_0, \xi_0)) x(t), \quad x(\tau_0) = \mathrm{id} \in \mathscr{L}(\mathbb{R}^d, \mathbb{R}^d),$$

   where D_x here means the partial derivative with respect to the state variable.
(b) *For $k \geq 2$ and $j \in \{2, \ldots, k\}$, the derivative $\mathrm{D}_\xi^j \lambda(t; \tau_0, \xi_0)$, as a function of t, is the unique solution of the initial value problem*

$$\dot{x}(t) = \mathrm{D}_x f(t, \lambda(t; \tau_0, \xi_0)) x(t) + R_j(t, \tau_0, \xi_0),$$

$$x(\tau_0) = 0 \in \mathscr{L}^j(\mathbb{R}^d, \mathbb{R}^d)$$

   with

$$R_j(t, \tau_0, \xi_0) := \sum_{i=1}^{j-1} \binom{j-1}{i} \mathrm{D}_\xi^i [\mathrm{D}_x f(t, \lambda(t; \tau_0, \xi_0))] \cdot \mathrm{D}_\xi^{j-i} \lambda(t; \tau_0, \xi_0).$$

Theorem 1.1. *Assume that the vector fields $f_0, f_1, \ldots, f_m$ of the control-affine system Σ are of class $\mathscr{C}^k$, $k \geq 1$. Then the k-th derivative $\mathrm{D}^k \varphi_{t,\omega} : \mathbb{R}^d \to \mathscr{L}^k(\mathbb{R}^d, \mathbb{R}^d)$ exists and the map $(t, x, \omega) \mapsto \mathrm{D}^k \varphi_{t,\omega}(x)$, $\mathbb{R} \times \mathbb{R}^d \times \mathscr{U} \to \mathscr{L}^k(\mathbb{R}^d, \mathbb{R}^d)$, is continuous.*

Proof. First consider the case $k = 1$. Proposition 1.4 guarantees that $\varphi_{t,\omega}$ is of class $\mathscr{C}^1$. The derivative $\mathrm{D}\varphi_{t,\omega}(x)$ (regarded as a function of t) satisfies the variational equation

$$\dot{z}(t) = \mathrm{D}f_0(\varphi(t, x, \omega)) z(t) + \sum_{i=1}^{m} \omega_i(t) \mathrm{D}f_i(\varphi(t, x, \omega)) z(t).$$

Let $V := \mathscr{L}(\mathbb{R}^d, \mathbb{R}^d)$, $P := \mathbb{R}^d \times \mathscr{U}$, and $g_i(t, z, p) := \mathrm{D}f_i(\varphi(t, p)) z$, $g_i : \mathbb{R} \times V \times P \to V$ for $i = 0, 1, \ldots, m$. Then P obviously is a locally compact metric space (using Proposition 1.14) and the g_i are continuous, since φ is continuous and the f_i are continuously differentiable. Moreover, each g_i is linear with respect to z and hence continuously differentiable with a derivative which depends continuously on (t, z). It thus follows from Proposition 1.17 that the derivative $\mathrm{D}\varphi_{t,\omega}(x)$ is continuous as a function of (t, x, ω). Indeed, using the notation $\phi(t; z, \omega, p)$ for the solution of $\dot{z}(t) = g_0(t, z(t), p) + \sum_{i=1}^{m} \omega_i(t) g_i(t, z(t), p)$, we have the equality $\mathrm{D}\varphi_{t,\omega}(x) = \phi(t; \mathrm{id}, \omega, (x, \omega))$ with the identity $\mathrm{id} \in \mathscr{L}(\mathbb{R}^d, \mathbb{R}^d)$. Continuity of the higher order derivatives can now be proved by an inductive argument. By Proposition 1.18, the assumption that the f_i are of class $\mathscr{C}^k$ ($k \geq 2$) guarantees that $\varphi_{t,\omega}$ is k times continuously differentiable and the k-th derivative $\mathrm{D}^k\varphi_{t,\omega}(x)$ is the solution of the initial value problem

$$\dot{z}(t) = g_0(t, z, p) + \sum_{i=1}^{m} \omega_i(t) g_i(t, z, p), \quad p = (x, \omega), \quad z(0) = 0 \in \mathscr{L}^j(\mathbb{R}^d, \mathbb{R}^d),$$

where the g_i are given by

$$g_i(t, z, (x, \omega)) = \mathrm{D}f_i(\varphi(t, x, \omega))z$$
$$+ \sum_{j=1}^{k-1} \binom{k-1}{j} \mathrm{D}^j \left[\mathrm{D}f_i(\varphi(t, x, \omega)) \right] \cdot \mathrm{D}^{k-j}\varphi_{t,\omega}(x).$$

Indeed, one easily checks that the assumptions of Proposition 1.18 are satisfied, maybe except for the global boundedness assumption for the derivatives up to order k. However, this boundedness assumption is satisfied after multiplying the right-hand side of the control-affine system by a cut-off function of class $\mathscr{C}^k$ with compact support. Putting $V := \mathscr{L}^j(\mathbb{R}^d, \mathbb{R}^d)$ and $P := \mathscr{U} \times \mathbb{R}^d$, we can apply Proposition 1.17 again, using continuity of φ and the induction hypothesis that $\mathrm{D}^j\varphi_{t,\omega}(x)$ is continuous in (t, x, ω) for $j = 1, \ldots, k-1$. This concludes the proof. $\square$

Control Sets

In the following, we are going to discuss control sets and chain control sets. Let $\Sigma = (\mathbb{T}, X, U, \mathscr{U}, \varphi)$ be a topological time-invariant system. To rule out trivialities and to avoid ad hoc assumptions, we additionally assume that X has no isolated points.

Definition 1.12. A nonempty set $D \subset X$ is called a *control set* of Σ if

(1) D is controlled invariant, that is, for every $x \in D$ there is $\omega \in \mathscr{U}$ with $\varphi(\mathbb{T}_+, x, \omega) \subset D$;

(2) For every $x \in D$ one has $D \subset \mathrm{cl}\, \mathcal{O}^+(x)$, that is, on D approximate controllability holds;

(3) D is maximal with properties (1) and (2), that is, if $D' \supset D$ satisfies (1) and (2), then $D' = D$.

A control set D is called an *invariant control set* if $\mathrm{cl}\, D = \mathrm{cl}\, \mathcal{O}^+(x)$ for all $x \in D$. All other control sets are called *variant*.

Proposition 1.19. *Let D be a maximal set with the property that for all $x \in D$ one has $D \subset \mathrm{cl}\, \mathcal{O}^+(x)$ and suppose that for an element $x \in D$ there are $\tau > 0$ and $\omega \in \mathcal{U}$ with $\varphi(\tau, x, \omega) \in D$. Then $\varphi(t, x, \omega) \in D$ for all $t \in [0, \tau]$.*

Proof. By the maximality property, it suffices to show that for $t \in [0, \tau]$ and every $y \in D$ one has $y \in \mathrm{cl}\, \mathcal{O}^+(\varphi(t, x, \omega))$ and $\varphi(t, x, \omega) \in \mathrm{cl}\, \mathcal{O}^+(y)$. The first assertion follows from $y \in \mathrm{cl}\, \mathcal{O}^+(\varphi(\tau, x, \omega)) \subset \mathrm{cl}\, \mathcal{O}^+(\varphi(t, x, \omega))$. In order to see the second one, let W be a neighborhood of $\varphi(t, x, \omega)$. By continuity of $\varphi(t, \cdot, \omega)$, there is a neighborhood V of x with $\varphi_{t,\omega}(V) \subset W$. There are $\omega_1 \in \mathcal{U}$ and $\tau_1 > 0$ with $\varphi(\tau_1, y, \omega_1) \in V$, because $x \in D$. By the axioms of systems, there exists an admissible control function ω_2 with

$$
\omega_2(t) = \begin{cases} \omega_1(t) & \text{for } t \in [0, \tau_1), \\ \omega(t - \tau_1) & \text{for } t \in [\tau_1, \tau_1 + t). \end{cases}
$$

This control function satisfies

$$
\varphi(\tau_1 + t, y, \omega_2) = \varphi(t, \varphi(\tau_1, y, \omega_1), \omega) \in \varphi_{t,\omega}(V) \subset W,
$$

showing that $\varphi(t, x, \omega) \in \mathrm{cl}\, \mathcal{O}^+(y)$. $\square$

The only control sets which are interesting for us are those with nonempty interior, which can be characterized as follows.

Proposition 1.20. *Let $D \subset X$ be a set which is maximal with the property that for all $x \in D$ one has $D \subset \mathrm{cl}\, \mathcal{O}^+(x)$ and assume that the interior of D is nonempty. Then D is a control set.*

Proof. It suffices to show that for every $x \in D$ there is $\omega \in \mathcal{U}$ with $\varphi(\mathbb{T}_+, x, \omega) \subset D$. Let $x \in D$. Then there are $\tau_0 \in \mathbb{T} \cap (0, \infty)$ and $\omega_0 \in \mathcal{U}$ with $y := \varphi(\tau_0, x, \omega_0) \in \mathrm{int}\, D$. From the assumption that X has no isolated points it follows that there exists $z \in \mathrm{int}\, D$ with $z \neq y$. Let V_z be an open neighborhood of z with $V_z \subset \mathrm{int}\, D$ and $y \notin \mathrm{cl}\, V_z$. Since $z \in \mathrm{cl}\, \mathcal{O}^+(y)$, there are $\omega_1 \in \mathcal{U}$ and $\tau_1 \in \mathbb{T} \cap (0, \infty)$ with $\varphi(\tau_1, y, \omega_1) \in V_z$. By continuity of $\varphi(\tau_1, \cdot, \omega_1)$ there is an open neighborhood $V_y \subset \mathrm{int}\, D$ of y with $V_y \cap V_z = \emptyset$ and $\varphi(\tau_1, V_y, \omega_1) \subset V_z$. By approximate controllability on D one can steer back from V_z into V_y. In this manner, one constructs a trajectory starting at x which first goes to y in time τ_0 using the control function ω_0, then from y into V_z in time τ_1 using ω_1 and back to V_y using some other control function. Then again, ω_1 is used to get to V_z in time τ_1, and this process is repeated. Since the time it takes to get from V_y into V_z is $\tau_1 > 0$ each time, this defines a trajectory defined on the time interval $\mathbb{T}_+$. By Proposition 1.19 the image of the whole trajectory is contained in D, which proves the assertion. $\square$

An important property of control sets with nonempty interior is the no-return property, defined as follows.

Definition 1.13. A set $A \subset X$ has the *no-return property* if $x \in A$, $\tau \in \mathbb{T}_+$, and $\omega \in \mathscr{U}$ with $\varphi(\tau, x, \omega) \in A$ implies $\varphi([0, \tau], x, \omega) \subset A$. That is, trajectories cannot leave the set A and then return.

Corollary 1.1. *A control set D with nonempty interior has the no-return property.*

Proof. The assertion follows from Proposition 1.19 if we can show that D is a maximal set with the property that $D \subset \operatorname{cl} \mathscr{O}^+(x)$ for all $x \in D$. Assume that this does not hold. Then D is a proper subset of a set D' with this property. Due to Proposition 1.20, D' is a control set, which implies the contradiction that D is not maximal with properties (1) and (2) of control sets. □

Proposition 1.21. *Let $D_0 \subset X$ be a set satisfying properties (1) and (2) in the definition of control sets. Then D_0 is contained in a control set.*

Proof. Define D as the union of all sets satisfying properties (1) and (2) and containing D_0. Then D is a control set. It is obvious that D is controlled invariant. For each pair of points $x, y \in D$, there is $z \in D_0$ with $y \in \operatorname{cl} \mathscr{O}^+(z)$ and $z \in \operatorname{cl} \mathscr{O}^+(x)$. By continuity of φ with respect to the state variable, one finds that $y \in \operatorname{cl} \mathscr{O}^+(x)$. The maximality property is obviously satisfied. □

With the same transitivity argument as in the proof above one sees that control sets are pairwisely disjoint. Indeed, if there were two control sets D_1 and D_2 with $D_1 \cap D_2 \neq \emptyset$, then $D_1 \cup D_2$ would satisfy properties (1) and (2) of control sets and hence would be contained in a control set $D \supset D_1 \cup D_2$, which contradicts maximality of both D_1 and D_2.

In the following, we assume that Σ additionally satisfies the following three properties:

(A) Continuity in time: $\mathbb{T} = \mathbb{R}$;
(B) The maps $\varphi_{t,\omega}$ are open, that is, they map open sets to open sets;
(C) For every $x \in X$ and every neighborhood V of x there is $\tau > 0$ such that
$\mathscr{O}_{\leq \tau}^+(x) \subset V$ and $\mathscr{O}_{\leq \tau}^-(x) \subset V$.

These three properties are satisfied in particular for smooth systems given by differential equations with bounded control range.

Proposition 1.22. *Let $\Sigma = (\mathbb{R}, M, \mathbb{R}^m, \mathscr{U}, \varphi)$ be a smooth system given by differential equations with bounded control range $\Omega \subset \mathbb{R}^m$. Then Σ satisfies (A)–(C).*

Proof. It is obvious that Σ satisfies (A) and (B). To prove (C), let M be endowed with some Riemannian metric and denote by ϱ the associated Riemannian distance. Fix $x \in M$ and a neighborhood V of x. Without loss of generality, we may assume that there exists a slightly larger neighborhood $W \supset V$ which has a compact closure. Let $\theta : M \to [0, 1]$ be a cut-off function of class $\mathscr{C}^1$ such that $\theta(x) \equiv 1$

on V and supp $\theta \subset \text{cl } W$. Define a new right-hand side by $\tilde{F} := \theta F$ and denote the corresponding trajectories by $\tilde{\varphi}(t, x, \omega)$. Then for all $\omega \in \mathcal{U}$ and $t \geq 0$ we find

$$\varrho(\tilde{\varphi}(t, x, \omega), x) \leq \mathcal{L}(\tilde{\varphi}(\cdot, x, \omega)|_{[0,t]}) = \int_0^t \left| \tilde{F}(\tilde{\varphi}(s, x, \omega), \omega(s)) \right| \, ds$$

$$\leq t \cdot \underbrace{\sup_{(y,u) \in W \times \Omega} \left| \tilde{F}(y, u) \right|}_{=:B}.$$

Boundedness of Ω and relative compactness of W guarantee that B is finite. This shows that for sufficiently small τ we have $\tilde{\varphi}([0, \tau], x, \omega) \subset V$ and hence $\varphi([0, \tau], x, \omega) \subset V$ for all $\omega \in \mathcal{U}$. Consequently, $\mathcal{O}^+_{\leq \tau}(x) \subset V$. The analogous statement for the backward orbit is proved in a similar way.   □

Lemma 1.2. *If Σ is locally accessible, then* $\text{cl } \mathcal{O}^+(x) = \text{cl int } \mathcal{O}^+(x)$ *for all $x \in M$.*

Proof. From property (C) and local accessibility it follows that for every neighborhood N of a point $y \in X$ there is an open set $V \subset \text{int } \mathcal{O}^+(y) \cap N$. Now fix $x \in X$ and choose $y \in \mathcal{O}^+(x)$. Observe that $\mathcal{O}^+(y) \subset \mathcal{O}^+(x)$. Then we find a sequence $x_n \to y$ with $x_n \in \text{int } \mathcal{O}^+(y) \subset \text{int } \mathcal{O}^+(x)$, which proves that int $\mathcal{O}^+(x)$ is dense in $\mathcal{O}^+(x)$ and hence cl int $\mathcal{O}^+(x) = \text{cl } \mathcal{O}^+(x)$.   □

Before we formulate our main results about control sets of systems satisfying properties (A)–(C), we introduce the *domain of attraction* of a control set D which is the set

$$A(D) := \left\{ x \in X : \text{cl } \mathcal{O}^+(x) \cap D \neq \emptyset \right\}.$$

We further remark that the domain of attraction of a control set with nonempty interior is an open neighborhood of the control set (but in general not of its closure). Indeed, if $x \in A(D)$, then there exist a sequence x_n in $\mathcal{O}^+(x)$ and a point $y \in D$ with $x_n \to y$. From y it is possible to get into the interior of D. By continuity, a whole neighborhood of y can be steered into the interior of D which implies that one can get from x into the interior of D. Then, using continuity again, a whole neighborhood of x can be steered into the interior, so x is an interior point of $A(D)$.

Proposition 1.23. *Let D be a control set with nonempty interior for the system Σ. Then the following assertions hold:*

  (i) *If Σ is locally accessible on X, then D is connected and cl int $D = \text{cl } D$.*
 (ii) *If Σ is locally accessible from $y \in \text{int } D$, then $y \in \mathcal{O}^+(x)$ for all $x \in D$.*
(iii) *If Σ is locally accessible from all $y \in \text{int } D$, then int $D \subset \mathcal{O}^+(x)$ for all $x \in D$, and for every $y \in \text{int } D$ one has*

$$D = \text{cl } \mathcal{O}^+(y) \cap \mathcal{O}^-(y).$$

In particular, on int D *exact controllability holds. That is, for any two states* $x, y \in$ int D *there are* $\omega \in \mathcal{U}$ *and* $\tau \geq 0$ *with* $\varphi(\tau, x, \omega) = y$.

(iv) *If* $\varphi(\cdot, x, \omega)$ *is a periodic trajectory, that is,* $\varphi(t + \tau, x, \omega) = \varphi(t, x, \omega)$ *for some* $\tau > 0$ *and all* $t \in \mathbb{R}$, *such that* $x \in$ int D, *then* $\varphi(t, x, \omega) \in$ int D *for all* $t \in \mathbb{R}$.

(v) *Let* $K_1 \subset A(D)$ *and* $K_2 \subset$ int D *be compact sets. Then, under the assumption of local accessibility on* int D, *there exists a time* $T = T(K_1, K_2) < \infty$ *such that for all* $x \in K_1$ *and* $y \in K_2$ *it holds that*

$$h(x, y) := \inf \{t \geq 0 \mid \exists \omega \in \mathcal{U} : \varphi(t, x, \omega) = y\} \leq T.$$

Proof. (i) For the proof of the equality cl int $D = $ cl D it suffices to show that $D \subset$ cl int D. To this end, let $x \in D$. By approximate controllability, there are an open neighborhood N of x and $t > 0$, $\omega \in \mathcal{U}$ with $\varphi_{t,\omega}(N) \subset$ int D. Since $x \in D \subset$ cl $\mathcal{O}^+(y) = $ cl int $\mathcal{O}^+(y)$ by Lemma 1.2, for every $y \in D$ the open set int $\mathcal{O}^+(y) \cap N$ is nonempty and because of the no-return property contained in D. Hence, there is a sequence x_n in int $\mathcal{O}^+(y) \cap N$ converging to x which implies $x \in$ cl int D. To show connectedness of D, assume to the contrary that D is the disjoint union of nonempty sets A and B which are open in D. Since cl $D = $ cl int D, both A and B have nonempty intersection with int D. Hence, we find $a \in$ int A and $b \in$ int B. Approximate controllability from a to b yields a contradiction (using that $\mathbb{T} = \mathbb{R}$).

(ii) Let $x \in D$. Then, using local accessibility from $y \in$ int D and property (C), one finds $t > 0$ such that $\emptyset \neq$ int $\mathcal{O}^-_{\leq t}(y) \subset$ int D. Choose $z \in$ int $\mathcal{O}^-_{\leq t}(y)$. Since $z \in$ cl $\mathcal{O}^+(x)$, one concludes that $y \in \mathcal{O}^+(x)$.

(iii) The first assertion is an immediate consequence of statement (ii). If $x \in D$ and $y \in$ int D, then x can approximately be reached from y and x can be steered to y by (ii). This implies $D \subset$ cl $\mathcal{O}^+(y) \cap \mathcal{O}^-(y)$. If, conversely, $x \in$ cl $\mathcal{O}^+(y) \cap \mathcal{O}^-(y)$ for some $y \in$ int D, then $D' := D \cup \{x\}$ satisfies the first two properties of control sets and hence $D' = D$, that is, $x \in D$.

(iv) Assume that $y = \varphi(t, x, \omega) \notin$ int D for some $t \in (0, \tau)$. From the no-return property of control sets with nonempty interior it follows that $y \in D \cap \partial D$. Then every neighborhood N_x of x is mapped by the open map $\varphi_{t,\omega}$ onto a neighborhood of y which necessarily contains elements of $X \backslash D$. On the other hand, points that are sufficiently close to y are mapped into the interior of D by $\varphi_{\tau-t,\Theta_t\omega}$. This leads to trajectories which start in int D, leave the control set D, and then return to int D in contradiction to the no-return property.

(v) The proof proceeds in three steps.

Step 1. For $x \in K_1$ and $y \in K_2$ we show that there is an open neighborhood N_x of x and a time $t = t(x, y)$ such that $h(z, y) \leq t < \infty$ for all $z \in N_x$. By local accessibility from y and property (C) there are $T > 0$ and $y_1 \in$ int $D \cap \mathcal{O}^-_{\leq T}(y)$ such that $N_{y_1} \subset$ int $D \cap \mathcal{O}^-_{\leq T}(y)$ for an open neighborhood N_{y_1} of y_1. By local accessibility from y_1 there exist $\omega \in \mathcal{U}$ and $t_1 > 0$ such that $\varphi(t_1, x, \omega) = y_1$. By continuity of φ with respect to the state variable, there is an open neighborhood N_x of x with $\varphi(t_1, z, \omega) \in N_{y_1}$ for all $z \in N_x$. Putting this together yields $N_x \subset \mathcal{O}^-_{\leq T + t_1}(y)$, hence $h(z, y) \leq t_1 + T$ for all $z \in N_x$.

Step 2. For $x \in K_1$ and $y \in K_2$ we show that there is an open neighborhood N_y of y and a time $t = t(x, y)$ such that $h(x, z) \leq t < \infty$ for all $z \in N_y$. To this end, let $x_1 \in K_2 \subset \operatorname{int} D$ and $\omega_1 \in \mathcal{U}$, $t_1 \geq 0$ such that $\varphi(t_1, x, \omega_1) = x_1$. By local accessibility on K_2 and property (C) we find $T > 0$ and $y_1 \in \operatorname{int} D \cap \mathcal{O}^+_{\leq T}(x_1)$ such that an open neighborhood N_{y_1} of y_1 is contained in $\operatorname{int} D \cap \mathcal{O}^+_{\leq T}(x_1)$. By local accessibility from y, there are $\omega_2 \in \mathcal{U}$ and $t_2 > 0$ with $\varphi(t_2, y_1, \omega_2) = y$. The open set N_{y_1} is mapped onto an open neighborhood N_y of y by the open map φ_{t_2, ω_2}. This implies $N_y \subset \mathcal{O}^+_{\leq T + t_1 + t_2}(x)$ and hence, $h(x, z) \leq T + t_1 + t_2$ for all $z \in N_y$.

Step 3. To prove the assertion, it suffices to show that for every pair $(x, y) \in K_1 \times K_2$ there are neighborhoods N_x of x and N_y of y such that h is bounded on $N_x \times N_y$. Then the assertion follows by compactness of $K_1 \times K_2$. Next to x and y, fix a further point $x_1 \in K_2$ as in Step 2. Then Step 1 gives a neighborhood N_x of x such that the time to get from N_x to x_1 is bounded. Step 2 shows that there is a neighborhood N_y of y such that the time to get from x_1 to N_y is bounded as well. This concludes the proof. □

For control-affine systems we additionally introduce the notion of chain control sets, which play an important role in Sects. 7.1 and 7.4. In the rest of this subsection, let $\Sigma = (\mathbb{R}, M, \mathbb{R}^m, \mathcal{U}, \varphi)$ be a control-affine system on a compact manifold M, and fix a metric ϱ on M. Note that Σ satisfies (A)–(C), since the control range of a control-affine system is assumed to be compact.

Definition 1.14. Let $x, y \in M$ and $\varepsilon, \tau > 0$. A *controlled (ε, τ)-chain* from x to y is given by $n \in \mathbb{N}$, $x_0, \ldots, x_n \in X$, $\omega_0, \ldots, \omega_{n-1} \in \mathcal{U}$, and $t_0, \ldots, t_{n-1} \geq \tau$ with $x_0 = x$, $x_n = y$, such that

$$\varrho(\varphi(t_j, x_j, \omega_j), x_{j+1}) < \varepsilon \quad \text{for all } j = 0, 1, \ldots, n - 1.$$

If for all $\varepsilon, \tau > 0$ there exists a controlled (ε, τ)-chain from x to y, we say that Σ is *chain controllable from x to y*.

Definition 1.15. A set $E \subset M$ is called a *chain control set* of Σ if it satisfies the following properties:

(1) For every $x \in E$ there is $\omega \in \mathcal{U}$ with $\varphi(\mathbb{R}, x, \omega) \subset E$;
(2) For all $x, y \in E$ and $\varepsilon, \tau > 0$, there is a controlled (ε, τ)-chain from x to y (the points in this chain are not necessarily elements of E);
(3) E is maximal with properties (1) and (2).

It is easy to see that the property of being a chain control set does not depend on the metric imposed on M. For chain control sets we have the following results.

Proposition 1.24. *The following assertions hold:*

(i) *Every chain control set E of Σ is closed.*

(ii) *Assume that Σ is locally accessible. Then every control set D with nonempty interior is contained in a chain control set E.*

(iii) *Different chain control sets of Σ are disjoint.*

(iv) *If E is a chain control set of Σ, then*

$$\mathcal{E} := \{(\omega, x) \in \mathcal{U} \times M \ : \ \varphi(\mathbb{R}, x, \omega) \subset E\}$$

is a maximal invariant chain transitive set for the control flow of Σ. On the other hand, if $\mathcal{E} \subset \mathcal{U} \times M$ is a maximal invariant chain transitive set for the control flow, then the projection of $\mathcal{E}$ to M is a chain control set.

Proof. (i) Let E be a chain control set. We prove the assertion by showing that cl E satisfies the properties (1) and (2) of chain control sets. By maximality, it then follows that $E = $ cl E. For every $x \in$ cl E there is a sequence $x_n \in E$ converging to x. By property (1), for every x_n there exists $\omega_n \in \mathcal{U}$ with $\varphi(\mathbb{R}, x_n, \omega_n) \subset E$. By weak*-compactness of $\mathcal{U}$ we may assume that $\omega_n \overset{*}{\rightharpoonup} \omega \in \mathcal{U}$. By continuity of φ this implies $\varphi(t, x, \omega) = \lim_{n \to \infty} \varphi(t, x_n, \omega_n) \in$ cl E for every $t \in \mathbb{R}$. Hence, cl E satisfies (1). Now let $x, y \in$ cl E and $\varepsilon, \tau > 0$. Let $\omega_0 \in \mathcal{U}$ be any control function such that $x_1 := \varphi(\tau, x, \omega_0) \in$ cl E. Then there are $x_2, \tilde{y} \in E$ with $\varrho(x_1, x_2) < \varepsilon$ and $\varrho(y, \tilde{y}) < \varepsilon/2$, and there is an $(\varepsilon/2, \tau)$-chain from x_2 to $\tilde{y}$. By adding the points x, x_1 at the beginning and y at the end of this chain, we obtain an (ε, τ)-chain from x to y. Hence, cl E satisfies (2).

(ii) It suffices to prove that cl D satisfies properties (1) and (2) of chain control sets. Obviously, (2) is satisfied. Hence, it remains to show that for every $x \in$ cl D there is $\omega \in \mathcal{U}$ with $\varphi(\mathbb{R}, x, \omega) \subset$ cl D. For every point $x \in$ int D we can construct a periodic trajectory starting and ending at x using local accessibility from x. Extending the corresponding control function periodically, one obtains a trajectory on $\mathbb{R}$ whose image is completely contained in int D. By Proposition 1.23 (i), for every $x \in \partial D$ we find a sequence $x_n \in$ int D converging to x. Then the same argument as already used in the proof of statement (i) shows that there exists $\omega \in \mathcal{U}$ with $\varphi(\mathbb{R}, x, \omega) \subset$ cl D.

(iii) It is easy to see that if E_1 and E_2 were two different chain control sets with nonempty intersection, the set $E := E_1 \cup E_2$ would be a chain control set as well, contradicting maximality.

(iv) Fix a metric $\varrho_{\mathcal{U}}$ on $\mathcal{U}$ as described in Proposition 1.14 by fixing a dense and countable subset $\{y_n\}_{n \geq 1}$ of $L^1(\mathbb{R}, \mathbb{R}^m)$. Let $(\omega, x), (\mu, y) \in \mathcal{E}$, pick $\varepsilon, \tau > 0$, and choose $N \in \mathbb{N}$ large enough so that

$$\sum_{n=N+1}^{\infty} 2^{-n} < \frac{\varepsilon}{2}.$$

Let $T_0 > 0$ be chosen so large that

$$\int_{\mathbb{R}\setminus[-T_0,T_0]} |y_i(\tau)|\,d\tau < \frac{\varepsilon}{2\,\mathrm{diam}\,\Omega} \quad \text{for } i = 1,\ldots,N.$$

We can assume without loss of generality that Ω consists of more than one point (which guarantees that $\mathrm{diam}\,\Omega > 0$), and also that $\tau > T_0$. Chain controllability from $\varphi(2\tau, x, \omega) \in E$ to $\varphi(-\tau, y, \mu) \in E$ yields the existence of $k \in \mathbb{N}$ and $x_0,\ldots,x_k \in M$, $\omega_0,\ldots,\omega_{k-1} \in \mathscr{U}$, $t_0,\ldots,t_{k-1} \geq \tau$ with $x_0 = \varphi(2\tau, x, \omega)$, $x_k = \varphi(-\tau, y, \mu)$, and

$$\varrho(\varphi(t_j, x_j, \omega_j), x_{j+1}) < \varepsilon \quad \text{for } j = 0,1,\ldots,k-1.$$

We now construct an (ε, τ)-chain from (ω, x) to (y, μ) in the following way. Define

$$t_{-2} = \tau, \qquad x_{-2} = x, \qquad \mu_{-2} = \omega,$$

$$t_{-1} = \tau, \qquad x_{-1} = \varphi(\tau, x, \omega), \qquad \mu_{-1}(t) = \begin{cases} \omega(t_{-2} + t) & \text{for } t \leq t_{-1} \\ \omega_0(t - t_{-1}) & \text{for } t > t_{-1} \end{cases}$$

and let the times $t_0,\ldots,t_{k-1}$ and the points $x_0,\ldots,x_k$ be as given before. Furthermore, put

$$t_k = \tau, \qquad x_{k+1} = y, \qquad \mu_{k+1} = \mu,$$

and define for $j = 0,1,\ldots,k-2$

$$\mu_j(t) = \begin{cases} \mu_{j-1}(t_{j-1} + t) & \text{for } t \leq 0 \\ \omega_j(t) & \text{for } 0 < t < t_j \\ \omega_{j+1}(t - t_j) & \text{for } t > t_j \end{cases},$$

$$\mu_{k-1}(t) = \begin{cases} \mu_{k-2}(t_{k-2} + t) & \text{for } t \leq 0 \\ \omega_{k-1}(t) & \text{for } 0 < t < t_{k-1} \\ \mu(t - t_{k-1} - \tau) & \text{for } t > t_{k-1} \end{cases},$$

$$\mu_k(t) = \begin{cases} \mu_{k-1}(t_{k-1} + t) & \text{for } t \leq 0 \\ \mu(t - \tau) & \text{for } t > 0 \end{cases}.$$

It is easy to see that

$$(\mu_{-2}, x_{-2}), \ (\mu_{-1}, x_{-1}), \ \ldots, (\mu_{k+1}, x_{k+1}) \quad \text{and} \quad t_{-2}, t_{-1}, \ldots, t_k \geq \tau$$

yield an (ε, τ)-chain from (ω, x) to (μ, y) provided that for $j = -2, -1, \ldots, k$
it holds that

$$\varrho_{\mathscr{U}}\left(\Theta_{t_j}\mu_j, \mu_{j+1}\right) < \varepsilon.$$

By the choice of τ and N, for all $\alpha_1, \alpha_2 \in \mathscr{U}$ we have

$$\varrho_{\mathscr{U}}(\alpha_1, \alpha_2) = \sum_{n=1}^{\infty} \frac{1}{2^n} \frac{|\int_{\mathbb{R}}\langle \alpha_1(t) - \alpha_2(t), y_n(t)\rangle \mathrm{d}t|}{1 + |\int_{\mathbb{R}}\langle \alpha_1(t) - \alpha_2(t), y_n(t)\rangle \mathrm{d}t|}$$

$$\leq \sum_{n=1}^{N} \frac{1}{2^n}\left[\left|\int_{\mathbb{R}\setminus[-\tau, \tau]}\langle \alpha_1(t) - \alpha_2(t), y_n(t)\rangle \mathrm{d}t\right|\right.$$

$$\left. + \left|\int_{-\tau}^{\tau}\langle \alpha_1(t) - \alpha_2(t), y_n(t)\rangle \mathrm{d}t\right|\right] + \frac{\varepsilon}{2}$$

$$< \varepsilon + \max_{n=1}^{N} \int_{-\tau}^{\tau} |\alpha_1(t) - \alpha_2(t)| \cdot |y_n(t)| \mathrm{d}t.$$

Hence, it suffices to show that for all considered pairs of control functions
the integrands vanish. This is immediate from the definition of μ_j, $j = -2, -1, \ldots, k+1$. Hence, chain transitivity of $\mathscr{E}$ is proved. Invariance of $\mathscr{E}$ is
immediate from its definition.

Now consider an invariant chain transitive set $\mathscr{E}$ for the control flow. For
$x \in \pi_M \mathscr{E}$ ($\pi_M : \mathscr{U} \times M \to M$, $(\omega, x) \mapsto x$) there exists $\omega \in \mathscr{U}$ with
$\varphi(t, x, \omega) \in \mathscr{E}$ for all $t \in \mathbb{R}$ by invariance. Now let $x, y \in \pi_M \mathscr{E}$ and choose
$\varepsilon, \tau > 0$. Then, by chain transitivity of $\mathscr{E}$, we can choose x_j, ω_j, t_j such
that the corresponding trajectories satisfy the required condition. The proof
is concluded by the observation that E is maximal if and only if $\mathscr{E}$ is maximal.

□

Remark 1.8. Statement (iv) in the preceding proposition establishes a relation
between the control-theoretic properties of the control-affine system and the dynamical properties of the associated control flow. This result is an important link between
control theory and classical dynamical systems theory. A related result asserts that
the control flow is topologically transitive on the lifts of control sets with nonempty
interior and compact closure, provided that local accessibility holds (cf. Colonius
and Kliemann [25, Proposition 4.3.3]).

1.5   Linearization and Regular Trajectories

A smooth system given by differential equations with continuously differentiable
right-hand side can be linearized along a trajectory both in the state and in the control
variable. This linearization can be regarded as a time-variant linear system. As one

would expect, global properties of the linear system imply corresponding local properties of the nonlinear system in the vicinity of the given trajectory. In particular, controllability of the linearization gives local controllability. Such relations between the linearization and the nonlinear system are used in Chap. 5. In this section, we introduce the necessary notions and give proofs of some basic results.

The Linearization Along Controlled Trajectories

Consider a smooth system $\Sigma = (\mathbb{R}, M, \mathbb{R}^m, \mathcal{U}, \varphi)$ given by differential equations

$$\dot{x}(t) = F(x(t), \omega(t)), \quad \omega \in \mathcal{U}, \tag{1.4}$$

with a continuously differentiable right-hand side $F : M \times \mathbb{R}^m \to TM$, that is, $F \in \mathcal{C}^1(M \times \mathbb{R}^m, TM)$. Moreover, we assume that (M, g) is a Riemannian $\mathcal{C}^3$-manifold. Then ϱ denotes the induced Riemannian distance on M.

In the following, we often have to consider partial derivatives of maps which depend on two arguments. For simplicity, we write D_1 and D_2 for the derivatives with respect to the first and second component, respectively. If the argument is an element of an infinite-dimensional Banach space, the corresponding derivative is the Fréchet derivative.

Definition 1.16. A pair $(\varphi(\cdot, x, \omega), \omega(\cdot))$ of a trajectory and the corresponding control function is called a *controlled trajectory*. Given a controlled trajectory, we consider the system

$$\frac{Dz}{dt}(t) = A(t)z(t) + B(t)\mu(t), \quad \mu \in L^\infty(\mathbb{R}, \mathbb{R}^m), \tag{1.5}$$

of differential equations, where

$$A(t) := \nabla F_{\omega(t)}(\varphi(t, x, \omega)) \quad \text{and} \quad B(t) := D_2 F(\varphi(t, x, \omega), \omega(t)).$$

The derivative on the left-hand side of (1.5) is the covariant derivative of $z(\cdot)$ along $\varphi(\cdot, x, \omega)$. We call system (1.5) the *linearization of Σ along the controlled trajectory* $(\varphi(\cdot, x, \omega), \omega(\cdot))$. A solution of (1.5) corresponding to $\mu \in L^\infty(\mathbb{R}, \mathbb{R}^m)$ with initial value $\lambda \in T_x M$ is a locally absolutely continuous vector field[4] $z : \mathbb{R} \to TM$ along $\varphi(\cdot, x, \omega)$ with $z(0) = \lambda$, satisfying the differential equation (1.5) for almost all $t \in \mathbb{R}$. If such a solution exists and is unique, we denote it by $\phi^{x,\omega}(\cdot, \lambda, \mu)$.

---

[4]Here we mean that z is locally absolutely continuous as a curve in TM, cf. Sect. A.3.

In Sontag [102, Theorem 1], the linearization of a system on Euclidean space is analyzed. The following proposition is a reduced version of the main result presented by Sontag. The system Σ in this proposition may be incomplete, that is, trajectories may only be defined on finite time intervals.

Proposition 1.25. *Let M be an open subset of $\mathbb{R}^d$ and $F : M \times \mathbb{R}^m \to \mathbb{R}^d$ a $\mathscr{C}^1$-mapping. Consider the system $(\mathbb{R}, \mathbb{R}^d, \mathbb{R}^m, \mathscr{U}, \varphi)$ given by*

$$\dot{x}(t) = F(x(t), \omega(t)), \quad \omega \in \mathscr{U} := L^\infty(\mathbb{R}, \mathbb{R}^m).$$

For fixed $\tau > 0$ define

$$\mathscr{D}_\tau := \{(x, \omega) \in M \times L^\infty([0, \tau], \mathbb{R}^m) \ : \ \tau \in I_{\max}(\omega, x)\},$$

where $I_{\max}(\omega, x)$ is the maximal interval of existence for $\varphi(\cdot, x, \omega)$. Then $\mathscr{D}_\tau$ is open in $M \times L^\infty([0, \tau], \mathbb{R}^m)$ and the mapping

$$\varphi_\tau : \mathscr{D}_\tau \to M, \quad (x, \omega) \mapsto \varphi(\tau, x, \omega),$$

is of class $\mathscr{C}^1$. For fixed $(x_0, \omega_0) \in M \times L^\infty([0, \tau], \mathbb{R}^m)$ and $(\lambda, \mu) \in \mathbb{R}^d \times L^\infty([0, \tau], \mathbb{R}^m)$, the curve

$$\xi(t) := D\varphi_\tau(x_0, \omega_0)(\lambda, \mu), \quad \xi : [0, \tau] \to \mathbb{R}^d,$$

is the unique solution of the Carathéodory differential equation

$$\dot{\xi}(t) = D_1 F(\varphi(t, x_0, \omega_0), \omega_0(t))\xi(t) + D_2 F(\varphi(t, x_0, \omega_0), \omega_0(t))\mu(t)$$

with initial value λ.

The following proposition describes the properties of the linearization in the general situation.

Proposition 1.26. *Let $(\varphi(\cdot, x, \omega), \omega(\cdot))$ be a controlled trajectory with corresponding linearization* (1.5). *Then the following statements hold:*

(i) For every $\tau > 0$ the mapping

$$\varphi_\tau : M \times L^\infty([0, \tau], \mathbb{R}^m) \to M, \quad (x, \omega) \mapsto \varphi(\tau, x, \omega),$$

is continuously (Fréchet) differentiable.

(ii) For every initial value $\lambda \in T_x M$ and every $\mu \in L^\infty(\mathbb{R}, \mathbb{R}^m)$ there exists a unique solution $\phi^{x,\omega}(\cdot, \lambda, \mu) : \mathbb{R} \to TM$ of (1.5) *satisfying*

$$\phi^{x,\omega}(t, \lambda, \mu) = D\varphi_t(x, \omega)(\lambda, \mu) \tag{1.6}$$

for all $t \in \mathbb{R}$ *and* $(\lambda, \mu) \in T_x M \times L^\infty(\mathbb{R}, \mathbb{R}^m)$, *where* D *stands for the total derivative of* $\varphi_t : M \times L^\infty(\mathbb{R}, \mathbb{R}^m) \to M$, *which consists of the derivative* $\mathrm{d}_x \varphi_t(\cdot, \omega) : T_x M \to T_{\varphi(t,x,\omega)} M$ *in the first, and the Fréchet derivative of* $\varphi_t(x, \cdot) : L^\infty(\mathbb{R}, \mathbb{R}^m) \to T_{\varphi(t,x,\omega)} M$ *in the second component.*

(iii) For every $\tau > 0$ *the mapping*

$$\phi^{x,\omega}(\tau, \cdot, \cdot) : T_x M \times L^\infty([0, \tau], \mathbb{R}^m) \to T_{\varphi(\tau,x,\omega)} M$$

is linear and continuous.

(iv) For each $t \in \mathbb{R}$ *let*

$$\phi_t^{x,\omega} := \phi^{\varphi(t,x,\omega), \Theta_t \omega}.$$

Then for all $t, s, r \in \mathbb{R}$ *and* $\lambda \in T_{\varphi(r,x,\omega)} M$ *we have*

$$\phi_{r+s}^{x,\omega}(t, \phi_r^{x,\omega}(s, \lambda, \mathbf{0}), \mathbf{0}) = \phi_r^{x,\omega}(t + s, \lambda, \mathbf{0}). \tag{1.7}$$

More general, for all $t, s \in \mathbb{R}$, $\lambda \in T_x M$, *and* $\mu \in L^\infty(\mathbb{R}, \mathbb{R}^m)$,

$$\phi_s^{x,\omega}(t, \phi^{x,\omega}(s, \lambda, \mu), \Theta_s \mu) = \phi^{x,\omega}(t + s, \lambda, \mu). \tag{1.8}$$

Proof. Statements (i) and (ii) follow from Proposition 1.25 by writing everything in local coordinates. The actual proof is lengthy and technical, so we do not go into details here. Statement (iii) follows immediately from (ii), which says that $\phi^{x,\omega}(\tau, \cdot, \cdot)$ is the Fréchet derivative of a $\mathscr{C}^1$-mapping. The identity (1.7) follows from (1.8) by replacing x with $\varphi(r, x, \omega)$ and ω with $\Theta_r \omega$, and taking the constant control function $\mu(t) := \mathbf{0}(t) \equiv 0$. In order to prove (1.8), write

$$X(t) := \phi_s^{x,\omega}(t, \phi^{x,\omega}(s, \lambda, \mu), \Theta_s \mu), \quad Y(t) := \phi^{x,\omega}(t + s, \lambda, \mu).$$

Both X and Y are locally absolutely continuous vector fields along the curve $\varphi(\cdot + s, x, \omega)$ and we have $X(0) = \phi^{x,\omega}(s, \lambda, \mu) = Y(0)$. Letting

$$A_X(t) := \nabla F_{\Theta_s \omega(t)}(\varphi(t, \varphi(s, x, \omega), \Theta_s \omega)) = \nabla F_{\omega(t+s)}(\varphi(t + s, x, \omega)),$$

$$A_Y(t) := \nabla F_{\omega(t)}(\varphi(t, x, \omega)),$$

$$B_X(t) := \mathrm{D}_2 F(\varphi(t, \varphi(s, x, \omega), \Theta_s \omega), \Theta_s \omega(t))$$

$$= \mathrm{D}_2 F(\varphi(t + s, x, \omega), \omega(t + s)),$$

$$B_Y(t) := \mathrm{D}_2 F(\varphi(t, x, \omega), \omega(t)),$$

we find that

$$\frac{\mathrm{D}X}{\mathrm{d}t}(t) = A_X(t)X(t) + B_X(t)\Theta_s\mu(t),$$

$$\frac{\mathrm{D}Y}{\mathrm{d}t}(t) = A_Y(t+s)Y(t) + B_Y(t+s)\mu(t+s)$$

$$= A_X(t)Y(t) + B_X(t)\Theta_s\mu(t).$$

Hence, X and Y are solutions of the same initial value problem, which gives $X = Y$.   □

Regular Trajectories

In the following, we introduce the notion of regularity for controlled trajectories.

Definition 1.17. Consider some $(x, \omega, \tau) \in M \times \mathcal{U} \times (0, \infty)$ and let $y := \varphi(\tau, x, \omega)$. Then we call the linearization along $(\varphi(\cdot, x, \omega), \omega(\cdot))$ *controllable on* $[0, \tau]$ if for each $\lambda_1 \in T_x M$ and $\lambda_2 \in T_y M$ there exists $\mu \in L^\infty([0, \tau], \mathbb{R}^m)$ with

$$\phi^{x,\omega}(\tau, \lambda_1, \mu) = \lambda_2.$$

In this case, we say that ω is *regular for x on* $[0, \tau]$, and the controlled trajectory $(\varphi(\cdot, x, \omega), \omega(\cdot))$ is called *regular on* $[0, \tau]$.

Proposition 1.27. *Consider the linearization along a controlled trajectory* $(\varphi(\cdot, x, \omega), \omega(\cdot))$. *Fix* $\tau > 0$ *and let* $y := \varphi(\tau, x, \omega)$. *Then the following statements are equivalent:*

 (i) The controlled trajectory $(\varphi(\cdot, x, \omega), \omega(\cdot))$ *is regular on* $[0, \tau]$.
 (ii) For each $\lambda \in T_x M$ *there is* $\mu \in L^\infty([0, \tau], \mathbb{R}^m)$ *with* $\phi^{x,\omega}(\tau, \lambda, \mu) = 0_y$.
 (iii) For each $\lambda \in T_y M$ *there is* $\mu \in L^\infty([0, \tau], \mathbb{R}^m)$ *with* $\phi^{x,\omega}(\tau, 0_x, \mu) = \lambda$.
 (iv) The linear operator $\mathrm{D}_2\varphi_\tau(x, \omega) : L^\infty([0, \tau], \mathbb{R}^m) \to T_y M$ *is surjective.*

Proof. Clearly, (i) implies (ii). Assume that (ii) holds and let $\lambda \in T_y M$. Define

$$\tilde{\lambda} := \phi_\tau^{x,\omega}(-\tau, -\lambda, \mathbf{0}) \in T_x M.$$

Then there is $\mu \in L^\infty([0, \tau], \mathbb{R}^m)$ with $\phi^{x,\omega}(\tau, \tilde{\lambda}, \mu) = 0_y$. Using Proposition 1.26, we conclude

$$\phi^{x,\omega}(\tau, 0_x, \mu) = \phi^{x,\omega}(\tau, \tilde{\lambda}, \mu) - \phi^{x,\omega}(\tau, \tilde{\lambda}, \mathbf{0})$$

$$= 0_y - \phi_0^{x,\omega}(\tau, \phi_\tau^{x,\omega}(-\tau, -\lambda, \mathbf{0}), \mathbf{0}) = \lambda.$$

Hence, (ii) implies (iii). Statement (iii) says that the linear operator $\phi^{x,\omega}(\tau, 0_x, \cdot) : L^\infty([0, \tau], \mathbb{R}^m) \to T_y M$ is surjective. By Proposition 1.26, this operator coincides with $D_2\varphi_\tau(x,\omega)$, which proves that (iii) implies (iv). Finally, assume that (iv) holds. Then $\phi^{x,\omega}(\tau,0_x,\cdot)$ is surjective and we can show (i) with similar arguments as used in the proof of the implication "(ii) $\Rightarrow$ (iii)". □

Next, we will use the above proposition in order to show that for the regularity of a controlled trajectory it is sufficient that an arbitrary piece of this trajectory is regular. Recall from Sect. 1.1 the notation used for the concatenation of two functions: If $\omega \in L^\infty([0,\tau],\mathbb{R}^m)$ and $\mu \in L^\infty([\tau,\tau+\rho],\mathbb{R}^m)$,

$$(\omega\mu)(t) = \begin{cases} \omega(t) \text{ for } t \in [0,\tau), \\ \mu(t) \text{ for } t \in [\tau,\tau+\rho]. \end{cases}$$

Furthermore, if $\omega \in L^\infty([0,\tau],\mathbb{R}^m)$ and $\rho > 0$, the function $\omega^\rho \in L^\infty([\rho,\tau+\rho],\mathbb{R}^m)$ is defined by $\omega^\rho(t) :\equiv \omega(t-\rho)$.

Proposition 1.28. *Let $x,y \in M$, $\iota,\rho > 0$, and $\omega \in L^\infty([0,\tau],\mathbb{R}^m)$, $\mu \in L^\infty([0,\rho],\mathbb{R}^m)$ such that $\varphi(\tau,x,\omega) = y$. Then the following assertions hold:*

(i) If ω is regular for x on $[0,\tau]$, then $\omega\mu^\tau$ is regular for x on $[0,\tau+\rho]$.
(ii) If μ is regular for y on $[0,\rho]$, then $\omega\mu^\tau$ is regular for x on $[0,\tau+\rho]$.

Proof. Let us assume that ω is regular for x on $[0,\tau]$. Proposition 1.27 implies that for each $\lambda \in T_x M$ we find $\mu \in L^\infty([0,\tau],\mathbb{R}^m)$ with $\phi^{x,\omega}(\tau,\lambda,\mu) = 0_y$. Using the function $\tilde\mu \in L^\infty([0,\tau+\rho],\mathbb{R}^m)$, defined by

$$\tilde\mu(t) := \begin{cases} \mu(t) \text{ for } t \in [0,\tau) \\ 0 \quad \text{ for } t \in [\tau,\tau+\rho] \end{cases},$$

we obtain

$$\phi^{x,\omega\mu^\tau}(\tau+\rho,\lambda,\tilde\mu) = \phi_\tau^{x,\omega\mu^\tau}(\rho,\phi^{x,\omega\mu^\tau}(\tau,\lambda,\tilde\mu),\tilde\mu^{-\tau})$$
$$= \phi^{y,\mu}(\rho,\phi^{x,\omega}(\tau,\lambda,\mu),\mathbf{0})$$
$$= \phi^{y,\mu}(\rho,0_y,\mathbf{0}) = 0_{\varphi(\rho,y,\mu)}.$$

Hence, again using Proposition 1.27, we see that $\omega\mu^\tau$ is regular for x on $[0,\tau+\rho]$. Statement (ii) is proved with a similar argument. □

Proposition 1.29. *Consider the linearization along the controlled trajectory $(\varphi(\cdot,x,\omega),\omega(\cdot))$. Then for all $\tau,C > 0$ there exist $\delta > 0$ and a function $\zeta = \zeta_{\tau,C} : [0,\delta) \to \mathbb{R}_+$ with $\zeta(b) \to 0$ for $b \searrow 0$ such that*

$$\left|\exp_{\varphi(\tau,x,\omega)}^{-1}(\varphi(\tau,y,\mu)) - \phi^{x,\omega}(\tau,\exp_x^{-1}(y),\mu-\omega)\right| \le \zeta(b)b \qquad (1.9)$$

for all $y \in M$ with $\varrho(x,y) \le b$ and $\mu \in L^\infty([0,\tau],\mathbb{R}^m)$ with $\|\omega - \mu\|_{[0,\tau]} \le Cb$, where $b \in [0,\delta)$ is small enough that $\exp_x^{-1}(y)$ and $\exp_{\varphi(\tau,x,\omega)}^{-1}(\varphi(\tau,y,\mu))$

are defined (that is, y and $\varphi(\tau, y, \mu)$ are contained in the codomains of the local diffeomorphisms, defined by restriction of $\exp_x$ and $\exp_{\varphi(\tau,x,\omega)}$ to appropriate open neighborhoods of 0_x and $0_{\varphi(\tau,x,\omega)}$, respectively).[5]

Proof. Let $\tau, C > 0$. Throughout the proof, we write $\|\cdot\|$ for the L^∞-norm of an essentially bounded function $\mu : [0, \tau] \to \mathbb{R}^m$, that is, $\|\mu\| = \operatorname{ess\,sup}_{t \in [0,\tau]} |\mu(t)|$. For given $\tau > 0$, consider the mappings

$$\alpha : M \times L^\infty([0, \tau], \mathbb{R}^m) \to M, \quad (y, \mu) \mapsto \varphi(\tau, y, \mu),$$

and

$$\tilde{\alpha} : T_x M \times L^\infty([0, \tau], \mathbb{R}^m) \supset \tilde{W} \to T_{\varphi(\tau,x,\omega)} M,$$

$$(y, \mu) \mapsto \exp_{\varphi(\tau,x,\omega)}^{-1}(\alpha(\exp_x(y), \mu)),$$

where $\tilde{W}$ is an open neighborhood of $(0_x, \omega)$, chosen small enough such that $\tilde{\alpha}$ is well-defined. Since $\alpha(\exp_x(0_x), \omega) = \varphi(\tau, x, \omega)$ and α is continuous (by Proposition 1.26 (i)), such $\tilde{W}$ exists. Differentiating $\tilde{\alpha}$ at $(0_x, \omega)$ by the chain rule yields

$$D\tilde{\alpha}(0_x, \omega) = d_{\varphi(\tau,x,\omega)} \exp_{\varphi(\tau,x,\omega)}^{-1} D\alpha(x, \omega) D(\exp_x \times \mathrm{id})(0_x, \omega).$$

Using that $D\alpha(x, \omega) = \phi^{x,\omega}(\tau, \cdot, \cdot)$ and that the derivative of the Riemannian exponential map at the origin is the identity, we obtain

$$D\tilde{\alpha}(0_x, \omega)(\lambda, \mu) = \phi^{x,\omega}(\tau, \lambda, \mu).$$

Thus, for all $(y, \mu) \in \tilde{W}$ we have

$$
\begin{aligned}
\exp_{\varphi(\tau,x,\omega)}^{-1}(\varphi(\tau, \exp_x(y), \omega)) &= \tilde{\alpha}(y, \mu) \\
&= \underbrace{\tilde{\alpha}(0_x, \omega)}_{=0_{\varphi(\tau,x,\omega)}} + D\tilde{\alpha}(0_x, \omega)(y, \mu - \omega) + r(y, \mu) \\
&= \phi^{x,\omega}(\tau, y, \mu - \omega) + r(y, \mu),
\end{aligned}
$$

where $r(y, \mu)$ satisfies

$$\lim_{(y,\mu) \to (0_x, \omega)} \frac{r(y, \mu)}{|y| + \|\mu - \omega\|} = 0_{\varphi(\tau,x,\omega)}. \tag{1.10}$$

---

[5]By $\|\cdot\|_{[0,\tau]}$ we denote the L^∞-norm on $L^\infty([0, \tau], \mathbb{R}^m)$.

Hence, we obtain

$$\left|\exp^{-1}_{\varphi(\tau,x,\omega)}(\varphi(\tau,\exp_x(y),\mu)) - \phi^{x,\omega}(\tau,y,\mu-\omega)\right| \equiv |r(y,\mu)|. \qquad (1.11)$$

Since $\tilde{W}$ is an open neighborhood of $(0_x,\omega)$, there exists $\delta = \delta(C) > 0$ such that $B(0_x,\delta) \times B(\omega,C\delta) \subset \tilde{W}$. Define $\zeta_{\tau,C} : [0,\delta) \to \mathbb{R}_+$ by

$$\zeta_{C,\tau}(b) := \begin{cases} b^{-1} \sup\limits_{\substack{|y|\le b, \\ \|\mu-\omega\|\le Cb}} |r(y,\mu)| & \text{for } b \in (0,\delta), \\ 0 & \text{for } b = 0. \end{cases}$$

Then from (1.11) we obtain (1.9). From (1.10) it follows that for every $\varepsilon > 0$ there is $b > 0$ such that $|y| \le b$ and $\|\mu-\omega\| \le Cb$ implies $|r(y,\mu)|/(|y|+\|\mu-\omega\|) \le \varepsilon$. Hence, from

$$\frac{|r(y,\mu)|}{|y|+\|\mu-\omega\|} = \frac{|r(y,\mu)|}{b}\frac{b}{|y|+\|\mu-\omega\|} \le \varepsilon$$

it follows that

$$\frac{|r(y,\mu)|}{b} \le \varepsilon\frac{|y|+\|\mu-\omega\|}{b} \le \varepsilon\frac{b(C+1)}{b} = \varepsilon(C+1).$$

For $b = b(\varepsilon)$ this implies

$$\zeta_{C,\tau}(b) = \sup_{\substack{|y|\le b, \\ \|\mu-\omega\|\le Cb}} \frac{|r(y,\mu)|}{b} \le \varepsilon(C+1),$$

which finishes the proof. □

A controlled trajectory $(\varphi(\cdot,x,\omega),\omega(\cdot))$ is called *periodic with period* $\tau > 0$ or τ-*periodic* if $(\varphi(t+\tau,x,\omega),\omega(t+\tau)) = (\varphi(t,x,\omega),\omega(t))$ for all $t \in \mathbb{R}$, or equivalently $\Theta_\tau\omega = \omega$ and $\varphi(\tau,x,\omega) = x$.

Proposition 1.30. *Let $(\varphi(\cdot,x,\omega),\omega(\cdot))$ be a τ-periodic controlled trajectory which is regular on $[0,\tau]$. Then there exists $C > 0$ such that for every $\lambda \in T_xM$ there is $\mu \in L^\infty([0,\tau],\mathbb{R}^m)$ with*

$$\phi^{x,\omega}(\tau,\lambda,\mu) = 0_x \quad and \quad \|\mu\|_{[0,\tau]} \le C|\lambda|,$$

where $\|\cdot\|_{[0,\tau]}$ denotes the L^∞-norm.

Proof. By regularity, for every $\lambda \in T_xM$ there exists at least one $\mu \in L^\infty([0,\tau],\mathbb{R}^m)$ such that $\phi^{x,\omega}(\tau,\lambda,\mu) = 0_x$, or equivalently

$$\phi^{x,\omega}(\tau,0_x,\mu) = \phi^{x,\omega}(\tau,-\lambda,\mathbf{0}).$$

Consider the automorphism $Q : T_x M \rightarrow T_x M$, $Q\lambda := \phi^{x,\omega}(\tau, -\lambda, \mathbf{0})$, and the continuous linear operator

$$L : L^\infty([0, \tau], \mathbb{R}^m) \rightarrow T_x M, \quad \mu \mapsto \phi^{x,\omega}(\tau, 0_x, \mu).$$

By Proposition 1.27, the regularity assumption is equivalent to L being surjective. Hence, by the *bounded inverse theorem* (cf. Bachman and Narici [6, Theorem 16.5]), there exists a constant $\tilde{C} > 0$ such that for all $\lambda \in T_x M$ there is $\mu \in L^\infty([0, \tau], \mathbb{R}^m)$ with $L\mu = Q\lambda$ and $\|\mu\|_{[0,\tau]} \leq \tilde{C}|Q\lambda|$. Thus, with $C := \tilde{C}\|Q\|$ the assertion holds.

$\square$

1.6   Comments and Bibliographical Notes

For control-affine systems it is usually not assumed that the control range is compact. However, we need this assumption to guarantee compactness of the set $\mathcal{U}$ of admissible control functions in the weak*-topology and continuity of the associated control flow. A comprehensive treatment of control sets for smooth systems given by differential equations can be found in the book of Colonius and Kliemann [25]. Control sets were first introduced by W. Kliemann in his Ph.D. thesis. For smooth discrete-time systems, control sets have been studied by Albertini and Sontag [2, 3] and Wirth [110, 112]. Control sets can also be defined for topological semigroup actions, see Patrão and San Martin [88] or San Martin and Tonelli [95]. The proof for continuity of the transition map of control-affine systems given in Sect. 1.4 is due to Fritz Colonius. However, it has not appeared before in the literature. The proofs that can be found in Colonius and Kliemann [25] and in Kloeden and Rasmussen [68] contain the same mistake which is related to the unjustified use of the Gronwall lemma.

Chapter 2
Introduction to Invariance Entropy

This chapter gives an introduction to the concept of invariance entropy. For topological time-invariant systems, different notions of invariance entropy are defined and investigated. We show that invariance entropy shares several elementary properties with the entropy notions in the classical theory of dynamical systems, that is, with metric and topological entropy. Section 2.3 provides a first nontrivial example for a system which allows for an explicit computation of the associated invariance entropies. The last two sections are concerned with the relations between invariance entropy and minimal data rates. In particular, the relation to the topological feedback entropy introduced by Nair et al. [85] is investigated in Sect. 2.4.

2.1 Definitions and Basic Properties

In the following, consider a topological time-invariant system $\Sigma = (\mathbb{T}, X, U, \mathscr{U}, \varphi)$. That is, the state space X is a metrizable topological space and for each admissible control function $\omega \in \mathscr{U}$ the map

$$\varphi_\omega : \mathbb{T}_+ \times X \to X$$

is assumed to be continuous. Recall from Chap. 1 that the transition map φ satisfies the following properties:

- For all $(x, \omega) \in X \times \mathscr{U}$ it holds that $\varphi(0, x, \omega) = x$;
- For all $t, s \in \mathbb{T}_+$, $x \in X$, and $\omega \in \mathscr{U}$ it holds that

$$\varphi(t + s, x, \omega) = \varphi(s, \varphi(t, x, \omega), \Theta_t \omega)$$

with the shift $\Theta : \mathbb{T} \times \mathscr{U} \to \mathscr{U}$, $(t, \omega(\cdot)) \mapsto \omega(t + \cdot)$, on the set of admissible control functions;

C. Kawan, *Invariance Entropy for Deterministic Control Systems*, Lecture Notes in Mathematics 2089, DOI 10.1007/978-3-319-01288-9_2,
© Springer International Publishing Switzerland 2013

- For each $t \in \mathbb{T}_+$ and $(x, \omega) \in X \times \mathcal{U}$, $\varphi(t, x, \omega)$ does not depend on the values of ω outside of $[0, t)$.

Invariance entropy is a nonnegative (possibly infinite) quantity which is assigned to a pair (K, Q) of subsets of X, which satisfies the properties described in the following definition.

Definition 2.1. A pair (K, Q) of nonempty subsets of X is called *admissible for* Σ if it satisfies the following properties:

(i) The set K is compact;
(ii) For each $x \in K$ there exists $\omega \in \mathcal{U}$ such that $\varphi(\mathbb{T}_+, x, \omega) \subset Q$ (in particular, $K \subset Q$).

Given an admissible pair (K, Q) and $\tau > 0$, a set $\mathcal{S} \subset \mathcal{U}$ is called (τ, K, Q)-*spanning* if

$$\forall x \in K \; \exists \omega \in \mathcal{S} : \; \varphi([0, \tau], x, \omega) \subset Q.$$

By $r_{\text{inv}}(\tau, K, Q)$ we denote the minimal number of elements such a set can have (if no finite (τ, K, Q)-spanning set exists, $r_{\text{inv}}(\tau, K, Q) := \infty$). If $K = Q$, we omit the argument K, that is, we write $r_{\text{inv}}(\tau, Q)$ and we speak of (τ, Q)-spanning sets.

The existence of (τ, K, Q)-spanning sets is guaranteed by property (ii); indeed, $\mathcal{U}$ is a (τ, K, Q)-spanning set for every $\tau > 0$. A pair of the form (Q, Q) is admissible if and only if Q is a compact and controlled invariant set (cf. Definition 1.9).

Definition 2.2. Given an admissible pair (K, Q), we define the *invariance entropy of* (K, Q) by

$$h_{\text{inv}}(K, Q) = h_{\text{inv}}(K, Q; \Sigma) := \limsup_{\tau \to \infty} \frac{1}{\tau} \log r_{\text{inv}}(\tau, K, Q).$$

Here, we use the convention that $\log = \log_2$ for discrete-time systems and $\log = \log_e = \ln$ for continuous-time systems. If $K = Q$, again we omit the argument K and write $h_{\text{inv}}(Q)$. Moreover, we let $\log \infty := \infty$.

Hence, the invariance entropy of (K, Q) measures the exponential growth rate of the minimal number of different control functions sufficient to stay in Q when starting in K, as time tends to infinity.

Example 2.1. Assume that an admissible pair (K, Q) satisfies $\mathcal{O}^+(x) \subset Q$ for all $x \in K$, that is, one cannot escape from Q when the initial state lies in K. Then $r_{\text{inv}}(\tau, K, Q) = 1$ for all τ and hence $h_{\text{inv}}(K, Q) = 0$. Another trivial example with $h_{\text{inv}}(K, Q) = 0$ is given if the set K is finite, since then $r_{\text{inv}}(\tau, K, Q) \leq \#K$ for all τ.

The following monotonicity properties are easy to see and hence we omit their proofs.

Proposition 2.1. *Let* (K, Q) *be an admissible pair.*

(i) *If* $\tau_1 < \tau_2$, *then* $r_{\mathrm{inv}}(\tau_1, K, Q) \leq r_{\mathrm{inv}}(\tau_2, K, Q)$.

(ii) *If* $Q \subset P$, *then also* (K, P) *is admissible and* $r_{\mathrm{inv}}(\tau, K, Q) \geq r_{\mathrm{inv}}(\tau, K, P)$ *for all* $\tau > 0$; *hence,* $h_{\mathrm{inv}}(K, Q) \geq h_{\mathrm{inv}}(K, P)$.

(iii) *If* $L \subset K$ *is closed in* X, *then also* (L, Q) *is admissible and* $r_{\mathrm{inv}}(\tau, L, Q) \leq r_{\mathrm{inv}}(\tau, K, Q)$ *for all* $\tau > 0$; *hence,* $h_{\mathrm{inv}}(L, Q) \leq h_{\mathrm{inv}}(K, Q)$.

(iv) *If* $\Sigma' = (\mathbb{T}, X, U, \mathscr{U}', \varphi')$ *is another system with* $\mathscr{U}' \supset \mathscr{U}$ *and* $\varphi'(t, x, \omega) = \varphi(t, x, \omega)$ *whenever* $\omega \in \mathscr{U}$, *then* (K, Q) *is also admissible for* Σ' *and* $h_{\mathrm{inv}}(K, Q; \Sigma') \leq h_{\mathrm{inv}}(K, Q; \Sigma)$.

Unfortunately, finite (τ, K, Q)-spanning sets for admissible pairs do not always exist (see Example 2.3). However, if Q is an open set, there is no problem.

Proposition 2.2. *If* Q *is open, then* $r_{\mathrm{inv}}(\tau, K, Q)$ *is finite for all* $\tau > 0$.

Proof. Let X be endowed with a metric ϱ. Fix $\iota \in \mathbb{T} \cap (0, \infty)$ and take $x \in K$. Then, by definition of admissible pairs, there exists $\omega_x \in \mathscr{U}$ with $\varphi(\mathbb{T}_+, x, \omega_x) \subset Q$. The set $[0, \tau] \times K$ is compact. Hence, the restriction of the continuous map $\varphi_{\omega_x} : \mathbb{T}_+ \times X \to X$ to this set is uniformly continuous. In particular, for each $\varepsilon > 0$ there exists $\delta > 0$ such that $\varrho(x, y) < \delta$ implies $\varrho(\varphi_{\omega_x}(t, x), \varphi_{\omega_x}(t, y)) < \varepsilon$ for all $t \in [0, \tau]$. By compactness of the finite-time orbit $\varphi_{\omega_x}([0, \tau] \times \{x\})$ and openness of Q, we find $\varepsilon > 0$ such that the ε-neighborhood of this orbit is still contained in Q. Taking $\delta_x = \delta(\varepsilon)$, we find that $\varphi_{\omega_x}(t, y) \in Q$ for all $t \in [0, \tau]$ and $y \in B(x, \delta_x)$. The balls $B(x, \delta_x)$, $x \in K$, form an open cover of K. By compactness we can choose a finite subcover $B(x_1, \delta_{x_1}), \ldots, B(x_n, \delta_{x_n})$. This implies that the corresponding set $\mathscr{S} := \{\omega_{x_1}, \ldots, \omega_{x_n}\}$ of control functions is (τ, K, Q)-spanning. $\qquad \square$

As will become clear later, the case $K = Q$ is of particular interest. Also in this case, minimal spanning sets need not be finite. But at least we have the following proposition.

Proposition 2.3. *Let* $Q \subset X$ *be a compact controlled invariant set. Then the following assertions hold:*

(i) *The number* $r_{\mathrm{inv}}(\tau, Q)$ *is either finite for all* $\tau > 0$ *or for none.*

(ii) *The function* $\tau \mapsto \log r_{\mathrm{inv}}(\tau, Q)$, $\mathbb{T} \cap (0, \infty) \to \mathbb{R}_+ \cup \{\infty\}$, *is subadditive and therefore*

$$h_{\mathrm{inv}}(Q) = \lim_{\tau \to \infty} \frac{1}{\tau} \log r_{\mathrm{inv}}(\tau, Q) = \inf_{\tau > 0} \frac{1}{\tau} \log r_{\mathrm{inv}}(\tau, Q).$$

Proof. (i) Assume that $r_{\mathrm{inv}}(\tau_0, Q) < \infty$ for some positive $\tau_0 \in \mathbb{T}$. By Proposition 2.1 (i) we have $r_{\mathrm{inv}}(\tau, Q) \leq r_{\mathrm{inv}}(\tau_0, Q) < \infty$ for all $\tau \in (0, \tau_0)$. Now let $\tau > \tau_0$. Fix an integer $k \geq 1$ with $k\tau_0 \geq \tau$ and let $\mathscr{S} = \{\omega_1, \ldots, \omega_n\}$, $n = r_{\mathrm{inv}}(\tau_0, Q)$, be a minimal (τ_0, Q)-spanning set. We may regard the elements of $\mathscr{S}$ as functions defined on $[0, \tau_0)$, that is, as elements of $\mathscr{U}[0, \tau_0)$. For every k-tuple $(i_0, \ldots, i_{k-1}) \in \{1, \ldots, n\}^k$ we define a function $\omega_{i_0, i_1, \ldots, i_{k-1}} \in \mathscr{U}[0, k\tau_0)$ by

$$\omega_{i_0,i_1,\ldots,i_{k-1}} := \omega_{i_0}\omega_{i_1}^{\tau_0}\omega_{i_2}^{2\tau_0}\cdots\omega_{i_{k-1}}^{(k-1)\tau_0}.$$

Since $\mathscr{U}$ is invariant with respect to shifts and finite concatenations, we can extend these functions to elements of $\mathscr{U}$. The set

$$\mathscr{S}_k := \left\{\omega_{i_0,i_1,\ldots,i_{k-1}} \ : \ (i_0,i_1,\ldots,i_{k-1}) \in \{1,\ldots,n\}^k\right\} \subset \mathscr{U},$$

obtained by this construction, has n^k elements. We claim that $\mathscr{S}_k$ is a $(k\tau_0, Q)$-spanning set. Indeed, take an arbitrary $x_0 \in Q$. Then there exists $\omega_{i_0} \in \mathscr{S}$ with $\varphi([0,\tau_0], x_0, \omega_{i_0}) \subset Q$. Let $x_1 := \varphi(\tau_0, x_0, \omega_{i_0})$. Then again, there exists $\omega_{i_1} \in \mathscr{S}$ with $\varphi([0,\tau_0], x_1, \omega_{i_1}) \subset Q$. Repeating this process, after k steps we have obtained control functions $\omega_{i_0}, \ldots, \omega_{i_{k-1}} \in \mathscr{S}$ such that

$$\varphi([0,\tau_0], x_j, \omega_{i_j}) \subset Q \quad \text{for } j = 0, 1, \ldots, k-1, \tag{2.1}$$

where $x_1, \ldots, x_k$ are defined recursively by

$$x_j := \varphi(\tau_0, x_{j-1}, \omega_{i_{j-1}}), \quad j = 1, \ldots, k.$$

Using the cocycle property of φ it follows inductively that

$$x_j = \varphi(j\tau_0, x_0, \omega_{i_0,i_1,\ldots,i_{k-1}}), \quad j = 0, 1, \ldots, k.$$

Taking an arbitrary $t \in [0, k\tau_0]$, we can write $t = j\tau_0 + r$ with an integer $j \in \{0, 1, \ldots, k\}$ and $r \in [0, \tau_0)$. Writing $\omega := \omega_{i_0,i_1,\ldots,i_{k-1}}$ and using (2.1), this gives

$$\varphi(t, x_0, \omega) = \varphi(r, \varphi(j\tau_0, x_0, \omega), \Theta_{j\tau_0}\omega) = \varphi(r, x_j, \Theta_{j\tau_0}\omega)$$
$$= \varphi(r, x_j, \omega_{i_j}) \in Q.$$

Hence, $\mathscr{S}_k$ is $(k\tau_0, Q)$-spanning, showing that

$$r_{\mathrm{inv}}(\tau, Q) \le r_{\mathrm{inv}}(k\tau_0, Q) \le n^k < \infty.$$

(ii) If $r_{\mathrm{inv}}(\tau, Q) = \infty$ for all τ, the assertion is trivial. Hence, by (i) we can assume that $r_{\mathrm{inv}}(\tau, Q) < \infty$ for all τ. Then it remains to show that

$$r_{\mathrm{inv}}(\tau_1 + \tau_2, Q) \le r_{\mathrm{inv}}(\tau_1, Q) \cdot r_{\mathrm{inv}}(\tau_2, Q) \quad \text{for all } \tau_1, \tau_2 > 0,$$

since the rest follows from the subadditivity lemma B.3. To this end, let $\mathscr{S}_j$ ($j = 1, 2$) be minimal (τ_j, Q)-spanning sets. Define control functions $\omega \in \mathscr{U}[0, \tau_1 + \tau_2)$ by

$$\omega := \omega_1\omega_2^{\tau_1}, \quad (\omega_1, \omega_2) \in \mathscr{S}_1 \times \mathscr{S}_2.$$

With similar arguments as in the proof of (i) it follows that the set of all these functions is $(\tau_1 + \tau_2, Q)$-spanning. Hence,

$$r_{\text{inv}}(\tau_1 + \tau_2, Q) \leq \#\mathscr{S}_1 \cdot \#\mathscr{S}_2 = r_{\text{inv}}(\tau_1, Q) \cdot r_{\text{inv}}(\tau_2, Q),$$

which concludes the proof.   □

Remark 2.1. The preceding proposition implies the equivalence of the following statements:

- $h_{\text{inv}}(Q)$ is finite;
- $r_{\text{inv}}(\tau, Q)$ is finite for some τ;
- $r_{\text{inv}}(\tau, Q)$ is finite for all τ.

The following examples show that both cases, $h_{\text{inv}}(Q) < \infty$ and $h_{\text{inv}}(Q) = \infty$, are possible.

Example 2.2. Consider a system of the form $\Sigma = (\mathbb{R}, \mathbb{R}, U, \mathscr{U}, \varphi)$, that is, the system is continuous in time and its state space is the real line. Moreover, assume that the maps $\varphi_{t,\omega}$ are invertible. Let $Q = [a, b]$ be a compact and controlled invariant interval. We claim that in this case $h_{\text{inv}}(Q) < \infty$. By the preceding remark, it suffices to show that $r_{\text{inv}}(\tau, Q) < \infty$ for some $\tau > 0$. If $a = b$, this is trivial, because then $r_{\text{inv}}(\tau, Q) = 1$ for all $\tau > 0$. Hence, we may assume that there exists $c \in (a, b)$. Since Q is controlled invariant, there are $\omega_a, \omega_b \in \mathscr{U}$ with $\varphi(\mathbb{R}_+, a, \omega_a) \subset Q$ and $\varphi(\mathbb{R}_+, b, \omega_b) \subset Q$. By continuity of φ in t we can choose $\tau > 0$ small enough that $\varphi([0, \tau], c, \omega_a) \subset Q$ and $\varphi([0, \tau], c, \omega_b) \subset Q$. By continuity of φ in x the sets $\varphi_{t,\omega_a}([a, c])$ and $\varphi_{t,\omega_b}([c, b])$, $t \in [0, \tau]$, are compact intervals, and by invertibility, these intervals are given by $[\varphi_{t,\omega_a}(a), \varphi_{t,\omega_a}(c)]$ and $[\varphi_{t,\omega_b}(c), \varphi_{t,\omega_b}(b)]$. Hence, the set $\mathscr{S} := \{\omega_a, \omega_b\}$ is (τ, Q)-spanning.

Note that the argument used in the preceding example does not work for a one-dimensional discrete-time system.

Example 2.3. Consider the smooth system given by the differential equations

$$\begin{pmatrix} \dot{x}(t) \\ \dot{y}(t) \end{pmatrix} = \left(\frac{x(t)}{(x(t)^2 + y(t)^2)^{1/2}} - \omega(t) \right)^2 \begin{pmatrix} x(t) \\ y(t) \end{pmatrix}, \quad \omega \in \mathscr{U},$$

with state space $M := \mathbb{R}^2 \backslash \{(0, 0)\}$ and control range $\Omega := [-1, 1]$. For every $z = (x, y) \in M$ there exists a constant control function $\omega_z \in \mathscr{U}$ such that (ω_z, z) is an equilibrium pair, namely

$$\omega_z(t) :\equiv \frac{x}{(x^2 + y^2)^{1/2}}.$$

Hence, every subset of M is controlled invariant. We define

$$Q := \left\{ (x, y) \in M : \frac{1}{2} \leq (x^2 + y^2)^{1/2} \leq 1 \right\},$$

that is, Q is the compact annulus with inner radius $1/2$ and outer radius 1. Obviously, from every state $z \in M$ one can only steer the system along the line through $(0,0)$ and z away from the origin. Hence, a point z on the outer boundary $S_1 := \{(x, y) : x^2 + y^2 = 1\} \subset Q$ can only be kept in Q for some positive time $\tau > 0$ by using the constant control function ω_z. Since one needs infinitely many of these control functions for all points on S_1, no finite (τ, Q)-spanning set exists. This shows that $h_{\mathrm{inv}}(Q) = \infty$.

Now we introduce another quantity associated with an admissible pair whose definition requires a metric.

Definition 2.3. Given an admissible pair (K, Q) such that Q is closed in X, and a metric ϱ on X, we define the *outer invariance entropy of* (K, Q) by

$$h_{\mathrm{inv,out}}(K, Q) := h_{\mathrm{inv,out}}(K, Q; \varrho; \Sigma) := \lim_{\varepsilon \searrow 0} h_{\mathrm{inv}}(K, N_\varepsilon(Q))$$

$$= \sup_{\varepsilon > 0} h_{\mathrm{inv}}(K, N_\varepsilon(Q)),$$

where $N_\varepsilon(Q) = \{y \in X : \exists x \in Q \text{ with } \varrho(x, y) < \varepsilon\}$ denotes the ε-neighborhood of Q.

The definition of $h_{\mathrm{inv,out}}(K, Q)$ is correct, that is, the limit for $\varepsilon \searrow 0$ exists and equals the supremum over $\varepsilon > 0$, since from Proposition 2.1 it follows that the pairs $(K, N_\varepsilon(Q))$ are admissible and that $\varepsilon_1 < \varepsilon_2$ implies $h_{\mathrm{inv}}(K, N_{\varepsilon_1}(Q)) \geq h_{\mathrm{inv}}(K, N_{\varepsilon_2}(Q))$. Hence, the supremum over $\varepsilon > 0$ is attained as the limit for $\varepsilon \searrow 0$.

The simple proof of the following proposition is omitted.

Proposition 2.4. *For every admissible pair* (K, Q), *where* Q *is closed, it holds that*

$$0 \leq h_{\mathrm{inv,out}}(K, Q) \leq h_{\mathrm{inv}}(K, Q) \leq \infty.$$

If ϱ_1 and ϱ_2 are two metrics on X which are uniformly equivalent on Q, in the sense that

$$\forall \varepsilon > 0 \; \exists \delta > 0 \; \forall x \in Q \; \forall y \in X :$$

$$\varrho_i(x, y) < \delta \quad \Rightarrow \quad \varrho_j(x, y) < \varepsilon \qquad (2.2)$$

for $(i, j) = (1, 2)$ and $(i, j) = (2, 1)$, then the value of the outer invariance entropy of (K, Q) is the same for both metrics.

Proposition 2.5. *Let* (K, Q) *be an admissible pair such that* Q *is closed in* X. *If* ϱ_1, ϱ_2 *are two metrics on* X *which are uniformly equivalent on* Q, *then* $h_{\mathrm{inv,out}}(K, Q; \varrho_1) = h_{\mathrm{inv,out}}(K, Q; \varrho_2)$. *If* Q *is compact, this is automatically satisfied, and hence, in this case, the outer invariance entropy is independent of the metric.*

Proof. From (2.2) it follows that for given $\varepsilon > 0$ and $\delta = \delta(\varepsilon)$ the δ-neighborhood of Q with respect to ϱ_1 is contained in the ε-neighborhood of Q with respect to ϱ_2 which gives $h_{\mathrm{inv}}(K, N_\delta(Q); \varrho_1) \geq h_{\mathrm{inv}}(K, N_\varepsilon(Q); \varrho_2)$ and thus $h_{\mathrm{inv,out}}(K, Q; \varrho_1) \geq h_{\mathrm{inv,out}}(K, Q; \varrho_2)$. The same holds with the roles of ϱ_1 and ϱ_2 interchanged. If Q is compact, the identity maps $\mathrm{id}_X : (X, \varrho_1) \to (X, \varrho_2)$ and $\mathrm{id}_X : (X, \varrho_2) \to (X, \varrho_1)$ are uniformly continuous on Q which implies uniform equivalence of ϱ_1 and ϱ_2 on Q. $\qquad\square$

An important question is under which conditions the two quantities $h_{\mathrm{inv,out}}(K, Q)$ and $h_{\mathrm{inv}}(K, Q)$ coincide. This becomes clearer in the following chapters, where we will see that $h_{\mathrm{inv,out}}(K, Q)$ is easier accessible to computation than $h_{\mathrm{inv}}(K, Q)$. In particular, it is easier to derive upper bounds for $h_{\mathrm{inv,out}}(K, Q)$. On the other hand, the quantity $h_{\mathrm{inv}}(Q)$ is the more interesting one in terms of its interpretation as a measure for the minimal data rate for invariance of Q, as we will see in Sect. 2.5.

2.2   Elementary Properties

In this section, we prove elementary properties of the invariance entropy, all of which are analogous to well-known properties of the classical dynamical entropy notions. In particular, we discuss the behavior of invariance entropy with respect to finite covers of K and Q, the entropy of power and product systems, and invariance under (time-variant) conjugacies. Moreover, conditions are given under which the limit in ε in the definition of $h_{\mathrm{inv,out}}(K, Q)$ becomes superfluous, and different conditions which imply that the limit superior in this definition can be replaced by a limit inferior.

Time Discretization

The following proposition shows that for the computation of the invariance entropy it is sufficient to consider the system at times which are integer multiples of some fixed time step $\tau > 0$. The only property used in the proof is that the function $\tau \mapsto r_{\mathrm{inv}}(\tau, K, Q)$ is non-decreasing.

Proposition 2.6. *Let Σ be a topological time-invariant system and (K, Q) an admissible pair. Then for all $\tau \in \mathbb{T} \cap (0, \infty)$ we have*

$$h_{\mathrm{inv}}(K, Q) = \limsup_{\mathbb{N} \ni n \to \infty} \frac{1}{n\tau} \log r_{\mathrm{inv}}(n\tau, K, Q). \qquad (2.3)$$

Proof. Obviously, the left-hand side of (2.3) is not less than the right-hand side. In order to show the converse, let $(\tau_k)_{k \geq 1}$, $\tau_k \in \mathbb{T} \cap (0, \infty)$, be an arbitrary sequence converging to ∞. Then for every $k \geq 1$ there exists $n_k \geq 1$ such that $n_k \tau \leq \tau_k \leq (n_k + 1)\tau$, and $n_k \to \infty$ for $k \to \infty$. By Proposition 2.1 we have

$$r_{\mathrm{inv}}(\tau_k, K, Q) \leq r_{\mathrm{inv}}((n_k + 1)\tau, K, Q)$$

and consequently

$$\frac{1}{\tau_k} \log r_{\mathrm{inv}}(\tau_k, K, Q) \leq \frac{1}{n_k \tau} \log r_{\mathrm{inv}}((n_k + 1)\tau, K, Q).$$

This yields

$$\limsup_{k \to \infty} \frac{1}{\tau_k} \log r_{\mathrm{inv}}(\tau_k, K, Q) \leq \limsup_{k \to \infty} \frac{1}{n_k \tau} \log r_{\mathrm{inv}}((n_k + 1)\tau, K, Q).$$

Since

$$\frac{1}{n_k \tau} \log r_{\mathrm{inv}}((n_k + 1)\tau, K, Q) = \frac{n_k + 1}{n_k} \frac{1}{(n_k + 1)\tau} \log r_{\mathrm{inv}}((n_k + 1)\tau, K, Q)$$

and $(n_k + 1)/n_k \to 1$ for $k \to \infty$, we obtain

$$\limsup_{k \to \infty} \frac{1}{\tau_k} \log r_{\mathrm{inv}}(\tau_k, K, Q) \leq \limsup_{k \to \infty} \frac{1}{n_k \tau} \log r_{\mathrm{inv}}(n_k \tau, K, Q)$$

$$\leq \limsup_{n \to \infty} \frac{1}{n \tau} \log r_{\mathrm{inv}}(n \tau, K, Q),$$

which yields the result. $\square$

Subsets

For a classical dynamical system, the topological entropy can only become smaller if one restricts the system to a subset of the state space. Moreover, if one considers a finite cover of the state space and the associated restrictions, the entropy of the dynamical system equals the maximum of the entropies of these restrictions. In the following, we prove analogous results for the invariance entropy.

Lemma 2.1. *For any functions $f_1, \ldots, f_N : \mathbb{T} \cap (0, \infty) \to \mathbb{R}$ ($\mathbb{T} \in \{\mathbb{Z}, \mathbb{R}\}$) it holds that*

$$\limsup_{\tau \to \infty} \frac{1}{\tau} \log \sum_{i=1}^{N} f_i(\tau) \leq \max_{i=1,\ldots,N} \limsup_{\tau \to \infty} \frac{1}{\tau} \log f_i(\tau).$$

Proof. For brevity we write

$$\lambda(f) := \limsup_{\tau \to \infty} \frac{1}{\tau} \log f(\tau)$$

for any function $f : \mathbb{T} \cap (0, \infty) \to \mathbb{R}$. We define $g : \mathbb{T} \cap (0, \infty) \to \mathbb{R}$ by

$$g(\tau) := \max_{i=1,\dots,N} f_i(\tau).$$

Then

$$\lambda\left(\sum_{i=1}^N f_i\right) \leq \lambda(Ng) = \limsup_{\tau \to \infty} \frac{1}{\tau} (\log N + \log g(\tau)) = \lambda(g).$$

Thus, it suffices to show that $\lambda(g) \leq \max_{i=1,\dots,N} \lambda(f_i)$. To this end, let $(\tau_k)_{k \geq 1}$, $\tau_k \in \mathbb{T} \cap (0, \infty)$, be a sequence with $\tau_k \to \infty$ and

$$\lambda(g) = \lim_{k \to \infty} \frac{1}{\tau_k} \log \max_{i=1,\dots,N} f_i(\tau_k).$$

Obviously, there exists $i_0 \in \{1, \dots, N\}$ such that $f_{i_0}(\tau_k) = \max_{i=1,\dots,N} f_i(\tau_k)$ for infinitely many k. Let $(\tau_{n_k})_{k \geq 1}$ be a corresponding subsequence. Then

$$\lambda(g) = \lim_{k \to \infty} \frac{1}{\tau_{n_k}} \log f_{i_0}(\tau_{n_k}) \leq \lambda(f_{i_0}) \leq \max_{i=1,\dots,N} \lambda(f_i),$$

which finishes the proof.   □

Proposition 2.7. *Let* $\Sigma = (\mathbb{T}, X, U, \mathscr{U}, \varphi)$ *be a topological time-invariant system and* (K, Q) *an admissible pair. Assume that* $K = \bigcup_{i=1}^N K_i$ *with finitely many compact sets* $K_1, \dots, K_N$. *Then each pair* (K_i, Q), $i = 1, \dots, N$, *is admissible and*

$$h_{\mathrm{inv}}(K, Q) = \max_{i=1,\dots,N} h_{\mathrm{inv}}(K_i, Q).$$

Proof. By Proposition 2.1 the pairs (K_i, Q) are admissible. If $\mathscr{S}$ is a minimal (τ, K, Q)-spanning set, then $\mathscr{S}$ is also (τ, K_i, Q)-spanning for each $i \in \{1, \dots, N\}$ and hence, $r_{\mathrm{inv}}(\tau, K_i, Q) \leq r_{\mathrm{inv}}(\tau, K, Q)$. This implies

$$\max_{i=1,\dots,N} h_{\mathrm{inv}}(K_i, Q) \leq h_{\mathrm{inv}}(K, Q). \qquad (2.4)$$

On the other hand, if $\mathscr{S}_i$ is a minimal (τ, K_i, Q)-spanning set, $i = 1, \dots, N$, then $\mathscr{S} := \bigcup_{i=1}^N \mathscr{S}_i$ is (τ, K, Q)-spanning, which implies

$$r_{\text{inv}}(\tau, K, Q) \leq \#\mathscr{S} \leq \sum_{i=1}^{N} \#\mathscr{S}_i = \sum_{i=1}^{N} r_{\text{inv}}(\tau, K_i, Q).$$

Using Lemma 2.1, we get

$$h_{\text{inv}}(K, Q) \leq \limsup_{\tau \to \infty} \frac{1}{\tau} \log \sum_{i=1}^{N} r_{\text{inv}}(\tau, K_i, Q) \leq \max_{i=1,\ldots,N} h_{\text{inv}}(K_i, Q),$$

which concludes the proof.   □

It is clear that the analogous result holds for the outer invariance entropy. In Sect. 7.2, we show by a counter-example that this result cannot be generalized to the case of a countable cover of K, that is, it is in general not true that $h_{\text{inv,out}}(K, Q) = \sup_{\alpha \in A} h_{\text{inv,out}}(K_\alpha, Q)$ if $\{K_\alpha\}_{\alpha \in A}$ is a countable cover of K consisting of compact sets (cf. Remark 7.3).

Proposition 2.8. *Let* $\Sigma = (\mathbb{T}, X, U, \mathscr{U}, \varphi)$ *be a topological time-invariant system and* $Q \subset X$ *a compact controlled invariant set. Assume that* $Q = \bigcup_{i=1}^{N} Q_i$ *with finitely many compact controlled invariant sets* $Q_1, \ldots, Q_N$. *Then*

$$h_{\text{inv}}(Q) \leq \max_{i=1,\ldots,N} h_{\text{inv}}(Q_i).$$

Proof. For every $i \in \{1, \ldots, N\}$ let $\mathscr{S}_i$ be a minimal (τ, Q_i)-spanning set and define $\mathscr{S} := \bigcup_{i=1}^{N} \mathscr{S}_i$. Then $\mathscr{S}$ is obviously (τ, Q)-spanning, which implies

$$r_{\text{inv}}(\tau, Q) \leq \#\mathscr{S} \leq \sum_{i=1}^{N} \#\mathscr{S}_i = \sum_{i=1}^{N} r_{\text{inv}}(\tau, Q_i).$$

By Lemma 2.1 we obtain

$$h_{\text{inv}}(Q) \leq \limsup_{\tau \to \infty} \frac{1}{\tau} \log \sum_{i=1}^{N} r_{\text{inv}}(\tau, Q_i) \leq \max_{i=1,\ldots,N} h_{\text{inv}}(Q_i),$$

which yields the result.   □

Power Rule

A well-known property of both metric and topological entropy of classical dynamical systems is that the entropy of the time-t-map is t times the entropy of the time-one-map. Sometimes this is called the *power rule* (cf. Downarowicz [39]).

In this subsection, we prove an analogous result for the invariance entropy of an admissible pair. However, in the formulation of this result we cannot treat discrete- and continuous-time systems simultaneously. We start with the simpler case of continuous-time systems.

Proposition 2.9. *Let $\Sigma = (\mathbb{R}, X, U, \mathscr{U}, \varphi)$ be a continuous-time topological time-invariant system. Then, for every real number $s > 0$, also $\Sigma_s := (\mathbb{R}, X, U, \mathscr{U}_s, \varphi^s)$ with*

$$\mathscr{U}_s := \left\{ \omega \in U^{\mathbb{R}} \ : \ \exists \mu_\omega \in \mathscr{U} \text{ with } \omega(t) \equiv \mu_\omega(st) \right\}$$

and

$$\varphi^s(t, x, \omega) :\equiv \varphi(st, x, \mu_\omega)$$

is a topological time-invariant system. If (K, Q) is an admissible pair for Σ, then it is also admissible for Σ_s and

$$h_{\mathrm{inv}}(K, Q; \Sigma_s) = s \cdot h_{\mathrm{inv}}(K, Q; \Sigma).$$

Proof. We leave the relatively easy proof that Σ_s is a system to the reader. To show that (K, Q) is admissible for Σ_s, let $x \in K$. Then there exists $\omega \in \mathscr{U}$ such that $\varphi(\mathbb{R}_+, x, \omega) \subset Q$. This implies

$$\varphi^s(t, x, \omega(s \cdot)) = \varphi(st, x, \omega) \in Q \quad \text{for all } t \in \mathbb{R}_+.$$

If $\mathscr{S} \subset \mathscr{U}$ is an $(s\tau, K, Q)$-spanning set for Σ, then $\mathscr{S}_s := \{\omega(s \cdot)\}_{\omega \in \mathscr{S}}$ is a (τ, K, Q)-spanning set for Σ_s with the same number of elements. Analogously, every (τ, K, Q)-spanning set for Σ_s gives an $(s\tau, K, Q)$-spanning set for Σ with the same number of elements. This proves that

$$r_{\mathrm{inv}}(s\tau, K, Q; \Sigma) = r_{\mathrm{inv}}(\tau, K, Q; \Sigma_s) \quad \text{for all } \tau > 0.$$

Therefore, we obtain

$$
\begin{aligned}
h_{\mathrm{inv}}(K, Q; \Sigma_s) &= \limsup_{\tau \to \infty} \frac{1}{\tau} \log r_{\mathrm{inv}}(\tau, K, Q; \Sigma_s) \\
&= \limsup_{\tau \to \infty} \frac{1}{\tau} \log r_{\mathrm{inv}}(s\tau, K, Q; \Sigma) \\
&= s \limsup_{\tau \to \infty} \frac{1}{s\tau} \log r_{\mathrm{inv}}(s\tau, K, Q; \Sigma) = s \cdot h_{\mathrm{inv}}(K, Q; \Sigma),
\end{aligned}
$$

concluding the proof.   □

For a smooth system given by differential equations with right-hand side F, the system Σ_s in the preceding proposition is also given by differential equations with corresponding right-hand side $s \cdot F$.

Proposition 2.10. *Let* $\Sigma = (\mathbb{R}, M, \mathbb{R}^m, \mathscr{U}, \varphi)$ *be a smooth system given by differential equations*

$$\dot{x}(t) = F(x(t), \omega(t)), \quad \omega \in \mathscr{U}.$$

Then, for every $s > 0$, *the system* Σ_s *defined in Proposition 2.9 is given by the differential equations*

$$\dot{x}(t) = s \cdot F(x(t), \omega(t)), \quad \omega \in \mathscr{U}. \tag{2.5}$$

Proof. It is clear that $\mathscr{U}_s = \mathscr{U}$. Moreover, for all $(x, \omega) \in M \times \mathscr{U}$ and $\tilde{\omega}(t) := \omega(t/s)$ we have

$$\frac{\mathrm{d}}{\mathrm{d}t}\varphi(st, x, \tilde{\omega}) = s \cdot F(\varphi(st, x, \tilde{\omega}), \omega(t))$$

for almost all $t \in \mathbb{R}$, which shows that the unique solution of (2.5) for the initial value x and the control function ω is given by $t \mapsto \varphi(st, x, \tilde{\omega})$. This proves the assertion. □

Finally, we formulate and prove the discrete-time power rule.

Proposition 2.11. *Let* $\Sigma = (\mathbb{Z}, X, U, \mathscr{U}, \varphi)$ *be a discrete-time topological time-invariant system. Then, for every integer* $k \geq 1$, *also* $\Sigma_k := (\mathbb{Z}, X, U_k, \mathscr{U}_k, \varphi^k)$ *with* $U_k = U \times \cdots \times U$ *(k factors),*

$$\mathscr{U}_k := \left\{ \omega \in U_k^{\mathbb{Z}} : \begin{array}{c} \exists \mu_\omega \in \mathscr{U} \text{ with} \\ \omega(n) \equiv (\mu_\omega(kn), \mu_\omega(kn+1), \ldots, \mu_\omega(kn+k-1)) \end{array} \right\}$$

and $\varphi^k : \mathbb{Z}_+ \times X \times \mathscr{U}_k \to X$ *given by*

$$\varphi^k(n, x, \omega) := \begin{cases} x & \text{if } n = 0, \\ \varphi_{k, \omega_{n-1}} \circ \cdots \circ \varphi_{k, \omega_0}(x) & \text{if } n \geq 1, \end{cases}$$

is a topological time-invariant system. If (K, Q) *is an admissible pair for* Σ, *then it is also admissible for* Σ_k *and*

$$h_{\mathrm{inv}}(K, Q; \Sigma_k) = k \cdot h_{\mathrm{inv}}(K, Q; \Sigma).$$

Proof. Let us first show that Σ_k is a system: The set $\mathscr{U}_k$ is shift-invariant, since for every $\omega \in \mathscr{U}_k$ we find

$$(\Theta_l \omega)(n) = (\Theta_{kl}\mu_\omega(kn), \ldots, \Theta_{kl}\mu_\omega(k(n+1)-1)),$$

and $\Theta_{kl}\mu_\omega \in \mathcal{U}$. The identity axiom $\varphi^k(0, x, \omega) = x$ is satisfied by definition. The cocycle property is proved by

$$\varphi^k(n + m, \cdot, \omega) = \left(\varphi_{k,\omega_{n+m-1}} \circ \cdots \circ \varphi_{k,\omega_m}\right) \circ \left(\varphi_{k,\omega_{m-1}} \circ \cdots \circ \varphi_{k,\omega_0}\right)$$

$$= \varphi^k_{n,\Theta_m\omega} \circ \varphi^k_{m,\omega} = \varphi^k(n, \varphi^k(m, \cdot, \omega), \Theta_m\omega).$$

By definition $\varphi^k(n, x, \omega)$ only depends on the restriction of ω to $\{0, 1, \ldots, n-1\}$. It is clear that Σ_k is a topological time-invariant system. To show that (K, Q) is admissible for Σ_k, let $x \in K$. Then there is $\mu \in \mathcal{U}$ with $\varphi(\mathbb{Z}_+, x, \mu) \subset Q$. Let $\omega \in \mathcal{U}_k$ be defined by $\omega(n) := (\mu(kn), \ldots, \mu(k(n+1)-1))$ for all $n \in \mathbb{Z}$. Then

$$\varphi^k(n, x, \omega) = \varphi_{k,\omega_{n-1}} \circ \cdots \circ \varphi_{k,\omega_0}(x)$$

$$= \varphi_{k,\Theta_{(n-1)k}\mu} \circ \cdots \circ \varphi_{k,\mu}(x) = \varphi(nk, x, \mu) \in Q$$

for all $n \geq 0$. Finally, the assertion about the invariance entropies is proved analogously to the continuous-time case. $\square$

Product Rule

The product rule for topological entropy asserts that $h_{\text{top}}(f \times g) = h_{\text{top}}(f) + h_{\text{top}}(g)$ for continuous maps $f : X \to X$ and $g : Y \to Y$ on compact Hausdorff spaces X, Y (cf. Adler et al. [1] and Goodwyn [51]). With Bowen's definition for topological entropy of uniformly continuous maps on non-compact metric spaces (see Sect. B.3), it can only be shown that $h_{\text{top}}(f \times g) \leq h_{\text{top}}(f) + h_{\text{top}}(g)$. Also for the invariance entropy, only the corresponding inequality can be proved.

Proposition 2.12. *Let* $\Sigma_i = (\mathbb{T}, X_i, U_i, \mathcal{U}_i, \varphi_i)$, $i = 1, 2$, *be two topological time-invariant systems with associated admissible pairs* (K_i, Q_i). *Then also* $\Sigma := (\mathbb{T}, X_1 \times X_2, U_1 \times U_2, \mathcal{U}_1 \times \mathcal{U}_2, \varphi_1 \times \varphi_2)$ *with*

$$\varphi_1 \times \varphi_2 : \mathbb{T}_+ \times (X_1 \times X_2) \times (\mathcal{U}_1 \times \mathcal{U}_2) \to X_1 \times X_2,$$

$$(t, (x_1, x_2), (\omega_1, \omega_2)) \mapsto (\varphi_1(t, x_1, \omega_1), \varphi_2(t, x_2, \omega_2)),$$

is a topological time-invariant system and $(K_1 \times K_2, Q_1 \times Q_2)$ *is an admissible pair for* Σ *such that*

$$h_{\text{inv}}(K_1 \times K_2, Q_1 \times Q_2; \Sigma) \leq h_{\text{inv}}(K_1, Q_1; \Sigma_1) + h_{\text{inv}}(K_2, Q_2; \Sigma_2). \quad (2.6)$$

Proof. The proof that Σ is a topological time-invariant system and $(K_1 \times K_2, Q_1 \times Q_2)$ is admissible for Σ is straightforward. (Note that $X_1 \times X_2$ is endowed with the product topology.) To show the formula for the entropies, let $\mathcal{S}_i$ be minimal

56

2 Introduction to Invariance Entropy

(τ, K_i, Q_i)-spanning sets, $i = 1, 2$. Then $\mathscr{S} := \mathscr{S}_1 \times \mathscr{S}_2 \subset \mathscr{U}_1 \times \mathscr{U}_2$ obviously is $(\tau, K_1 \times K_2, Q_1 \times Q_2)$-spanning and thus

$$r_{\text{inv}}(\tau, K_1 \times K_2, Q_1 \times Q_2; \Sigma) \le r_{\text{inv}}(\tau, K_1, Q_1; \Sigma_1) \cdot r_{\text{inv}}(\tau, K_2, Q_2; \Sigma_2),$$

which implies (2.6). □

It is not clear if the converse inequality in (2.6) holds. However, this seems improbable due to the fact that the definition of invariance entropy involves the limit superior (instead of the limit inferior). But also in the case $K = Q$, where we have shown that the limit superior is a limit, it is not clear how the other inequality could be proved.

Invariance Under Conjugacy

The importance of entropy in classical dynamical systems in particular is based on the fact that it is a dynamical invariant, that is, conjugate systems (in the corresponding category, measure-theoretic or topological) have the same entropy. In this subsection, we describe an appropriate notion of topological conjugacy for control systems, which preserves the invariance entropy of an admissible pair.

Definition 2.4. Consider two topological time-invariant systems $\Sigma_1 = (\mathbb{T}, X_1, U_1, \mathscr{U}_1, \varphi_1)$ and $\Sigma_2 = (\mathbb{T}, X_2, U_2, \mathscr{U}_2, \varphi_2)$. Let $\pi : \mathbb{T}_+ \times X_1 \to X_2$, $(t, x) \mapsto \pi_t(x)$, be a continuous map and $h : \mathscr{U}_1 \to \mathscr{U}_2$ a map such that

$$\pi_t(\varphi_1(t, x, \omega)) = \varphi_2(t, \pi_0(x), h(\omega)) \tag{2.7}$$

holds for all $t \in \mathbb{T}_+$, $x \in X_1$, and $\omega \in \mathscr{U}_1$. Then (π, h) is called a *time-variant semi-conjugacy from Σ_1 to Σ_2*. If π is independent of $t \in \mathbb{T}_+$, we can regard π as a map from X_1 to X_2 and we say that (π, h) is a *(time-invariant) semi-conjugacy from Σ_1 to Σ_2*. If each of the maps $\pi_t : X_1 \to X_2$ is a homeomorphism and $h : \mathscr{U}_1 \to \mathscr{U}_2$ is invertible, we call (π, h) a *time-variant conjugacy from Σ_1 to Σ_2*.

Remark 2.2. It is easy to see that if (π, h) is a time-variant conjugacy from Σ_1 to Σ_2, then (ψ, h^{-1}) with $\psi_t(y) := \pi_t^{-1}(y)$ is a time-variant conjugacy from Σ_2 to Σ_1.

Remark 2.3. For smooth systems with the same set $\mathscr{U}$ of admissible control functions, one usually speaks of a *state equivalence* or *state-space equivalence* if there is a diffeomorphism π such that $(\pi, \text{id}_{\mathscr{U}})$ is a topological conjugacy.

Proposition 2.13. *Consider two systems $\Sigma_1 = (\mathbb{T}, X_1, U_1, \mathscr{U}_1, \varphi_1)$ and $\Sigma_2 = (\mathbb{T}, X_2, U_2, \mathscr{U}_2, \varphi_2)$ and let (π, h) be a time-variant semi-conjugacy from Σ_1 to Σ_2. Further assume that (K, Q) is an admissible pair for Σ_1 and*

$$\pi_t(Q) \subset \pi_0(Q) \quad \text{for all } t > 0. \tag{2.8}$$

Then $(\pi_0(K), \pi_0(Q))$ is an admissible pair for system Σ_2 and

$$h_{\mathrm{inv}}(K, Q; \Sigma_1) \geq h_{\mathrm{inv}}(\pi_0(K), \pi_0(Q); \Sigma_2).$$

Moreover, if Q is compact and the family $\{\pi_t\}_{t \in \mathbb{T}_+}$ is pointwise equicontinuous, then

$$h_{\mathrm{inv,out}}(K, Q; \Sigma_1) \geq h_{\mathrm{inv,out}}(\pi_0(K), \pi_0(Q); \Sigma_2).$$

Proof. Since π is continuous, the set $\pi_0(K)$ is compact. Let $y \in \pi_0(K)$. Then there exists $x \in K$ with $y = \pi_0(x)$. Since (K, Q) is admissible for Σ_1, there is $\omega \in \mathcal{U}_1$ with $\varphi_1(\mathbb{T}_+, x, \omega) \subset Q$. Using the semi-conjugacy equation (2.7) and assumption (2.8), this implies

$$\varphi_2(t, y, h(\omega)) = \pi_t(\varphi_1(t, x, \omega)) \in \pi_t(Q) \subset \pi_0(Q) \quad \text{for all } t \in \mathbb{T}_+.$$

Therefore, $(\pi_0(K), \pi_0(Q))$ is admissible for Σ_2.

Now let $\mathcal{S} \subset \mathcal{U}_1$ be a minimal (τ, K, Q)-spanning set. With the same arguments as above we find that $h(\mathcal{S}) \subset \mathcal{U}_2$ is $(\tau, \pi_0(K), \pi_0(Q))$-spanning. Hence,

$$r_{\mathrm{inv}}(\tau, \pi_0(K), \pi_0(Q)) \leq \# h(\mathcal{S}) \leq \# \mathcal{S} = r_{\mathrm{inv}}(\tau, K, Q),$$

implying that $h_{\mathrm{inv}}(\pi_0(K), \pi_0(Q); \Sigma_2) \leq h_{\mathrm{inv}}(K, Q; \Sigma_1)$.

Now assume that Q is compact. Let ϱ_1 denote a metric on X_1 and ϱ_2 a metric on X_2. By equicontinuity of $\{\pi_t\}$ we have that

$$\forall x \in X \; \forall \varepsilon > 0 \; \exists \delta > 0 \; \forall t \in \mathbb{T}_+ : \; \varrho_1(x, y) < \delta \Rightarrow \varrho_2(\pi_t(x), \pi_t(y)) < \varepsilon.$$

As for a single map, the equicontinuity is uniform on the compact set Q, that is,

$$\forall \varepsilon > 0 \; \exists \delta > 0 \; \forall t \in \mathbb{T}_+ \; \forall x \in Q, \; y \in X :$$
$$\varrho_1(x, y) < \delta \Rightarrow \varrho_2(\pi_t(x), \pi_t(y)) < \varepsilon.$$

We claim that if $\mathcal{S} \subset \mathcal{U}_1$ is a $(\tau, K, N_\delta(Q))$-spanning set with $\delta = \delta(\varepsilon)$ as above, then $h(\mathcal{S}) \subset \mathcal{U}_2$ is $(\tau, \pi_0(K), N_\varepsilon(\pi_0(Q)))$-spanning. Indeed, take $y \in \pi_0(K)$ and $x \in K$ with $y = \pi_0(x)$. Let $\omega \in \mathcal{S}$ with $\varphi_1([0, \tau], x, \omega) \subset N_\delta(Q)$. Then for each $t \in [0, \tau]$ there exists $x_t \in Q$ with $\varrho_1(x_t, \varphi_1(t, x, \omega)) < \delta$. This implies

$$\varrho_2(\varphi_2(t, y, h(\omega)), \pi_t(x_t)) = \varrho_2(\pi_t(\varphi_1(t, x, \omega)), \pi_t(x_t)) < \varepsilon$$

for all $t \in [0, \tau]$. Since $\pi_t(x_t) \in \pi_t(Q) \subset \pi_0(Q)$, this proves the claim. Hence, we obtain $r_{\mathrm{inv}}(\tau, \pi_0(K), N_\varepsilon(\pi_0(Q))) \leq r_{\mathrm{inv}}(\tau, K, N_\delta(Q))$, which implies the assertion about the outer invariance entropies. $\qquad \square$

For two smooth systems given by differential equations, a sufficient condition for the existence of a topological conjugacy can be formulated in terms of the right-hand sides of these systems.

Proposition 2.14. *Let $\Sigma_i = (\mathbb{R}, M_i, \mathbb{R}^{m_i}, \mathscr{U}_i, \varphi_i)$ $(i = 1, 2)$ be two smooth systems given by differential equations*

$$\dot{x}(t) = F_i(x(t), \omega(t)), \quad \omega \in \mathscr{U}_i, \quad i = 1, 2,$$

with corresponding control ranges $\Omega_i \subset \mathbb{R}^{m_i}$, which are bounded. Assume that $\pi : \mathbb{R}_+ \times M_1 \to M_2$ is a $\mathscr{C}^1$-map and that $H : \mathbb{R} \times \mathbb{R}^{m_1} \to \mathbb{R}^{m_2}$ is continuous with $H(t, \Omega_1) \subset \Omega_2$ for all $t \in \mathbb{R}$ and

$$\frac{\partial \pi}{\partial t}(t, x) + \frac{\partial \pi}{\partial x}(t, x) F_1(x, u) = F_2(\pi(t, x), H(t, u)) \tag{2.9}$$

for all $t \in \mathbb{R}_+$, $x \in M_1$ and $u \in \mathbb{R}^{m_1}$. Let

$$h : \mathscr{U}_1 \to \mathscr{U}_2, \quad h(\omega)(t) :\equiv H(t, \omega(t)). \tag{2.10}$$

Then (π, h) is a time-variant semi-conjugacy from Σ_1 to Σ_2.

Proof. First we have to show that h is well-defined, that is, $h(\mathscr{U}_1) \subset \mathscr{U}_2$. It is clear that $h(\omega)$ is measurable for each $\omega \in \mathscr{U}_1$. Since Ω_2 is bounded, it follows from $H(t, \Omega_1) \subset \Omega_2$ that $h(\omega)$ is essentially bounded and $h(\omega)(t) \in \Omega_2$ for almost all $t \in \mathbb{R}$, which proves that $h(\omega) \in \mathscr{U}_2$.

In order to show the semi-conjugacy identity (2.7), we fix $(x, \omega) \in M_1 \times \mathscr{U}_1$ and consider the curves

$$\xi(t) :\equiv \pi(t, \varphi_1(t, x, \omega)), \quad \eta(t) :\equiv \varphi_2(t, \pi_0(x), h(\omega)).$$

We have $\xi(0) = \pi(0, x) = \pi_0(x) = \eta(0)$. Differentiating ξ gives (for almost all $t \in \mathbb{R}_+$)

$$
\begin{aligned}
\dot{\xi}(t) &= \frac{\partial \pi}{\partial t}(t, \varphi_1(t, x, \omega)) + \frac{\partial \pi}{\partial x}(t, \varphi_1(t, x, \omega)) \frac{\partial}{\partial t} \varphi_1(t, x, \omega) \\
&= \frac{\partial \pi}{\partial t}(t, \varphi_1(t, x, \omega)) + \frac{\partial \pi}{\partial x}(t, \varphi_1(t, x, \omega)) F_1(\varphi_1(t, x, \omega), \omega(t)) \\
&\overset{(2.9)}{=} F_2\left(\pi(t, \varphi_1(t, x, \omega)), H(t, \omega(t))\right) \\
&\overset{(2.10)}{=} F_2\left(\xi(t), h(\omega)(t)\right).
\end{aligned}
$$

By differentiating η we obtain

$$\dot{\eta}(t) = F_2(\varphi_2(t, \pi_0(x), h(\omega)), h(\omega)(t)) = F_2(\eta(t), h(\omega)(t)).$$

Hence, by uniqueness of solutions, $\xi(t) = \eta(t)$ for all $t \in \mathbb{R}_+$, which proves the assertion. □

Example 2.4. Consider a linear system $\Sigma = (\mathbb{T}, X, U, \mathscr{U}, \varphi)$. Let $X = Y \oplus Z$ be a decomposition into linear subspaces Y and Z which are invariant under each of the maps $\varphi_{t,0} : X \to X$, that is, $\varphi_{t,0} Y \subset Y$ and $\varphi_{t,0} Z \subset Z$ for all $t \in \mathbb{T}_+$. Let $\pi \in \mathscr{L}(X, X)$ be the projection onto Y along Z. Then one obtains another linear system $\Sigma' = (\mathbb{T}, Y, U, \mathscr{U}, \varphi')$ with

$$\varphi'(t, y, \omega) := \pi \varphi(t, y, \omega) \quad \text{for all } (t, y, \omega) \in \mathbb{T}_+ \times Y \times \mathscr{U}.$$

We only show the cocycle property of φ' and leave the rest to the reader. By linearity of $\varphi(t, \cdot, \cdot)$, we obtain

$$\varphi'(t + s, y, \omega) = \pi \varphi(t + s, y, \omega) = \pi \varphi(s, \varphi(t, y, \omega), \Theta_t \omega)$$
$$= \pi \varphi(s, \pi \varphi(t, y, \omega), \Theta_t \omega) + \pi \underbrace{\varphi(s, (\mathrm{id} - \pi) \varphi(t, y, \omega), \mathbf{0})}_{\in \ker \pi = Z}$$
$$= \varphi'(s, \varphi'(t, y, \omega), \Theta_t \omega).$$

We claim that $(\pi, \mathrm{id}_{\mathscr{U}})$ is a (time-invariant) semi-conjugacy from Σ to Σ'. To show this, first note that

$$\pi \varphi(t, x, \mathbf{0}) = \pi \Big(\underbrace{\varphi(t, \pi x, \mathbf{0})}_{\in Y = \mathrm{im}\,\pi} + \underbrace{\varphi(t, (\mathrm{id} - \pi) x, \mathbf{0})}_{\in Z = \ker \pi} \Big) = \varphi(t, \pi x, \mathbf{0}).$$

This implies

$$\pi \varphi(t, x, \omega) = \pi \varphi(t, x, \mathbf{0}) + \pi \varphi(t, 0, \omega) = \varphi(t, \pi x, \mathbf{0}) + \pi \varphi(t, 0, \omega)$$
$$= \pi \varphi(t, \pi x, \omega) = \varphi'(t, \pi x, \omega).$$

Since π is continuous, this proves the claim. In Chap. 3, we use a semi-conjugacy of this form to compute the outer invariance entropy of a linear system.

Example 2.5. Consider two planar linear systems $\Sigma_i = (\mathbb{R}, \mathbb{R}^2, \mathbb{R}^{m_i}, \mathscr{U}_i, \varphi_i)$ given by differential equations corresponding to matrix pairs (A_i, B_i) with

$$A_1 = \begin{bmatrix} 1 & -1 \\ 1 & 1 \end{bmatrix}, \quad B_1 \in \mathbb{R}^{2 \times m_1} \text{ arbitrary,}$$

and

$$A_2 = \begin{bmatrix} 1 & 0 \\ 0 & 1 \end{bmatrix}, \quad B_2 = \begin{bmatrix} 1 & 0 \\ 0 & 1 \end{bmatrix}.$$

Let the corresponding control ranges be given by $\Omega_1 = B(0, 1) \subset \mathbb{R}^{m_1}$ and $\Omega_2 = B(0, \|B_1\|) \subset \mathbb{R}^2$, where $\|\cdot\|$ denotes the operator norm induced by the standard Euclidean norm in $\mathbb{R}^2$. Define transformations

$$\pi(t, x) := \begin{bmatrix} \cos(t) & \sin(t) \\ -\sin(t) & \cos(t) \end{bmatrix} x, \quad \pi : \mathbb{R}_+ \times \mathbb{R}^2 \to \mathbb{R}^2,$$

$$H(t, u) := \begin{bmatrix} \cos(t) & \sin(t) \\ -\sin(t) & \cos(t) \end{bmatrix} B_1 u, \quad H : \mathbb{R} \times \mathbb{R}^{m_1} \to \mathbb{R}^2.$$

Then we obtain

$$\frac{\partial \pi}{\partial t}(t, x) + \frac{\partial \pi}{\partial x}(t, x)(A_1 x + B_1 u)$$

$$= \begin{bmatrix} -\sin(t) & \cos(t) \\ -\cos(t) & -\sin(t) \end{bmatrix} x$$

$$+ \begin{bmatrix} \cos(t) & \sin(t) \\ -\sin(t) & \cos(t) \end{bmatrix} \left(\begin{bmatrix} 1 & -1 \\ 1 & 1 \end{bmatrix} x + B_1 u \right)$$

$$= \begin{bmatrix} \cos(t) & \sin(t) \\ -\sin(t) & \cos(t) \end{bmatrix} x + \begin{bmatrix} \cos(t) & \sin(t) \\ -\sin(t) & \cos(t) \end{bmatrix} B_1 u$$

$$= A_2 \pi(t, x) + B_2 H(t, u).$$

Moreover, for every $t \in \mathbb{R}$ and $u \in \Omega_1$ we have $|u| \leq 1$, which implies $|H(t, u)| \leq \|B_1\| \|u\| \leq \|B_1\|$ and hence $H(t, u) \in \Omega_2$. By Proposition 2.14, we have thus proved that (π, h) with $h(\omega)(t) = H(t, \omega(t))$ is a time-variant semi-conjugacy from Σ_1 to Σ_2. If $m_1 = 2$ and $B_1 \in O(2)$ (the orthogonal group in dimension 2), then (π, h) is a topological conjugacy. Indeed, in this case $\|B_1\| = 1$ and hence $\Omega_2 = \Omega_1$. Furthermore, $H(t, \Omega_1) = \Omega_2$ for every $t \in \mathbb{R}$ and $H(t, \cdot)$ is invertible, which implies that also $h : \mathcal{U}_1 \to \mathcal{U}_2$ has an inverse. In this case, every sufficiently small compact ball Q centered at $0 \in \mathbb{R}^2$ is controlled invariant for Σ_1, since for each $x \in Q$ we have the equilibrium pair (u_x, x) with $u_x := -B_1^{-1} A_1 x \in \Omega_1$. Moreover, $\pi_t(Q) = Q$ for all $t \in \mathbb{R}_+$ which implies that $h_{\mathrm{inv}}(Q; \Sigma_1) = h_{\mathrm{inv}}(Q; \Sigma_2)$. In particular, this shows that $h_{\mathrm{inv}}(Q; \Sigma_1)$ does not depend on the imaginary parts of the eigenvalues of A_1.

Isolated Sets

A property of topological entropy observed by Keynes and Robertson [67] and Bowen [11] is that for expansive dynamical systems the limit for ε tending to zero in the definition via spanning or separated sets can be omitted. In this subsection, we formulate a somehow similar condition for an admissible pair (K, Q) of a topological system Σ which guarantees that the same holds true for the outer invariance entropy of (K, Q).

We assume that $\Sigma = (\mathbb{T}, X, U, \mathcal{U}, \varphi)$ is a topological time-invariant system with the following additional properties:

- The state space X is locally compact;
- The set $\mathcal{U}$ of admissible control functions is endowed with a topology that makes it a sequentially compact space, that is, every sequence in $\mathcal{U}$ has a convergent subsequence;
- The transition map $\varphi : \mathbb{T}_+ \times X \times \mathcal{U} \to X$ is continuous when $\mathcal{U}$ is endowed with the above topology.

These properties are satisfied in particular for a control-affine system. Then the appropriate topology on $\mathcal{U}$ is the weak*-topology of $L^\infty(\mathbb{R}, \mathbb{R}^m) = L^1(\mathbb{R}, \mathbb{R}^m)^*$ (cf. Sect. 1.4).

Definition 2.5. Let ϱ denote a fixed metric on X. A compact set $Q \subset X$ is called *isolated* if there exists $\delta_0 > 0$ such that for all $(x, \omega) \in \mathrm{cl}\, N_{\delta_0}(Q) \times \mathcal{U}$ the following implication holds:

$$\varphi(\mathbb{T}_+, x, \omega) \subset \mathrm{cl}\, N_{\delta_0}(Q) \quad \Rightarrow \quad \varphi(\mathbb{T}_+, x, \omega) \subset Q. \qquad (2.11)$$

Proposition 2.15. *Let (K, Q) be an admissible pair such that Q is compact and isolated with constant δ_0. Then it holds that*

$$h_{\mathrm{inv,out}}(K, Q) = h_{\mathrm{inv}}(K, N_\varepsilon(Q)) \quad \text{for all } \varepsilon \in (0, \delta_0].$$

Proof. By local compactness of X and Lemma A.3 we may assume that δ_0 is small enough that $\mathrm{cl}\, N_{\delta_0}(Q)$ is compact, since the assumption (2.11) is also satisfied for smaller δ_0. Then first we show the following:

$$\forall \rho > 0 \ \forall \varepsilon \in (0, \delta_0] \ \exists n \in \mathbb{N} \ \forall (x, \omega) \in \mathrm{cl}\, N_{\delta_0}(Q) \times \mathcal{U} :$$

$$\max_{t \in [0,n]} \mathrm{dist}(\varphi(t, x, \omega), Q) \le \varepsilon \quad \Rightarrow \quad \mathrm{dist}(x, Q) < \rho.$$

To this end, we assume that the opposite is true, that is,

$$\exists \rho > 0 \ \exists \varepsilon \in (0, \delta_0] \ \forall n \in \mathbb{N} \ \exists (x_n, \omega_n) \in \mathrm{cl}\, N_{\delta_0}(Q) \times \mathcal{U} :$$

$$\max_{t \in [0,n]} \mathrm{dist}(\varphi(t, x_n, \omega_n), Q) \le \varepsilon \quad \text{and} \quad \mathrm{dist}(x_n, Q) \ge \rho.$$

By compactness of $\mathrm{cl}\, N_{\delta_0}(Q)$ and $\mathcal{U}$ we may assume that $(x_n, \omega_n) \to (x, \omega) \in \mathrm{cl}\, N_{\delta_0}(Q) \times \mathcal{U}$. By continuity of $\mathrm{dist}(\cdot, Q)$ (see Lemma A.2) we obtain

$$\mathrm{dist}(x, Q) = \lim_{n \to \infty} \mathrm{dist}(x_n, Q) \ge \rho \quad \Rightarrow \quad x \notin Q.$$

For arbitrary $t_0 \in \mathbb{T}_+$ we have

$$
\begin{aligned}
\operatorname{dist}(\varphi(t_0, x, \omega), Q) &= \lim_{n \to \infty} \operatorname{dist}(\varphi(t_0, x_n, \omega_n), Q) \\
&\leq \limsup_{n \to \infty} \underbrace{\max_{t \in [0,t_0]} \operatorname{dist}(\varphi(t, x_n, \omega_n), Q)}_{\leq \varepsilon \text{ for } n \geq t_0} \leq \varepsilon \leq \delta_0.
\end{aligned}
$$

Hence, $\varphi(\mathbb{T}_+, x, \omega) \subset \operatorname{cl} N_{\delta_0}(Q)$ which implies $\varphi(\mathbb{T}_+, x, \omega) \subset Q$ in contradiction to $x \notin Q$.

Now let $0 < \varepsilon_1 < \varepsilon_2 \leq \delta_0$. Then, by what we have shown, there exists $n \in \mathbb{N}$ such that for all $(x, \omega) \in \operatorname{cl} N_{\delta_0}(Q) \times \mathscr{U}$ it holds that

$$
\max_{t \in [0,n]} \operatorname{dist}(\varphi(t, x, \omega), Q) \leq \varepsilon_2 \quad \Rightarrow \quad \operatorname{dist}(x, Q) < \varepsilon_1. \tag{2.12}
$$

For arbitrary $\tau \in \mathbb{T} \cap (0, \infty)$, let $\mathscr{S}$ be a minimal $(n + \tau, K, N_{\varepsilon_2}(Q))$-spanning set. Pick $x \in K$. Then there exists $\omega_x \in \mathscr{S}$ with

$$
\varphi([0, n + \tau], x, \omega_x) \subset N_{\varepsilon_2}(Q).
$$

For every $s \in [0, \tau]$ we have

$$
\max_{t \in [0,n]} \operatorname{dist}(\varphi(t, \varphi(s, x, \omega_x), \Theta_s \omega_x), Q) = \max_{t \in [0,n]} \operatorname{dist}(\varphi(t + s, x, \omega_x), Q) < \varepsilon_2.
$$

Hence, by (2.12) we have

$$
\operatorname{dist}(\varphi(s, x, \omega_x), Q) < \varepsilon_1 \text{ for all } s \in [0, \tau],
$$

which implies that $\mathscr{S}$ is a $(\tau, K, N_{\varepsilon_1}(Q))$-spanning set. Therefore,

$$
r_{\text{inv}}(\tau, K, N_{\varepsilon_1}(Q)) \leq r_{\text{inv}}(n + \tau, K, N_{\varepsilon_2}(Q)) \text{ for all } \tau > 0,
$$

which immediately gives

$$
h_{\text{inv}}(K, N_{\varepsilon_1}(Q)) \leq h_{\text{inv}}(K, N_{\varepsilon_2}(Q)).
$$

Together with the trivial inequality $h_{\text{inv}}(K, N_{\varepsilon_2}(Q)) \leq h_{\text{inv}}(K, N_{\varepsilon_1}(Q))$ (see Proposition 2.1) this implies the assertion. □

Example 2.6. Consider a continuous-time linear system given by the differential equations

$$
\dot{x}(t) = Ax(t) + B\omega(t), \quad \omega \in \mathscr{U},
$$

with $A \in \mathbb{R}^{d \times d}$ and $B \in \mathbb{R}^{d \times m}$ such that (A, B) is controllable and the control range $\Omega \subset \mathbb{R}^m$ is a compact and convex set with $0 \in \operatorname{int} \Omega$. Assume further that all eigenvalues of A have positive real parts. Then there exists a unique open control set $D \subset \mathbb{R}^d$ with compact closure Q which is given by $D = \mathcal{O}^-(0)$. To prove the latter, note that $0 \in \text{int } D$ (see Colonius and Kliemann [25, Example 3.2.16]) and hence $D = \text{cl } \mathcal{O}^+(0) \cap \mathcal{O}^-(0)$ (see Proposition 1.23 (iii) and note that controllability of (A, B) together with $0 \in \text{int } \Omega$ implies local accessibility). Since $\text{int } D \subset \mathcal{O}^+(0)$, we have

$$\varphi(\mathbb{R}_+, \text{int } D, \mathbf{0}) = \bigcup_{t \geq 0} e^{At} \text{ int } D \subset \mathcal{O}^+(0),$$

and since the linear semiflow $\{e^{At}\}_{t \in \mathbb{R}_+}$ is expanding, this implies $\mathcal{O}^+(0) = \mathbb{R}^d$. In particular, there are constants $c, \alpha > 0$ such that

$$|e^{At} x| \geq c e^{\alpha t} |x| \quad \text{for all } t \geq 0 \text{ and } x \in \mathbb{R}^d, \tag{2.13}$$

which follows from Lemma B.2. Now let $x \in \mathbb{R}^d \setminus Q$ and $\omega \in \mathcal{U}$. Define

$$\beta := \text{dist}(x, Q) > 0.$$

For given $\tau > 0$ define

$$y := -\int_0^\tau e^{-As} B\omega(s) \mathrm{d}s.$$

Then $y \in \mathcal{O}^-(0) = D$, since

$$\varphi(\tau, y, \omega) = e^{A\tau} \left(-\int_0^\tau e^{-As} B\omega(s) \mathrm{d}s \right) + \int_0^\tau e^{A(\tau - s)} B\omega(s) \mathrm{d}s = 0.$$

This implies

$$|\varphi(\tau, x, \omega)| = |\varphi(\tau, x, \omega) - \varphi(\tau, y, \omega)| = |e^{A\tau}(x - y)|$$

$$\overset{(2.13)}{\geq} c e^{\alpha\tau} \underbrace{|x - y|}_{\geq \text{dist}(x, Q)} \geq c e^{\alpha\tau} \beta.$$

Hence, $\varphi(\tau, x, \omega) \to \infty$ for $\tau \to \infty$, which implies that Q is isolated for any constant $\delta_0 > 0$. If $K \subset D$ is any compact set, then (K, Q) is admissible which follows from controlled invariance of D. Hence, $h_{\text{inv,out}}(K, Q) = h_{\text{inv,out}}(K, N_\varepsilon(Q))$ for all $\varepsilon > 0$.

Remark 2.4. For smooth systems given by differential equations the transition map is defined on $\mathbb{R} \times M \times \mathcal{U}$. Therefore, we can also introduce another notion of isolated sets for such systems by saying that $Q \subset X$ is isolated if there exists $\delta_0 > 0$ with

$$\varphi(\mathbb{R}, x, \omega) \subset \text{cl } N_{\delta_0}(Q) \quad \Rightarrow \quad \varphi(\mathbb{R}, x, \omega) \subset Q.$$

Then, if we assume that Q is compact and controlled invariant in forward and backward time (that is, for every $x \in Q$ there is $\omega \in \mathscr{U}$ with $\varphi(\mathbb{R}, x, \omega) \subset Q$), the assertion of Proposition 2.15 is valid as well, which can be shown with the same arguments.

Inner Control Sets

For topological entropy of maps on compact metric spaces, it is well-known that the limit superior in the definition via spanning or separated sets can be replaced by a limit inferior without changing the value of the entropy (see, for instance, Downarowicz [39, Chapter 6]). It is not clear at all if this is also true for the invariance entropy (except for the case $K = Q$). However, under certain controllability assumptions in a neighborhood of Q, we will see that it holds true for the outer invariance entropy. The notion of an *inner control set*, introduced next, makes these assumptions precise.

Definition 2.6. Consider a topological time-invariant system $\Sigma = (\mathbb{T}, X, U, \mathscr{U}, \varphi)$ whose state space X has no isolated points.[1] A set $A \subset X$ is called an *inner control set* if there exists a family of systems of the form

$$\Sigma_\rho = (\mathbb{T}, X, U, \mathscr{U}_\rho, \varphi_\rho), \quad \rho \in [0, 1],$$

and an associated family of control sets $D_\rho \subset X$ with nonempty interiors and compact closures such that the following conditions are satisfied:

(i) Any two of the transition maps φ_ρ coincide on the intersection of their domains (therefore we omit the index and just write φ for all of them);
(ii) $\mathscr{U}_{\rho_2} \subset \mathscr{U}_{\rho_1}$ whenever $\rho_1 < \rho_2$, and $\mathscr{U} = \mathscr{U}_0$;
(iii) $\operatorname{cl} D_{\rho_2} \subset \operatorname{int} D_{\rho_1}$ whenever $\rho_1 < \rho_2$, and $A = D_1$;
(iv) For every neighborhood W of $\operatorname{cl} A$ there is $\rho \in [0, 1)$ with $\operatorname{cl} D_\rho \subset W$.

Proposition 2.16. *Let Q be the closure of an inner control set A of a topological time-invariant system Σ. Then, for every compact set $K \subset A$, the pair (K, Q) is admissible for Σ. If K has nonempty interior, we have*

$$h_{\mathrm{inv,out}}(K, Q) = \lim_{\varepsilon \searrow 0} \liminf_{\tau \to \infty} \frac{1}{\tau} \log r_{\mathrm{inv}}(\tau, K, N_\varepsilon(Q)). \tag{2.14}$$

Proof. It is easy to see that (K, Q) is admissible. Indeed, since A is a control set for system Σ_1, for every $x \in K$ one finds $\omega \in \mathscr{U}_1 \subset \mathscr{U}_0 = \mathscr{U}$ such that $\varphi(\mathbb{T}_+, x, \omega) \subset A \subset Q$.

---

[1]This is our standard assumption for control sets, cf. Sect. 1.4.

Now let ϱ be a metric on X. From conditions (iii) and (iv) it follows that we can find a monotonically increasing sequence $(\rho_n)_{n \in \mathbb{N}}$ in $[0, 1)$ with $D_{\rho_n} \subset N_{1/n}(Q)$ for all $n \in \mathbb{N}$. Since $Q = \mathrm{cl}\, D_1 \subset \mathrm{int}\, D_{\rho_n}$ for all $n \in \mathbb{N}$, we can find a monotonically decreasing sequence $(\varepsilon_n)_{n \in \mathbb{N}}$ of positive real numbers with $\varepsilon_n \searrow 0$ such that $\mathrm{cl}\, N_{\varepsilon_n}(Q) \subset D_{\rho_n}$ for all $n \in \mathbb{N}$.

For each $n \in \mathbb{N}$ it is possible to steer from all states in $N_{\varepsilon_n}(Q)$ to K using only finitely many control functions taken from the set $\mathscr{U}_{\rho_n}$ (using compactness of $\mathrm{cl}\, N_{\varepsilon_n}(Q)$ and continuity of φ in x). By the no-return property of control sets with nonempty interior (see Corollary 1.1), the corresponding trajectories do not leave $D_{\rho_n} \subset N_{1/n}(Q)$. Let α_n be the minimal number of control functions which are necessary to do so. Then for every $\tau > 0$ and $m \in \mathbb{N}$ we can construct an $(m\tau, K, N_{1/n}(Q))$-spanning set with cardinality less than $\alpha_n^m r_{\mathrm{inv}}(\tau, K, N_\varepsilon(Q))^m$ for all $\varepsilon \in (0, \varepsilon_n]$ (by iterated concatenation of the control functions in a minimal $(\tau, K, N_\varepsilon(Q))$-spanning set and the control functions used to steer from $N_{\varepsilon_n}(Q)$ to K). Hence, we obtain

$$r_{\mathrm{inv}}\left(m\tau, K, N_{1/n}(Q)\right) \leq \alpha_n^m r_{\mathrm{inv}}(\tau, K, N_\varepsilon(Q))^m$$

for all $m \in \mathbb{N}$, $\tau > 0$, and $0 < \varepsilon \leq \varepsilon_n$. Using Proposition 2.6, this implies

$$h_{\mathrm{inv}}\left(K, N_{1/n}(Q)\right) = \limsup_{m \to \infty} \frac{1}{m\tau} \log r_{\mathrm{inv}}\left(m\tau, K, N_{1/n}(Q)\right)$$

$$\leq \limsup_{m \to \infty} \frac{1}{\tau} (\log \alpha_n + \log r_{\mathrm{inv}}(\tau, K, N_\varepsilon(Q)))$$

$$= \frac{1}{\tau} \log \alpha_n + \frac{1}{\tau} \log r_{\mathrm{inv}}(\tau, K, N_\varepsilon(Q)).$$

Therefore, we have

$$h_{\mathrm{inv}}\left(K, N_{1/n}(Q)\right) \leq \lim_{\varepsilon \searrow 0} \liminf_{\tau \to \infty} \left(\frac{1}{\tau} \log \alpha_n + \frac{1}{\tau} \log r_{\mathrm{inv}}(\tau, K, N_\varepsilon(Q))\right)$$

$$= \lim_{\varepsilon \searrow 0} \liminf_{\tau \to \infty} \frac{1}{\tau} \log r_{\mathrm{inv}}(\tau, K, N_\varepsilon(Q)).$$

Since this inequality holds for every $n \in \mathbb{N}$, the assertion follows. $\qquad \square$

Remark 2.5. Note that from (2.14) it does not necessarily follow that the limit $\lim_{\tau \to \infty}(1/\tau) \log r_{\mathrm{inv}}(\tau, K, N_\varepsilon(Q))$ exists for any $\varepsilon > 0$.

A simple example for a system with inner control sets would be a one-dimensional linear system as considered in the following section.

2.3 A One-Dimensional Linear Example

In this section, we give a first nontrivial example for an explicit computation of the invariance entropy. Consider the one-dimensional linear system given by the differential equations

$$\dot{x}(t) = ax(t) + \omega(t), \quad \omega \in \mathcal{U},$$

where $a > 0$. We assume that the control range Ω contains the interval $[-1, 1]$. By Proposition 1.5 the transition map of this system has the form

$$\varphi(t, x, \omega) = e^{at}x + \int_0^t e^{a(t-s)}\omega(s)\mathrm{d}s. \tag{2.15}$$

The compact interval $Q := [-1/a, 1/a]$ is controlled invariant, since for each $x \in Q$ the constant control function $\omega_x(t) \equiv -ax$ is admissible and (ω_x, x) is an equilibrium pair.

Proposition 2.17. $h_{\mathrm{inv}}(Q) = h_{\mathrm{inv,out}}(Q) = a.$

Proof. We subdivide the proof into two steps.

Step 1. We show the estimate $h_{\mathrm{inv}}(Q) \leq a$. To this end, note that for constant control functions formula (2.15) reduces to

$$\varphi(t, x, \omega) = e^{at}\left(x + \frac{\omega}{a}\right) - \frac{\omega}{a}. \tag{2.16}$$

Now we explicitly construct (τ_k, Q)-spanning sets for the times

$$\tau_k := k\frac{\log 2}{a}, \quad k \geq 1.$$

For each $k \geq 1$ we subdivide Q into 2^k subintervals of the same length:

$$Q_j := -\frac{1}{a} + \frac{2}{a}\left[\frac{j}{2^k}, \frac{j+1}{2^k}\right], \quad j = 0, 1, \ldots, 2^k - 1.$$

Then we associate to each Q_j a constant control function defined by

$$\omega_j(t) :\equiv 1 - \frac{2j}{2^k - 1}.$$

These control functions are admissible, since their values are contained in $[-1, 1]$. Now we apply the control function ω_j to the interval Q_j. For each $t \in [0, \tau_k]$, using (2.16), we obtain

$$\varphi\left(t, -\frac{1}{a} + \frac{2}{a}\frac{j}{2^k}, \omega_j\right) = \frac{1}{a}\left(2j e^{at}\left(\frac{1}{2^k} - \frac{1}{2^k - 1}\right) - \left(1 - \frac{2j}{2^k - 1}\right)\right)$$

$$\geq \frac{1}{a}\left(2j e^{a\tau_k}\left(\frac{1}{2^k} - \frac{1}{2^k - 1}\right) - \left(1 - \frac{2j}{2^k - 1}\right)\right)$$

$$= \frac{1}{a}\left(2j\,2^k\left(\frac{1}{2^k} - \frac{1}{2^k - 1}\right) - \left(1 - \frac{2j}{2^k - 1}\right)\right)$$

$$= \frac{1}{a}\left(-\frac{2j}{2^k - 1} - \left(1 - \frac{2j}{2^k - 1}\right)\right) = -\frac{1}{a}.$$

For the lower estimate we used that $1/2^k - 1/(2^k - 1) < 0$. For the right endpoint of Q_j we get

$$\varphi\left(t, -\frac{1}{a} + \frac{2}{a}\frac{j+1}{2^k}, \omega_j\right) = \frac{1}{a}\left(2e^{at}\left(\frac{j+1}{2^k} - \frac{j}{2^k - 1}\right) - \left(1 - \frac{2j}{2^k - 1}\right)\right)$$

$$\leq \frac{1}{a}\left(2e^{a\tau_k}\left(\frac{j+1}{2^k} - \frac{j}{2^k - 1}\right) - \left(1 - \frac{2j}{2^k - 1}\right)\right)$$

$$= \frac{1}{a}\left(2\left(j + 1 - \frac{2^k j}{2^k - 1}\right) - \left(1 - \frac{2j}{2^k - 1}\right)\right)$$

$$= \frac{1}{a}\left(\left(2 - \frac{2j}{2^k - 1}\right) - \left(1 - \frac{2j}{2^k - 1}\right)\right) = \frac{1}{a}.$$

For the upper estimate we used that $(j + 1)/2^k - j/(2^k - 1) \geq 0$ which is equivalent to $j \leq 2^k - 1$. Since each of the maps φ_{t,ω_j}, $t \in [0, \tau_k]$, maps Q_j onto a compact interval without reversing the left and right endpoints, we have shown that the set $\mathscr{S}_k := \{\omega_0, \ldots, \omega_{2^k - 1}\}$ is (τ_k, Q)-spanning, which implies

$$r_{\mathrm{inv}}(\tau_k, Q) \leq 2^k \quad \text{for all } k \geq 1.$$

Using Proposition 2.6, this gives

$$h_{\mathrm{inv}}(Q) \leq \limsup_{k \to \infty} \underbrace{\frac{1}{\tau_k} \log(2^k)}_{=a} = a,$$

which finishes the first step of the proof.

Step 2. Since $h_{\mathrm{inv,out}}(Q) \leq h_{\mathrm{inv}}(Q)$, it suffices to show that $h_{\mathrm{inv,out}}(Q) \geq a$ to conclude the proof. Denote the one-dimensional Lebesgue measure on $\mathbb{R}$ by λ. For some fixed $\varepsilon > 0$, assume that $\mathscr{S}$ is a minimal (and hence finite) $(\tau, Q, N_\varepsilon(Q))$-spanning set and define

$$Q_\omega := \{x \in Q \,:\, \varphi(\tau, x, \omega) \in N_\varepsilon(Q)\}, \quad \omega \in \mathscr{S}.$$

Then from the definition of spanning sets it follows that $Q = \bigcup_{\omega \in \mathcal{S}} Q_\omega$. Moreover, each Q_ω is Lebesgue measurable, since it is the intersection of the compact set Q and the preimage $\varphi_{\tau,\omega}^{-1}(N_\varepsilon(Q))$. Since $\varphi_{\tau,\omega}(Q_\omega) \subset N_\varepsilon(Q)$, we have

$$\lambda\left(\varphi_{\tau,\omega}(Q_\omega)\right) = e^{a\tau}\lambda(Q_\omega) \leq \lambda(N_\varepsilon(Q)),$$

which implies

$$\lambda(Q) \leq \sum_{\omega \in \mathcal{S}} \lambda(Q_\omega) \leq \#\mathcal{S} \cdot \max_{\omega \in \mathcal{S}} \lambda(Q_\omega) \leq \#\mathcal{S} \cdot \frac{\lambda(N_\varepsilon(Q))}{e^{a\tau}}.$$

Hence, we obtain $r_{\text{inv}}(\tau, Q, N_\varepsilon(Q)) \geq c_\varepsilon e^{a\tau}$ for all $\tau > 0$ with $c_\varepsilon := \lambda(Q)/\lambda(N_\varepsilon(Q)) \in (0, \infty)$. This implies $h_{\text{inv,out}}(Q) \geq h_{\text{inv}}(Q, N_\varepsilon(Q)) \geq a$. $\square$

The idea of explicitly constructing spanning sets to obtain upper bounds, as in Step 1 of the above proof, is in general not feasible, since either no analytic descriptions of the transition map and the set Q are available, or they are too complicated. However, the volume argument in Step 2, used to obtain a lower bound for the outer invariance entropy, is one of the key ideas in this book. Variations of this idea are used in the Chaps. 3, 4, 6, and 7.

2.4 Relations to Topological Feedback Entropy

In this section, the concept of topological feedback entropy introduced by Nair, Evans, Mareels, and Moran in their seminal work [85] is explained and compared to invariance entropy. In particular, we present one of the main results of [85] which relates the topological feedback entropy to the minimal data rate necessary for rendering a compact subset of the state space invariant by means of a coder-controller pair. We also introduce another slightly different version of invariance entropy for discrete-time systems, which is shown to coincide with the topological feedback entropy. In particular, this proves that up to technical assumptions invariance entropy and topological feedback entropy are the same quantity or that one is an alternative characterization of the other, respectively.

The Definition of Topological Feedback Entropy

Consider a discrete-time time-invariant system which is given by difference equations

$$x_{k+1} = F(x_k, u_k), \quad u_k \in U, \quad k \in \mathbb{Z}_+, \tag{2.17}$$

where $F : X \times U \to X$ is a map, X a topological space, and U a nonempty set. Moreover, we assume that for each $u \in U$ the map $F_u : X \to X$, $x \mapsto F(x, u)$, is continuous. In our general notation, this system is written as $\Sigma = (\mathbb{Z}, X, U, U^{\mathbb{Z}}, \varphi)$ with

$$\varphi(k, x, \omega) = \begin{cases} x & \text{for } k = 0, \\ F_{\omega_{k-1}} \circ \cdots \circ F_{\omega_1} \circ F_{\omega_0}(x) & \text{for } k \geq 1. \end{cases}$$

In [85], for a compact set $Q \subset X$ with nonempty interior two invariance conditions of increasing strength are considered:[2]

- Q is called *weakly invariant* if there exists a time $n \in \mathbb{N}$ such that for every $x \in Q$ there is $\omega \in U^{\mathbb{Z}}$ with $\varphi(n, x, \omega) \in \operatorname{int} Q$.
- Q is called *strongly invariant* if for every $x \in Q$ there is $u \in U$ with $F(x, u) \in \operatorname{int} Q$.

Depending on which of these invariance conditions is imposed on Q, the *weak* or *strong topological feedback entropy* of Q is defined. For simplicity, we only consider the strong version in the following, that is, we assume that Q is strongly invariant.

A triple $(\mathscr{A}, \tau, G)$ is called an *invariant open cover* of Q if it satisfies the following properties:

- $\mathscr{A}$ is an open cover of Q;
- τ is a positive integer;
- G is a finite sequence of maps $G_k : \mathscr{A} \to U$, $k = 0, 1, \ldots, \tau - 1$, such that for every $A \in \mathscr{A}$ it holds that

$$\varphi(k, A, G(A)) \subset \operatorname{int} Q \quad \text{for } k = 1, \ldots, \tau,$$

that is, if the initial value $x \in Q$ lies in the set $A \in \mathscr{A}$, then any control sequence ω with $\omega_k = G_k(A)$ for $k = 0, 1, \ldots, \tau - 1$ yields $\varphi(k, x, \omega) \in \operatorname{int} Q$ for $k = 1, \ldots, \tau$.[3]

Existence of invariant open covers is seen as follows: Given $\tau \in \mathbb{N}$, strong invariance guarantees that for every $x \in Q$ we find $u_0, u_1, \ldots, u_{\tau-1} \in U$ with $\varphi(k, x, (u_i)) \in \operatorname{int} Q$ for $k = 1, \ldots, \tau$. Let $\omega_x \in U^{\mathbb{Z}}$ be a sequence with $\omega_x(k) = u_k$ for $k = 0, 1, \ldots, \tau - 1$. Since $\operatorname{int} Q$ is open, for every $k \in \{1, \ldots, \tau\}$ there is an open neighborhood W_k of $\varphi(k, x, \omega_x)$ with $W_k \subset \operatorname{int} Q$. By continuity, there are

---

[2]In fact, in Nair et al. [85], the two invariance conditions are a little bit stronger, since they require a compact set $Q' \subset \operatorname{int} Q$ such that one can steer into $\operatorname{int} Q'$ from Q. For simplicity, we slightly weaken these assumptions.

[3]Here and in the following, we write, with a little abuse of notation, $G(A)$ for the finite sequence $(G_0(A), \ldots, G_{\tau-1}(A))$.

corresponding open neighborhoods $V_1, \ldots, V_\tau$ of x with $\varphi(k, V_k, \omega_x) \subset W_k$. Then also $A_x := \bigcap_{k=1}^{\tau} V_k$ is an open neighborhood of x and

$$\varphi(k, A_x, \omega_x) \subset W_k \subset \text{int } Q \quad \text{for } k = 1, \ldots, \tau.$$

Now we can define an invariant open cover $(\mathscr{A}, \tau, G)$ by $\mathscr{A} := \{A_x\}_{x \in Q}$ and $G_k(A_x) := \omega_x(k)$ for all $x \in Q$ and $k = 0, 1, \ldots, \tau - 1$.

Given an arbitrary invariant open cover $(\mathscr{A}, \tau, G)$, for any sequence $\alpha = (A_i)_{i \in \mathbb{Z}_+}$ of sets in $\mathscr{A}$ we define the control sequence

$$\omega(\alpha) := (u_0, u_1, \ldots) \text{ with } (u_l)_{l=(i-1)\tau}^{i\tau-1} = G(A_{i-1}) \text{ for all } i \geq 1. \qquad (2.18)$$

We further define for each $j \in \mathbb{N}$ the set

$$B_j(\alpha) := \{x \in X \mid \varphi(i\tau, x, \omega(\alpha)) \in A_i \text{ for } i = 0, 1, \ldots, j-1\}. \qquad (2.19)$$

Then $B_j(\alpha)$ is an open set, since it can be written as the finite intersection of preimages of open sets under continuous maps, namely

$$B_j(\alpha) = \bigcap_{i=0}^{j-1} \{x \in X \mid \varphi(i\tau, x, \omega(\alpha)) \in A_i\} = \bigcap_{i=0}^{j-1} \varphi_{i\tau, \omega(\alpha)}^{-1}(A_i).$$

Furthermore, for each $j \in \mathbb{N}$, letting α run through all sequences of elements in $\mathscr{A}$, the family

$$\mathscr{B}_j = \mathscr{B}_j(\mathscr{A}, \tau, G) := \{B_j(\alpha) : \alpha \in \mathscr{A}^{\mathbb{Z}_+}\}$$

is an open cover of Q. Let $N(\mathscr{B}_j|Q)$ denote the minimal number of elements in a finite subcover of $\mathscr{B}_j$, and define the (strong) topological feedback entropy $h_{\text{fb}}(Q)$ by

$$h_{\text{fb}}(\mathscr{A}, \tau, G) := \lim_{j \to \infty} \frac{1}{j\tau} \log N(\mathscr{B}_j|Q) = \inf_{j \geq 1} \frac{1}{j\tau} \log N(\mathscr{B}_j|Q),$$

$$h_{\text{fb}}(Q) := \inf_{(\mathscr{A}, \tau, G)} h_{\text{fb}}(\mathscr{A}, \tau, G),$$

where the infimum is taken over all invariant open covers $(\mathscr{A}, \tau, G)$ of Q, and $\log = \log_2$. The quantity $h_{\text{fb}}(\mathscr{A}, \tau, G)$ is simply called the entropy of $(\mathscr{A}, \tau, G)$. Existence of the limit follows from the subadditivity lemma B.3. For the sake of completeness, we give a proof for this fact.

Lemma 2.2. *For each invariant open cover $(\mathscr{A}, \tau, G)$ of Q the sequence $j \mapsto \log N(\mathscr{B}_j|Q)$ is subadditive.*

Proof. For $k, l \in \mathbb{N}$, the open cover $\mathscr{B}_{k+l}$ consists of all sets of the form

$$\bigcap_{i=0}^{k+l-1} \varphi_{i\tau,\omega(\alpha)}^{-1}(A_i) = \bigcap_{i=0}^{l-1} \varphi_{(i+k)\tau,\omega(\alpha)}^{-1}(A_{i+k}) \cap \bigcap_{i=0}^{k-1} \varphi_{i\tau,\omega(\alpha)}^{-1}(A_i)$$

$$= \varphi_{k\tau,\omega(\alpha)}^{-1} \left[\bigcap_{i=0}^{l-1} \varphi_{i\tau,\Theta_{k\tau}\omega(\alpha)}^{-1}(A_{i+k}) \right] \cap \bigcap_{i=0}^{k-1} \varphi_{i\tau,\omega(\alpha)}^{-1}(A_i),$$

where α runs through all sequences of elements in $\mathscr{A}$. From the above expression one sees that these are exactly the sets

$$\varphi_{k\tau,\omega(\alpha)}^{-1}(B) \cap C, \quad B \in \mathscr{B}_l, \ C \in \mathscr{B}_k, \ \alpha = \alpha(C).$$

If $\tilde{\mathscr{B}}_l$ and $\tilde{\mathscr{B}}_k$ are minimal (and hence finite) subcovers of $\mathscr{B}_l$ and $\mathscr{B}_k$, respectively, then the sets

$$\varphi_{k\tau,\omega(\alpha)}^{-1}(B) \cap C, \quad B \in \tilde{\mathscr{B}}_l, \ C \in \tilde{\mathscr{B}}_k, \ \alpha = \alpha(C), \tag{2.20}$$

form a (not necessarily minimal) subcover of $\mathscr{B}_{k+l}$. To see this, let $x \in Q$. Since $\tilde{\mathscr{B}}_k$ is a cover of Q, we have $x \in C$ for some $C \in \tilde{\mathscr{B}}_k$. Let $\alpha = \alpha(C) = (A_i)_{i \geq 0}$ be the corresponding sequence of elements in $\mathscr{A}$. Then, by definition,

$$\varphi(i\tau, x, \omega(\alpha)) \in A_i \quad \text{for } i = 0, 1, \ldots, k-1,$$

and thus,

$$\varphi(k\tau, x, \omega(\alpha)) = \varphi\left(\tau, \varphi((k-1)\tau, x, \omega(\alpha)), \Theta_{(k-1)\tau}\omega(\alpha)\right)$$

$$\in \varphi\left(\tau, A_{k-1}, \Theta_{(k-1)\tau}\omega(\alpha)\right)$$

$$= \varphi\left(\tau, A_{k-1}, G(A_{k-1})\right) \subset \operatorname{int} Q \subset Q.$$

Since $\tilde{\mathscr{B}}_l$ is a cover of Q, this implies $x \in \varphi_{k\tau,\omega(\alpha)}^{-1}(B)$ for some $B \in \tilde{\mathscr{B}}_l$, and hence $x \in \varphi_{k\tau,\omega(\alpha)}^{-1}(B) \cap C$. Since there are exactly $\#\tilde{\mathscr{B}}_k \cdot \#\tilde{\mathscr{B}}_l$ sets of the form (2.20), we obtain

$$N(\mathscr{B}_{k+l}|Q) \leq N(\mathscr{B}_k|Q) \cdot N(\mathscr{B}_l|Q),$$

which implies the assertion. $\qquad\square$

Topological Feedback Entropy and Minimal Data Rates

The importance of the quantity $h_{\mathrm{fb}}(Q)$ lies in the fact that it characterizes the smallest average data rate above which it is possible to render the set Q invariant by a causal coding and control law which is realized as follows: Suppose that a sensor, which is connected to a controller by a noiseless digital channel, measures the state at discrete sampling times τ_k, $k \geq 0$, say $\tau_k = k$. At time τ_k, one discrete-valued symbol s_k from a finite coding alphabet S_k of time-varying size is transmitted. The *transmission data rate* of the channel is defined as the asymptotic average bit rate

$$R\left(\{S_k\}_{k\in\mathbb{Z}_+}\right) := \liminf_{k\to\infty} \frac{1}{k} \sum_{j=0}^{k-1} \log \#S_j.$$

For a time-independent alphabet $S_k \equiv S$ this reduces to $\log \#S$. Each symbol transmitted by the coder may depend on all past and present states and past symbols, that is, we have a *coder mapping*

$$\gamma_k : X^{k+1} \times S_0 \times \cdots \times S_{k-1} \to S_k,$$

$$(x_0, \ldots, x_k, s_0, \ldots, s_{k-1}) \mapsto s_k = \gamma_k (x_0, \ldots, x_k, s_0, \ldots, s_{k-1}).$$

Assuming that the digital channel is errorless, at time τ_k the controller has $s_0, \ldots, s_k$ available and generates a control value $u_k = \delta_k(s_0, \ldots, s_k)$, where δ_k is the *controller mapping*

$$\delta_k : S_0 \times \cdots \times S_k \to U.$$

We define the *coder-controller* as the triple

$$(S, \gamma, \delta) = \left(\{S_k\}_{k\in\mathbb{Z}_+}, \{\gamma_k\}_{k\in\mathbb{Z}_+}, \{\delta_k\}_{k\in\mathbb{Z}_+}\right).$$

We say that the coder-controller is *periodic with period r* if it satisfies

$$S_k = S_{k \bmod r},$$

$$\gamma_k \left(\{x_i\}_{i=0}^{k}, \{s_i\}_{i=0}^{k-1}\right) = \gamma_{k \bmod r} \left(\{x_i\}_{i=r\lfloor k/r\rfloor}^{k}, \{s_i\}_{i=r\lfloor k/r\rfloor}^{k-1}\right),$$

$$\delta_k \left(\{s_i\}_{i=0}^{k}\right) = \delta_{k \bmod r} \left(\{s_i\}_{i=r\lfloor k/r\rfloor}^{k}\right).$$

Now let $Q \subset X$ be a strongly invariant set. We say that a coder-controller (S, γ, δ) *renders Q (strongly) invariant* if for every $x_0 \in Q$ the sequence of states $(x_k)_{k\in\mathbb{Z}_+}$ generated by the coder-controller satisfies $x_k \in \text{int}\, Q$ for all $k \geq 1$. In [85, Theorem 1] we find the following result.[4]

Theorem 2.1. *Consider the discrete-time system given by (2.17). Assume that $Q \subset X$ is a compact set with nonempty interior which is strongly invariant. Then*

---

[4]In [85] the result is formulated for the weak topological feedback entropy, which is more general.

$$h_{\text{fb}}(Q) = \inf_{(S,\gamma,\delta)} R(S),$$

where the infimum is taken over all coder-controllers (S, γ, δ) that render Q strongly invariant.

Proof. The proof is subdivided into two steps.

Step 1. Let (S, γ, δ) be an arbitrary coder-controller that renders Q invariant. From the definition of $R(S)$ it follows that for every $\varepsilon > 0$ there are infinitely many $l \in \mathbb{N}$ with

$$\frac{1}{l} \sum_{j=0}^{l-1} \log \# S_j < R(S) + \varepsilon. \tag{2.21}$$

For every such l we define a periodic coder-controller $(S^P, \gamma^P, \delta^P)$ with period l by

$$S_k^P := S_{k \bmod l},$$

$$s_k = \gamma_k^P \left(\{x_i\}_{i=0}^{k}, \{s_i\}_{i=0}^{k-1} \right)$$

$$:= \gamma_{k \bmod l} \left(\{x_i\}_{i=l\lfloor k/l \rfloor}^{k}, \{s_i\}_{i=l\lfloor k/l \rfloor}^{k-1} \right),$$

$$u_k = \delta_k^P \left(\{s_i\}_{i=0}^{k} \right) := \delta_{k \bmod l} \left(\{s_i\}_{i=l\lfloor k/l \rfloor}^{k} \right).$$

By construction, this new coder-controller also renders Q invariant. Writing each $k \in \mathbb{N}$ as $k = p_k l + q_k$ with $p_k \in \mathbb{Z}_+$ and $q_k \in \{0, \dots, l-1\}$, the associated transmission data rate $R(S^P)$ can be computed as

$$R(S^P) = \liminf_{k\to\infty} \frac{1}{k} \sum_{j=0}^{k-1} \log \# S_j^P = \liminf_{k\to\infty} \frac{1}{k} \sum_{j=0}^{k-1} \log \# S_{j \bmod l}$$

$$= \liminf_{k\to\infty} \frac{1}{p_k l + q_k} \left(\sum_{j=0}^{p_k l - 1} \log \# S_{j \bmod l} + \sum_{j=p_k l}^{p_k l + q_k - 1} \log \# S_{j \bmod l} \right)$$

$$= \liminf_{k\to\infty} \left(\frac{p_k}{p_k l + q_k} \sum_{j=0}^{l-1} \log \# S_j + \frac{1}{p_k l + q_k} \sum_{j=0}^{q_k - 1} \log \# S_j \right)$$

$$= \frac{1}{l} \sum_{j=0}^{l-1} \log \# S_j.$$

With (2.21) this implies $R(S^P) < R(S) + \varepsilon$. Each sequence of symbols in $S_0 \times \cdots \times S_{l-1}$ defines a coding region in X which is defined as the set of all initial states x which force the coder to generate this sequence. The total number n of nonempty and disjoint coding regions is less than or equal to $\prod_{j=0}^{l-1} \#S_j$. Let $C_1, \ldots, C_n$ denote these coding regions and note that $Q \subset \bigcup_{i=1}^n C_i$. From $(S^P, \gamma^P, \delta^P)$ we can now construct an invariant open cover $(\mathscr{A}, \tau, G)$ of Q as follows: The time τ is set to l. For every x_0 in one of the coding regions $C_i = C_i(c_0, \ldots, c_{l-1})$ there exists an open neighborhood $N(x_0)$ such that for every $y_0 \in N(x_0)$ the same sequence $(c_0, \ldots, c_{l-1})$ of symbols gives $y_1, \ldots, y_l \in$ int Q, due to continuity of the transition map with respect to the state variable. Thus, we can "blow up" the sets C_i by setting $A_i := \bigcup_{x_0 \in C_i} N(x_0)$. This defines the open cover $\mathscr{A} := \{A_1, \ldots, A_n\}$ of Q. Finally, the mapping sequence G is defined by $G(A_i) :=$ the symbol sequence $(c_0, \ldots, c_{l-1})$ corresponding to the coding region C_i. By construction, it is clear that $(\mathscr{A}, \tau, G)$ is an invariant open cover. The entropy of $(\mathscr{A}, \tau, G)$ can be estimated by

$$h_{\text{fb}}(\mathscr{A}, \tau, G) = \lim_{j \to \infty} \frac{1}{j\tau} \log \underbrace{N(\mathscr{B}_j | Q)}_{\leq n^j} \leq \frac{1}{l} \log \prod_{j=0}^{l-1} \#S_j = R(S^P).$$

Therefore, we obtain

$$h_{\text{fb}}(Q) \leq h_{\text{fb}}(\mathscr{A}, \tau, G) \leq R(S^P) < R(S) + \varepsilon.$$

Since ε can be chosen arbitrarily, this gives $h_{\text{fb}}(Q) \leq R(S)$.

Step 2. It remains to show that data rates arbitrarily close to $h_{\text{fb}}(Q)$ can be achieved. To this end, let $(\mathscr{A}, \tau, G)$ be an invariant open cover of Q and define a periodic coder-controller rendering Q invariant as follows: Let

$$H := h_{\text{fb}}(\mathscr{A}, \tau, G) = \lim_{j \to \infty} \frac{\log N(\mathscr{B}_j | Q)}{j\tau}.$$

Then it follows that for every $\varepsilon > 0$ there is $j \in \mathbb{N}$ with

$$\frac{1}{j\tau} \log N(\mathscr{B}_j | Q) \leq H + \frac{\varepsilon}{2}.$$

Let us fix such $j = j(\varepsilon)$, and let $\{D_1, \ldots, D_m\}$ be a minimal subcover of $\mathscr{B}_j$, where $m = N(\mathscr{B}_j | Q)$ by definition. We construct a periodic coding law using these possibly overlapping sets via the rule

$$s_k := \begin{cases} \min\{i : x_k \in D_i\} & \text{when } k \in (j\tau)\mathbb{Z}_+, \\ 1 & \text{otherwise.} \end{cases}$$

The size of the coding alphabet S_k is $\#S_k = m$ if k is a multiple of $j\tau$ and $\#S_k = 1$ otherwise. Hence, the average data rate of this coder is

$$
\begin{aligned}
R(S) &= \liminf_{k\to\infty} \frac{1}{k} \sum_{i=0}^{k-1} \log \#S_i \\
&= \liminf_{k\to\infty} \frac{1}{k} \sum_{i=0}^{\lfloor k/(j\tau)\rfloor-1} \log \#S_{i(j\tau)} \\
&= \liminf_{k\to\infty} \frac{1}{k} \left\lfloor \frac{k}{j\tau} \right\rfloor \log m \\
&= \frac{\log m}{j\tau} = \frac{\log N(\mathscr{B}_j|Q)}{j\tau} \leq H + \frac{\varepsilon}{2}.
\end{aligned}
$$

The controller is constructed as follows. By definition, for each set $B_j \in \mathscr{B}_j$ there are $A_0, \ldots, A_{j-1} \in \mathscr{A}$ such that (2.19) holds. Upon receiving the symbol $s_{l(j\tau)} = i$ which indexes an open set D_i in the minimal subcover of $\mathscr{B}_j$, the controller finds $\alpha = (A_0, \ldots, A_{j-1})$ with $D_i = B_j(\alpha)$ and then generates inputs via the periodic rule

$$
\{u_k\}_{k=(lj+q)\tau}^{(lj+q+1)\tau-1} = G(A_q) \quad \text{for all } l \in \mathbb{Z}_+, \ q \in \{0, \ldots, j-1\}.
$$

By definition of invariant open covers this yields $x_{(lj+q)\tau} \in \operatorname{int} Q$ and hence, the constructed coder-controller renders Q invariant. By definition of $h_{\mathrm{fb}}(Q)$ we find $(\mathscr{A}, \tau, G)$ with $h_{\mathrm{fb}}(\mathscr{A}, \tau, G) - h_{\mathrm{fb}}(Q) \leq \varepsilon/2$. With the associated coder-controller (S, γ, δ) constructed as above, this gives

$$
\begin{aligned}
R(S) - h_{\mathrm{fb}}(Q) &= (R(S) - h_{\mathrm{fb}}(\mathscr{A}, \tau, G)) + (h_{\mathrm{fb}}(\mathscr{A}, \tau, G) - h_{\mathrm{fb}}(Q)) \\
&\leq \frac{\varepsilon}{2} + \frac{\varepsilon}{2} = \varepsilon,
\end{aligned}
$$

which concludes the proof. $\square$

Inner Invariance Entropy

In the following, we introduce a modified version of invariance entropy for discrete-time systems, which turns out to coincide with the topological feedback entropy.

Definition 2.7. Consider a discrete-time topological time-invariant system of the form $\Sigma = (\mathbb{Z}, X, U, U^{\mathbb{Z}}, \varphi)$ and a compact set $Q \subset X$ with nonempty interior which has the following property: For every $x \in Q$ there is $u \in U$ with $\varphi(1, x, u) \in$ int Q. For $\tau \in \mathbb{N}$ a subset $\mathscr{S} \subset U^{\mathbb{Z}}$ is called $(\tau, Q, \text{int } Q)$-*spanning*[5] if

$$\forall x \in Q \, \exists \omega \in \mathscr{S} : \; \varphi([1, \tau], x, \omega) \subset \text{int } Q.$$

The minimal cardinality of such a set is denoted by $r_{\text{inv,int}}(\tau, Q)$, and the *inner invariance entropy* of Q is defined by

$$h_{\text{inv,int}}(Q) := \lim_{\tau \to \infty} \frac{1}{\tau} \log r_{\text{inv,int}}(\tau, Q).$$

Proposition 2.18. *The following assertions hold:*

(i) *For each $\tau \in \mathbb{N}$ it holds that $r_{\text{inv,int}}(\tau, Q) < \infty$.*
(ii) *The sequence $\tau \mapsto \log r_{\text{inv,int}}(\tau, Q)$ is subadditive and hence, the definition of $h_{\text{inv,int}}(Q)$ is correct and*

$$h_{\text{inv,int}}(Q) = \inf_{\tau \geq 1} \frac{1}{\tau} \log r_{\text{inv,int}}(\tau, Q).$$

Proof. (i) Let $\tau \in \mathbb{N}$ and pick an arbitrary $x \in Q$. By strong invariance we find $\omega \in U^{\mathbb{Z}}$ with $x_j := \varphi(j, x, \omega) \in \text{int } Q$ for $j = 1, \dots, \tau$. Each x_j has an open neighborhood $V_j \subset \text{int } Q$. By continuity, we find an open neighborhood W_x of x with $\varphi(j, W_x, \omega) \subset V_j$ for $j = 1, \dots, \tau$. By compactness of Q finitely many of such neighborhoods are sufficient to cover Q. The corresponding control sequences form a finite $(\tau, Q, \text{int } Q)$-spanning set.
(ii) Again, we apply the subadditivity lemma B.3. In order to show subadditivity, consider a $(\tau_1, Q, \text{int } Q)$-spanning set $\mathscr{S}_1$ and a $(\tau_2, Q, \text{int } Q)$-spanning set $\mathscr{S}_2$. Then define control sequences of length $\tau_1 + \tau_2$ by

$$\omega := (u_0, \dots, u_{\tau_1-1}, v_0, \dots, v_{\tau_2-1}) \in U^{\tau_1+\tau_2}$$

for each $\omega_1 = (u_0, \dots, u_{\tau_1-1}) \in \mathscr{S}_1$ and $\omega_2 = (v_0, \dots, v_{\tau_2-1}) \in \mathscr{S}_2$. The set of all such control sequences ω is a $(\tau_1 + \tau_2, Q, \text{int } Q)$-spanning set of cardinality $\#\mathscr{S}_1 \cdot \#\mathscr{S}_2$, which implies $\log r_{\text{inv,int}}(\tau_1+\tau_2, Q) \leq \log r_{\text{inv,int}}(\tau_1, Q) + \log r_{\text{inv,int}}(\tau_2, Q)$. $\square$

Using the above proposition we can prove the following theorem.

Theorem 2.2. *Consider a system $\Sigma = (\mathbb{Z}, X, U, U^{\mathbb{Z}}, \varphi)$ with a strongly invariant compact set $Q \subset X$ with nonempty interior. Then*

---

[5]Be aware that this terminology might cause some confusion, since the notion of a (τ, K, Q)-spanning set usually implies that $K \subset Q$. However, we do not use this notation again after this section.

$$h_{\mathrm{fb}}(Q) = h_{\mathrm{inv,int}}(Q).$$

Proof. The proof is subdivided into two steps.

Step 1. We show that $h_{\mathrm{fb}}(Q) \le h_{\mathrm{inv,int}}(Q)$: For a fixed $\tau \in \mathbb{N}$ let $\mathscr{S}$ be a minimal (and hence finite) $(\tau, Q, \operatorname{int} Q)$-spanning set. For each $\omega \in \mathscr{S}$ define

$$A_\omega := \{x \in X \mid \varphi(j, x, \omega) \in \operatorname{int} Q \text{ for } j = 1, \ldots, \tau\}.$$

It is clear that the sets A_ω form an open cover $\mathscr{A}$ of Q. Now define τ maps $G_k : \mathscr{A} \to U$ by

$$G_k(A_\omega) := \omega(k), \quad k = 0, \ldots, \tau - 1.$$

Clearly, $(\mathscr{A}, \tau, G)$ is an invariant open cover of Q. We have the trivial inequality $\#\mathscr{B}_j \le (\#\mathscr{A})^j$ which implies

$$h_{\mathrm{fb}}(Q) \le \lim_{j \to \infty} \frac{\log N(\mathscr{B}_j|Q)}{j\tau} \le \frac{\log \#\mathscr{A}}{\tau} = \frac{1}{\tau} \log r_{\mathrm{inv,int}}(\tau, Q).$$

Since this holds for every $\tau \in \mathbb{N}$, we obtain the desired inequality of entropies.

Step 2. To show the converse inequality, let $(\mathscr{A}, \tau, G)$ be an invariant open cover of Q. Choosing a finite subcover $\mathscr{A}'$ of $\mathscr{A}$ and restricting the maps G_k to that subcover, we obtain another invariant open cover $(\mathscr{A}', \tau, G')$ such that $h_{\mathrm{fb}}(\mathscr{A}', \tau, G') \le h_{\mathrm{fb}}(\mathscr{A}, \tau, G)$. Therefore, we may assume that $\mathscr{A}$ is finite. Then we can construct a $(j\tau, Q, \operatorname{int} Q)$-spanning set $\mathscr{S}_j$ for each $j \in \mathbb{N}$ with $N(\mathscr{B}_j|Q)$ elements as follows: Let $\tilde{\mathscr{B}}_j$ be a minimal subcover of $\mathscr{B}_j$. Each element of $\tilde{\mathscr{B}}_j$ corresponds to a particular sequence of elements in $\mathscr{A}$ and an associated control sequence $\omega(\alpha)$ as defined in (2.18). The set of these control sequences obviously forms a $(j\tau, Q, \operatorname{int} Q)$-spanning set. Hence, we obtain

$$r_{\mathrm{inv,int}}(j\tau, Q) \le N(\mathscr{B}_j|Q) \quad \text{for all } j \in \mathbb{N},$$

implying

$$h_{\mathrm{inv,int}}(Q) = \lim_{j \to \infty} \frac{1}{j\tau} \log r_{\mathrm{inv,int}}(j\tau, Q)$$

$$\le \lim_{j \to \infty} \frac{1}{j\tau} \log N(\mathscr{B}_j|Q) = h_{\mathrm{fb}}(\mathscr{A}, \tau, G).$$

Since this holds for every invariant open cover, the desired inequality $h_{\mathrm{inv,int}}(Q) \le h_{\mathrm{fb}}(Q)$ follows. $\square$

Finally, the following proposition relates the quantities $h_{\mathrm{inv}}(Q)$ and $h_{\mathrm{inv,int}}(Q)$.

Proposition 2.19. *Consider a system* $\Sigma = (\mathbb{Z}, X, U, U^{\mathbb{Z}}, \varphi)$ *with a strongly invariant compact set* $Q \subset X$ *with nonempty interior. Then* Q *is controlled invariant and*

$$h_{\text{inv}}(Q) \leq h_{\text{inv,int}}(Q) < \infty.$$

Proof. It is clear that Q is controlled invariant. If $\mathscr{S} \subset U^{\mathbb{Z}}$ is a $(\tau, Q, \text{int } Q)$-spanning set, then it is also (τ, Q)-spanning, which implies $r_{\text{inv}}(\tau, Q) \leq r_{\text{inv,int}}(\tau, Q)$ and hence $h_{\text{inv}}(Q) \leq h_{\text{inv,int}}(Q)$. Finiteness of $h_{\text{inv,int}}(Q)$ is a consequence of Proposition 2.18.   □

Remark 2.6. We will not introduce the notion of inner invariance entropy for continuous-time systems. For such systems, it is not a priori clear how inner invariance entropy could or should be defined. The requirement that trajectories with initial values in a compact set Q immediately enter the interior of Q might be too strict in many cases. One could assume that this happens after a finite time $\tau_0 > 0$. But then, it is not clear what we should require about the trajectory on the time interval $[0, \tau_0]$. There are several possibilities which we are not going to discuss at this point.

Remark 2.7. In Colonius [19] another notion of controlled invariance for a compact set Q with nonempty interior in the state space of a continuous-time system is considered. The set Q is called *locally controlled invariant* if for all $\varepsilon > 0$ and every $x \in Q$ there are $\tau > 0$ and $\omega \in \mathscr{U}$ with $\text{dist}(\varphi(t, x, \omega), Q) < \varepsilon$ for all $t \in [0, \tau]$ and $\varphi(\tau, x, \omega) \in \text{int } Q$. For such a set, a corresponding notion of invariance entropy can be defined and related to minimal data rates (cf. [19, Theorem 7]).

2.5   Relations to Minimal Data Rates

In this section, we give an alternative characterization of the quantity $h_{\text{inv}}(Q)$ which resembles the definition of topological feedback entropy and also allows to give an interpretation of $h_{\text{inv}}(Q)$ in terms of minimal data rates. Throughout, we assume that $\Sigma = (\mathbb{T}, X, U, \mathscr{U}, \varphi)$ is a topological time-invariant system and $Q \subset X$ a compact controlled invariant set.

Alternative Characterization via Invariant Covers

Recall the definition of invariant open covers used in the definition of topological feedback entropy. Without the requirement of "openness", we obtain the following notion.

Definition 2.8. An *invariant cover* of Q is a triple $(\mathscr{A}, \tau, v)$, where $\mathscr{A}$ is a finite cover[6] of Q, $\tau \in \mathbb{T} \cap (0, \infty)$ is a positive time, and $v : \mathscr{A} \rightarrow \mathscr{U}[0, \tau)$ a map, assigning to each set in $\mathscr{A}$ an admissible control function such that

$$\varphi([0, \tau], A, v(A)) \subset Q \quad \text{for all } A \in \mathscr{A}.$$

We also say that the triple $(\mathscr{A}, \tau, v)$ is *invariantly covering* the set Q.

The following proposition yields a first relation between the quantity $h_{\mathrm{inv}}(Q)$ and invariant covers of Q.

Proposition 2.20. *It holds that $h_{\mathrm{inv}}(Q) < \infty$ if and only if there exists an invariant cover $(\mathscr{A}, \tau, v)$ of Q.*

Proof. Assume that $h_{\mathrm{inv}}(Q) < \infty$. Then, by Proposition 2.3, there exists a finite (τ, Q)-spanning set $\mathscr{S} = \{\omega_1, \ldots, \omega_n\}$ for some $\tau > 0$. Define

$$A_j := \{x \in Q \mid \varphi([0, \tau], x, \omega_j) \subset Q\}, \quad j = 1, \ldots, n.$$

Let $\mathscr{A} := \{A_1, \ldots, A_n\}$ and let $v : \mathscr{A} \rightarrow \mathscr{U}[0, \tau)$ be given by $v(A_j) := \omega_j|_{[0,\tau)}$. Then obviously $(\mathscr{A}, \tau, v)$ is an invariant cover of Q. On the other hand, if $(\mathscr{A}, \tau, v)$ is an invariant cover of Q, the set $v(\mathscr{A}) \subset \mathscr{U}[0, \tau)$ is a finite (τ, Q)-spanning set for Q. Hence, by Proposition 2.3, $h_{\mathrm{inv}}(Q) < \infty$. □

To each invariant cover $\mathscr{C} = (\mathscr{A}, \tau, v)$ of Q we assign a number which we call the *entropy of $\mathscr{C}$*.

Definition 2.9. Let $\mathscr{C} = (\mathscr{A}, \tau, v)$ be an invariant cover of Q with $\mathscr{A} = \{A_1, \ldots, A_q\}$. We denote by ω_a the control function $v(A_a)$ for $a = 1, \ldots, q$, and we define for every word $[a_0, a_1, \ldots, a_{N-1}]$ ($N \in \mathbb{N}$) with $a_j \in \{1, \ldots, q\}$ a control function by

$$\omega_{a_0, a_1, \ldots, a_{N-1}} := \omega_{a_0} \omega_{a_1}^{\tau} \cdots \omega_{a_{N-1}}^{(N-1)\tau}, \quad \omega_{a_0, a_1, \ldots, a_{N-1}} \in \mathscr{U}[0, N\tau).$$

The word $[a_0, a_1, \ldots, a_{N-1}]$ is called *admissible* for $(Q, \mathscr{C})$ if there exists a point $x \in Q$ with

$$\varphi(j\tau, x, \omega_{a_0, a_1, \ldots, a_{N-1}}) \in A_{a_j} \quad \text{for } j = 0, 1, \ldots, N-1.$$

We write $\mathscr{W}_N(\mathscr{C}; Q)$ for the set of all admissible words of length N. The *entropy of $\mathscr{C}$* is then defined by

---

[6]Note that in the definition of topological feedback entropy it would have been sufficient to consider finite covers. Here we restrict ourselves to finite covers from the beginning.

$$h(\mathscr{C}; Q) := \lim_{N \to \infty} \frac{\log \#\mathscr{W}_N(\mathscr{C}; Q)}{N\tau} = \inf_{N \geq 1} \frac{\log \#\mathscr{W}_N(\mathscr{C}; Q)}{N\tau}. \tag{2.22}$$

The limit in (2.22) exists by subadditivity of $(\log \#\mathscr{W}_N(\mathscr{C}; Q))_{N \geq 1}$.

Proposition 2.21. *For every invariant cover $\mathscr{C}$ the sequence $N \mapsto \log \#\mathscr{W}_N(\mathscr{C}; Q)$ is subadditive.*

Proof. It suffices to show that

$$\#\mathscr{W}_{N_1+N_2}(\mathscr{C}; Q) \leq \#\mathscr{W}_{N_1}(\mathscr{C}; Q) \cdot \#\mathscr{W}_{N_2}(\mathscr{C}; Q) \quad \text{for all } N_1, N_2 \in \mathbb{N}.$$

To this end, we define an injective map

$$\alpha : \mathscr{W}_{N_1+N_2}(\mathscr{C}; Q) \to \mathscr{W}_{N_1}(\mathscr{C}; Q) \times \mathscr{W}_{N_2}(\mathscr{C}; Q),$$

which implies

$$\begin{aligned}
\#\mathscr{W}_{N_1+N_2}(\mathscr{C}; Q) &= \#\alpha\,(\mathscr{W}_{N_1+N_2}(\mathscr{C}; Q)) \\
&\leq \#\,(\mathscr{W}_{N_1}(\mathscr{C}; Q) \times \mathscr{W}_{N_2}(\mathscr{C}; Q)) \\
&= \#\mathscr{W}_{N_1}(\mathscr{C}; Q) \cdot \#\mathscr{W}_{N_2}(\mathscr{C}; Q).
\end{aligned}$$

Let $\mathscr{C} = (\mathscr{A}, \tau, v)$ with $\mathscr{A} = \{A_1, \ldots, A_q\}$, and let $[a_0, a_1, \ldots, a_{N_1+N_2-1}] \in \mathscr{W}_{N_1+N_2}(\mathscr{C}; Q)$. Then there exists $x \in Q$ with $\varphi(j\tau, x, \omega_{a_0,a_1,\ldots,a_{N_1+N_2-1}}) \in A_{a_j}$ for $j = 0, 1, \ldots, N_1 + N_2 - 1$. Let $y := \varphi(N_1\tau, x, \omega_{a_0,a_1,\ldots,a_{N_1+N_2-1}})$. Then $y \in Q$ and by the cocycle property of φ we obtain

$$\varphi\left(j\tau, y, \omega_{a_{N_1},a_{N_1+1},\ldots,a_{N_1+N_2-1}}\right) \in A_{a_{N_1+j}} \quad \text{for } j = 0, 1, \ldots, N_2 - 1.$$

This proves that $[a_{N_1}, a_{N_1+1}, \ldots, a_{N_1+N_2-1}]$ is an admissible word of length N_2. Hence, we can define α by

$$\alpha : [a_0, a_1, \ldots, a_{N_1+N_2-1}] \mapsto ([a_0, a_1, \ldots, a_{N_1-1}], [a_{N_1}, \ldots, a_{N_1+N_2-1}]).$$

Injectivity of α is obvious. $\square$

The entropy of an invariant cover is always an upper bound for $h_{\text{inv}}(Q)$.

Proposition 2.22. *For every invariant cover $\mathscr{C} = (\mathscr{A}, \tau, v)$ of Q it holds that*

$$h_{\text{inv}}(Q) \leq h(\mathscr{C}; Q) \leq \frac{\log \#\mathscr{A}}{\tau}.$$

Proof. Let $q = \#\mathscr{A}$. Since $\mathscr{W}_N(\mathscr{C}; Q) \subset \{1, \ldots, q\}^N$, we have $\#\mathscr{W}_N(\mathscr{C}; Q) \leq \#\{1, \ldots, q\}^N = q^N$ and thus

$$\frac{\log \#\mathscr{W}_N(\mathscr{C};Q)}{N\tau} \le \frac{\log q^N}{N\tau} = \frac{\log \#\mathscr{A}}{\tau} \quad \text{for all } N \in \mathbb{N}.$$

This implies $h(\mathscr{C};Q) \le (\log \#\mathscr{A})/\tau$. Now consider for every $N \in \mathbb{N}$ the set

$$\mathscr{S}_N := \{\omega_{a_0,a_1,\dots,a_{N-1}} \mid [a_0,a_1,\dots,a_{N-1}] \in \mathscr{W}_N(\mathscr{C};Q)\}.$$

Let $\mathscr{A} = \{A_1,\dots,A_q\}$ and $\omega_a = v(A_a)$ for $a = 1,\dots,q$. We want to show that $\mathscr{S}_N$ is $(N\tau,Q)$-spanning. To this end, pick $x_0 \in Q$ arbitrarily. Then there exists $a_0 \in \{1,\dots,q\}$ with $x_0 \in A_{a_0}$. This implies

$$\varphi([0,\tau],x_0,\omega_{a_0}) \subset \varphi([0,\tau],A_{a_0},v_{a_0}) \subset Q.$$

Let $x_1 := \varphi(\tau,x_0,\omega_{a_0})$. Then there is $a_1 \in \{1,\dots,q\}$ with $x_1 \in A_{a_1}$ and we obtain $\varphi([0,\tau],x_1,\omega_{a_1}) \subset Q$. Again, for $x_2 := \varphi(\tau,x_1,\omega_{a_1})$ we have $x_2 \in A_{a_2}$ for some a_2. Repeating this process, after N steps we have found an admissible word $[a_0,a_1,\dots,a_{N-1}]$ for $(Q,\mathscr{C})$, since the cocycle property of φ implies

$$\varphi(j\tau,x_0,\omega_{a_0,a_1,\dots,a_{N-1}}) \in A_{a_j} \quad \text{for } j = 0,1,\dots,N-1.$$

Hence, $\mathscr{S}_N$ is $(N\tau,Q)$-spanning and we obtain

$$r_{\mathrm{inv}}(N\tau,Q) \le \#\mathscr{W}_N(\mathscr{C};Q) \quad \text{for all } N \in \mathbb{N},$$

which implies

$$h_{\mathrm{inv}}(Q) = \lim_{N\to\infty} \frac{1}{N\tau} \log r_{\mathrm{inv}}(N\tau,Q)$$
$$\le \lim_{N\to\infty} \frac{\log \#\mathscr{W}_N(\mathscr{C};Q)}{N\tau} = h(\mathscr{C};Q).$$

This finishes the proof.   □

The following lemma shows that, in order to approximate $h_{\mathrm{inv}}(Q)$ by the quantities $h(\mathscr{C};Q)$, it is sufficient to consider invariant covers $(\mathscr{A},\tau,v)$, where $\mathscr{A}$ is a partition of Q.

Lemma 2.3. *For every invariant cover $\mathscr{C} = (\mathscr{A},\tau,v)$ of Q there exists another invariant cover $\tilde{\mathscr{C}} = (\tilde{\mathscr{A}},\tau,\tilde{v})$ such that $\tilde{\mathscr{A}}$ is a partition of Q with $\#\tilde{\mathscr{A}} = \#\mathscr{A}$ and*

$$h(\tilde{\mathscr{C}};Q) \le h(\mathscr{C};Q).$$

Proof. Let $\mathscr{A} = \{A_1,\dots,A_q\}$ and define sets $\tilde{A}_1,\dots,\tilde{A}_q$ by

$$\tilde{A}_1 := A_1, \quad \tilde{A}_j := A_j \setminus \bigcup_{i=1}^{j-1} A_i \quad \text{for } j = 2,\dots,q.$$

It can be checked easily that $\tilde{\mathscr{A}} := \{\tilde{A}_1, \ldots, \tilde{A}_q\}$ is a partition of Q. By setting $\tilde{v}(\tilde{A}_j) := v(A_j)$, $j = 1, \ldots, q$, we obtain an invariant cover $\tilde{\mathscr{C}} = (\tilde{\mathscr{A}}, \tau, \tilde{v})$, since $\tilde{A}_j \subset A_j$. Now let $[a_0, a_1, \ldots, a_{N-1}]$ be an admissible word of length N for $(Q, \tilde{\mathscr{C}})$. Then there exists $x \in Q$ with

$$\varphi\left(j\tau, x, \tilde{\omega}_{a_0, a_1, \ldots, a_{N-1}}\right) = \varphi\left(j\tau, x, \omega_{a_0, a_1, \ldots, a_{N-1}}\right) \in \tilde{A}_{a_j} \subset A_{a_j}$$

for $j = 0, 1, \ldots, N - 1$. This implies that $[a_0, a_1, \ldots, a_{N-1}]$ is also admissible for $(Q, \mathscr{C})$ and hence $\mathscr{W}_N(\tilde{\mathscr{C}}; Q) \subset \mathscr{W}_N(\mathscr{C}; Q)$ for all $N \in \mathbb{N}$, which yields the assertion. $\square$

Now we can formulate the alternative characterization of $h_{\mathrm{inv}}(Q)$ in terms of invariant covers.

Theorem 2.3. *For the compact and controlled invariant set Q it holds that*

$$h_{\mathrm{inv}}(Q) = \inf_{\mathscr{C}} h(\mathscr{C}; Q),$$

where the infimum[7] is taken over all invariant covers $\mathscr{C} = (\mathscr{A}, \tau, v)$ of Q such that $\mathscr{A}$ is a Borel measurable partition of Q. Moreover, it suffices to consider only times τ which are integer multiples of some $\tau_0 \in \mathbb{T} \cap (0, \infty)$.

Proof. The case $h_{\mathrm{inv}}(Q) = \infty$ is treated by Proposition 2.20. Hence, we may assume that $h_{\mathrm{inv}}(Q) < \infty$. By Proposition 2.22 it suffices to show that there exists a sequence $(\mathscr{C}_k)_{k \in \mathbb{N}}$, $\mathscr{C}_k = (\mathscr{A}_k, \tau_k, v_k)$, of invariant covers such that $\mathscr{A}_k$ is a measurable partition of Q, $\tau_k = k\tau_0$ for some $\tau_0 > 0$, and $h(\mathscr{C}_k; Q) \to h_{\mathrm{inv}}(Q)$ for $k \to \infty$. To this end, fix $\tau_0 > 0$ and let $\tau_k := k\tau_0$. For each $k \in \mathbb{N}$ let $\mathscr{S}_k = \{\omega_1^k, \ldots, \omega_{n_k}^k\}$ be a minimal $(k\tau_0, Q)$-spanning set and define the cover $\tilde{\mathscr{A}}_k = \{\tilde{A}_1, \ldots, \tilde{A}_{n_k}\}$ by

$$\tilde{A}_j := \left\{x \in Q \ : \ \varphi\left([0, k\tau_0], x, \omega_j^k\right) \subset Q\right\}, \quad j = 1, \ldots, n_k.$$

The set $\tilde{A}_j$ is a G_δ-set (that is, the intersection of countably many open sets) for every $j \in \{1, \ldots, n_k\}$, which follows from the identity

$$\tilde{A}_j = \bigcap_{n \in \mathbb{N}} \left\{x \in Q \ : \ \varphi\left([0, k\tau_0], x, \omega_j^k\right) \subset N_{1/n}(Q)\right\},$$

for any metric ϱ on X. Now we construct a measurable partition $\mathscr{A}_k$ from $\tilde{\mathscr{A}}_k$ by

$$A_1 := \tilde{A}_1, \quad A_j := \tilde{A}_j \setminus \bigcup_{i=1}^{j-1} \tilde{A}_i, \quad j = 2, \ldots, n_k.$$

---

[7] $\inf \emptyset$ is defined as ∞ (in case there is no invariant cover).

Let $v_k : \mathscr{A} \to \mathscr{U}[0, \tau_k)$ be given by $v_k(A_j) := \omega_j^k|_{[0,\tau_k)}$, $j = 1, \ldots, n_k$. Then $(\mathscr{A}_k, \tau_k, v_k)$ is obviously an invariant cover of Q and we have

$$h_{\mathrm{inv}}(Q) = \lim_{k \to \infty} \frac{1}{k\,\tau_0} \log n_k = \inf_{k \in \mathbb{N}} \frac{1}{k\,\tau_0} \log n_k.$$

Hence, for given $\varepsilon > 0$ we can choose $k_0 \in \mathbb{N}$ big enough such that $(1/(k\tau_0)) \log n_k - h_{\mathrm{inv}}(Q) < \varepsilon$ for all $k \geq k_0$. Together with Proposition 2.22 we obtain

$$h_{\mathrm{inv}}(Q) \leq h(\mathscr{C}_k; Q) \leq \frac{\log n_k}{k\,\tau_0} < h_{\mathrm{inv}}(Q) + \varepsilon \quad \text{for all } k \geq k_0,$$

which implies the assertion.   □

Remark 2.8. The characterization of $h_{\mathrm{inv}}(Q)$ in terms of invariant covers can serve as a starting point for a numerical algorithm for the computation of $h_{\mathrm{inv}}(Q)$. Indeed, if the cover $\mathscr{A}$ of an invariant cover $\mathscr{C} = (\mathscr{A}, \tau, v)$ is a partition, the quantity $h(\mathscr{C}; Q)$ can be approximated numerically by an algorithm developed by Froyland, Junge, and Ochs [47]. The main problem is then to find an invariant cover whose entropy is close to $h_{\mathrm{inv}}(Q)$.

The Data Rate Theorem

On the basis of Theorem 2.3 we can prove an analog to the data rate theorem 2.1.

Consider system Σ and suppose that a sensor, which is connected to a controller via a digital noiseless channel, measures its states at sampling times $k\tau$, $k \in \mathbb{Z}_+$, for some fixed time step $\tau > 0$ (see Fig. 2.1). The state at time $k\tau$ is coded using a finite coding alphabet S_k. We require that the sequence $(\#S_k)_{k \in \mathbb{Z}_+}$ satisfies

$$\liminf_{k \to \infty} \frac{1}{k} \sum_{j=0}^{k-1} \log \#S_j < \infty. \tag{2.23}$$

The coder transmits a symbol $s_k \in S_k$, which may depend on the present state and on all past states. The corresponding coder mapping is denoted by

$$\gamma_k : X^{k+1} \to S_k.$$

At time $k\tau$ the controller has $k+1$ symbols $s_0, s_1, \ldots, s_k$ available and generates an admissible finite-time control function $u_k \in \mathscr{U}[0, \tau)$. We denote the corresponding controller mapping by

$$\delta_k : S_0 \times S_1 \times \cdots \times S_k \to \mathscr{U}[0, \tau).$$

Fig. 2.1 A coder-controller

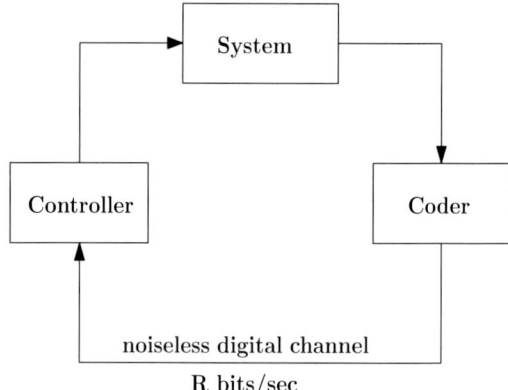

$$\text{R bits/sec}$$

Definition 2.10. The quadruple $\mathcal{H} := (S, \gamma, \delta, \tau)$, where $S = (S_k)_{k \in \mathbb{Z}_+}$, $\gamma = (\gamma_k)_{k \in \mathbb{Z}_+}$, and $\delta = (\delta_k)_{k \in \mathbb{Z}_+}$, is called a *coder-controller*.[8] We define the *transmission data rate* of $\mathcal{H}$ by

$$R(\mathcal{H}) := \liminf_{k \to \infty} \frac{1}{k\tau} \sum_{j=0}^{k-1} \log \# S_j,$$

which by (2.23) is a finite number. We say that $\mathcal{H}$ *renders Q invariant* if for all $x_0 \in Q$ the sequence $(x_k)_{k \in \mathbb{Z}_+}$, defined recursively by

$$x_k := \varphi\left(\tau, x_{k-1}, \omega_{k-1}\right)$$

with

$$\omega_{k-1} = \delta_{k-1}(\gamma_0(x_0), \gamma_1(x_0, x_1), \ldots, \gamma_{k-1}(x_0, x_1, \ldots, x_{k-1})),$$

satisfies

$$\varphi([0, \tau], x_k, \omega_k) \subset Q \quad \text{for all } k \in \mathbb{Z}_+.$$

That is, starting in Q at time $k = 0$ one stays in Q for all future times.

Theorem 2.4. *For the compact and controlled invariant set Q it holds that*

$$h_{\text{inv}}(Q) = \inf_{\mathcal{H}} R(\mathcal{H}),$$

where the infimum is taken over all coder-controllers $\mathcal{H}$ which render Q invariant.[9]

---

[8]Note that this definition differs a bit from the one used in Sect. 2.4.

[9]As in Theorem 2.3, $\inf \emptyset$ is defined as ∞ (in case there is no coder-controller rendering Q invariant).

Proof. The proof proceeds in three steps.

Step 1. Assume that $h_{\text{inv}}(Q) = \infty$ and let $\mathscr{H} = (S, \gamma, \delta, \tau)$ be a coder-controller rendering Q invariant. Consider the sets

$$A_s := \{x \in Q \mid \varphi([0, \tau], x, \delta_0(s)) \subset Q\}, \quad s \in S_0.$$

The family $\mathscr{A} := \{A_s\}_{s \in S_0}$ is a finite cover of Q, since $x \in A_{\delta_0(\gamma_0(x))}$ holds by Definition 2.10. With $v(A_s) := \delta_0(s)$, $s \in S_0$, one obtains an invariant cover $(\mathscr{A}, \tau, v)$ of Q. By Proposition 2.20, this contradicts $h_{\text{inv}}(Q) = \infty$. Hence, the assertion holds if $h_{\text{inv}}(Q) = \infty$.

Step 2. For an arbitrary coder-controller $\mathscr{H} = (S, \gamma, \delta, \tau)$, rendering Q invariant, we show that $R(\mathscr{H}) \geq h_{\text{inv}}(Q)$: It immediately follows from the definition of $R(\mathscr{H})$ that for given $\varepsilon > 0$ there exists $r \in \mathbb{N}$ such that

$$\frac{1}{r\tau} \sum_{k=0}^{r-1} \log \#S_k < R(\mathscr{H}) + \varepsilon. \tag{2.24}$$

For every tuple $(s_0, s_1, \ldots, s_{r-1}) \in S_0 \times S_1 \times \cdots \times S_{r-1}$ let

$$A_{s_0, s_1, \ldots, s_{r-1}} := \{x_0 \in Q \mid \gamma_j(x_0, \ldots, x_j) = s_j \text{ for } j = 0, 1, \ldots, r-1\},$$

where $x_0, x_1, \ldots, x_{r-1}$ are defined as in Definition 2.10. Then the family $\mathscr{A}$ of all the sets $A_{s_0, s_1, \ldots, s_{r-1}}$ obviously is a finite cover of Q, which can be extended to an invariant cover $\mathscr{C} = (\mathscr{A}, r\tau, v)$, where v assigns to the set $A_{s_0, s_1, \ldots, s_{r-1}}$ the control function given by concatenation of $\omega_0, \omega_1, \ldots, \omega_{r-1}$, which are defined as in Definition 2.10. By Proposition 2.22 we obtain

$$h_{\text{inv}}(Q) \leq h(\mathscr{C}; Q) \leq \frac{\log \#\mathscr{A}}{r\tau} = \frac{\log \prod_{k=0}^{r-1} \#S_k}{r\tau}$$

$$= \frac{\sum_{k=0}^{r-1} \log \#S_k}{r\tau} \overset{(2.24)}{<} R(\mathscr{H}) + \varepsilon.$$

Since this holds for every $\varepsilon > 0$, the assertion follows.

Step 3. We show that there exist coder-controllers rendering Q invariant whose transmission data rates come arbitrarily close to $h_{\text{inv}}(Q)$: By the proof of Theorem 2.3 there exists a sequence $(\mathscr{C}_n)_{n \in \mathbb{N}}$, $\mathscr{C}_n = (\mathscr{A}_n, \tau_n, v_n)$, $\mathscr{A}_n = \{A_1^n, \ldots, A_{q_n}^n\}$, of invariant covers of Q such that $\mathscr{A}_n$ is a partition of Q and

$$\frac{\log q_n}{\tau_n} \to h_{\text{inv}}(Q) \quad \text{for } n \to \infty. \tag{2.25}$$

Define the coder-controller $\mathscr{H}^n = (S^n, \gamma^n, \delta^n, \tau^n)$ by

$$\tau^n := \tau_n,$$

$$S_k^n := \{1, \ldots, q_n\},$$

$$\gamma_k^n(x_0, x_1, \ldots, x_k) := s_k \text{ if } x_k \in A_{s_k}^n,$$

$$\delta_k^n(s_0, s_1, \ldots, s_k) := v_n(A_{s_k}^n)|_{[0,\tau_n)}$$

for all $k \in \mathbb{Z}_+$ and for each $n \in \mathbb{N}$. Hence, in particular the alphabet of $\mathscr{H}^n$ is time-invariant and the coder and controller mappings depend only on the present state or symbol, respectively. From the definition of invariant covers it immediately follows that $\mathscr{H}^n$ renders Q invariant. For the corresponding transmission data rates we obtain

$$R(\mathscr{H}^n) = \liminf_{k \to \infty} \frac{1}{k\tau_n} \underbrace{\sum_{j=0}^{k-1} \log q_n}_{=k \log q_n} = \frac{\log q_n}{\tau_n} \overset{(2.25)}{\longrightarrow} h_{\mathrm{inv}}(Q).$$

This completes the proof.   □

Remark 2.9. In the data rate theorem for the topological feedback entropy, the symbols generated by the coder are allowed to depend on the past symbols. But since the past symbols can be generated from the past states, in the preceding theorem we only consider coders with inputs from the state space.

2.6 Comments and Bibliographical Notes

Most of the results of this chapter can be found in the thesis Kawan [62] and the paper Colonius and Kawan [23], but only for smooth systems given by differential equations. The definition of invariance entropy for admissible pairs (K, Q) is somewhat different to the definitions given in [23, 62], where it is assumed that Q is controlled invariant and compact. In the above literature, our notion of invariance entropy is called *strict invariance entropy* and the term *invariance entropy* refers to our notion of *outer invariance entropy*. The product theorem 2.12 for invariance entropy has not appeared before. The notion of inner invariance entropy and its relation to topological feedback entropy can be found in Colonius, Kawan, and Nair [28]. One topic not treated in this book can be found in Colonius and Helmke [22], where the outer invariance entropy for admissible pairs of (A, B)-invariant subspaces of linear systems given by differential equations is investigated. Another such topic is treated in Colonius and Kawan [24], where a version of invariance entropy for systems with outputs is investigated. (Here the control objective is to make a subset in the output space invariant.) In Colonius, Fukuoka, and Santana [27] one finds an extension of the notion of invariance entropy to topological semigroups acting on metric spaces. The paper Da Silva [32] treats an extension of invariance entropy to random control systems. Finally, in Hagihara and Nair [56] two extensions of topological feedback entropy are introduced, one for systems with an output map and one for systems with discontinuous right-hand sides. Here the considered control task is to steer the system into a target set. The probably most interesting questions left open in this chapter are the following: Under which conditions do the quantities $h_{\text{inv}}(K, Q)$ and $h_{\text{inv,out}}(K, Q)$ coincide? and: Looking at the characterization of $h_{\text{inv}}(Q)$ via invariant covers, are there conditions which guarantee the existence of "generators", that is, invariant covers $\mathscr{C}$ such that $h_{\text{inv}}(Q) = h(\mathscr{C}; Q)$? Note that the one-dimensional linear result of Proposition 2.17 provides a simple example where $h_{\text{inv,out}}(Q) = h_{\text{inv}}(Q)$.

Chapter 3
Linear and Bilinear Systems

After we have introduced the notion of invariance entropy and studied its elementary
properties, we now start to develop techniques for its computation or estimation.
Before we turn our attention to smooth nonlinear systems, we look at linear and
bilinear systems on Euclidean space.

Our main result for linear systems provides a formula for the outer invariance
entropy of an admissible pair (K, Q) such that K has positive Lebesgue measure
and Q is compact. In this case, it turns out that $h_{\mathrm{inv,out}}(K, Q)$ is given by the sum of
the logarithms of the unstable eigenvalues of the linear operator $\varphi_{1,0}$ corresponding
to time $t = 1$ and the constant zero control function. The proof of this result is based
on Bowen's classical theorem about the topological entropy of a linear operator
and similar volume growth arguments as already applied in Sect. 2.3 for the scalar
continuous-time case. We also provide a partial generalization of this theorem to the
general case without assumptions on K and Q except for the ones required in the
definition of $h_{\mathrm{inv,out}}(K, Q)$. Here, only an upper bound can be given which, next to
the eigenvalues, involves certain dimension characteristics of the set K.

Following the study of linear systems, we introduce the more general class of
inhomogeneous bilinear systems which are defined as generalizations of systems
given by differential equations of the form $\dot{x}(t) = A(\omega_1(t))x(t) + B\omega_2(t)$, where
$A(\omega) = A_0 + \sum_{i=1}^{m} \omega_i A_i$. Here, we cannot provide a formula for the invariance
entropy, but only a lower bound.[1] This bound is expressed in terms of volume
growth rates on invariant subbundles of the extended state space for the associated
homogeneous system. Using the theorem of Selgrade, one can choose subbundles
where these growth rates become maximal and relate the lower bound to the
Lyapunov exponents of the homogeneous system.

---

[1] An upper bound is given in Sect. 7.3 under additional controllability assumptions, using the
nonlinear techniques developed in Chap. 5.

C. Kawan, *Invariance Entropy for Deterministic Control Systems*, Lecture Notes
in Mathematics 2089, DOI 10.1007/978-3-319-01288-9_3,
© Springer International Publishing Switzerland 2013

3.1   Linear Systems

In this section, we consider a linear time-invariant system $\Sigma = (\mathbb{T}, X, U, \mathcal{U}, \varphi)$. That is, X is a d-dimensional and U an m-dimensional vector space over a field $\mathbb{K} \in \{\mathbb{R}, \mathbb{C}\}$ and Σ can be extended to a topological time-invariant system whose transition map is linear as a function of the last two arguments (see Definition 1.4). Since every finite-dimensional vector space over $\mathbb{C}$ can be identified with a real vector space of twice the dimension, we can assume that $\mathbb{K} = \mathbb{R}$. In the proofs of this section, some of the material about capacitive dimension and topological entropy, collected in Sect. B.3, is used.

Theorem 3.1. *Assume that (K, Q) is an admissible pair for Σ, where Q is compact and K has positive Lebesgue measure. Then*

$$h_{\mathrm{inv,out}}(K, Q) = \sum_{\lambda \in \sigma(\varphi_{1,0})} \max\{0, n_\lambda \log |\lambda|\}, \tag{3.1}$$

where $\varphi_{1,0} : X \to X$ is the linear operator corresponding to time $\tau = 1 \in \mathbb{T}$ and the constant control function $\mathbf{0}(t) \equiv 0 \in U$. Moreover, n_λ denotes the algebraic multiplicity of the eigenvalue λ.

Proof. Fix a norm $|\cdot|$ on X, let ϱ denote the corresponding metric and λ the induced Lebesgue measure. The proof now proceeds in three steps.

Step 1. Recall that $\varphi(t, \cdot, \cdot)$ is the restriction of a linear map for which we use the same notation. Hence, we can write

$$\varphi(t, x, \omega) = \varphi(t, x, \mathbf{0}) + \varphi(t, 0, \omega).$$

The map

$$\Phi : (t, x) \mapsto \varphi(t, x, \mathbf{0}), \quad \mathbb{T}_+ \times X \to X,$$

is a linear dynamical system on X which follows from linearity of $\varphi(t, \cdot, \cdot)$ and the cocycle property. In the case that $\mathbb{T} = \mathbb{R}$, Proposition 1.3 guarantees that $\Phi(t, x) = e^{At}x$ for some $A \in \mathscr{L}(X, X)$. We claim that the outer invariance entropy $h_{\mathrm{inv,out}}(K, Q)$ is bounded from above by the topological entropy of Φ, which by Sect. B.3 is given by

$$h_{\mathrm{top},\varrho}(\Phi) = \sum_{\lambda \in \sigma(\varphi_{1,0})} \max\{0, n_\lambda \log |\lambda|\}.$$

For fixed $\tau \in \mathbb{T} \cap (0, \infty)$ and $\varepsilon > 0$ let $E \subset K$ be a $(\tau, \varepsilon, \Phi)$-separated subset of K of maximal cardinality. Then E also $(\tau, \varepsilon, \Phi)$-spans the set K (see Sect. B.3 for an argument), which means that

$$\forall x \in K \; \exists y \in E : \; \max_{t \in [0,\tau]} \varrho(\varphi_{t,0}(x), \varphi_{t,0}(y)) < \varepsilon. \tag{3.2}$$

Since (K, Q) is admissible, for every $y \in E$ we find some $\omega_y \in \mathscr{U}$ with $\varphi(t, y, \omega_y) \in Q$ for all $t \in [0, \tau]$. We claim that $\mathscr{S} := \{\omega_y\}_{y \in E}$ is a $(\tau, K, N_\varepsilon(Q))$-spanning set. Indeed, for an arbitrary $x \in K$ take $y \in E$ as in (3.2). Then for all $t \in [0, \tau]$ we have

$$\left| \varphi(t, x, \omega_y) - \varphi(t, y, \omega_y) \right| = \left| \varphi(t, x - y, \mathbf{0}) \right| = \varrho(\varphi_{t,0}(x), \varphi_{t,0}(y)) < \varepsilon,$$

and hence $\varphi(t, x, \omega_y) \in B(\varphi(t, y, \omega_y), \varepsilon) \subset N_\varepsilon(Q)$. This implies

$$r_{\mathrm{inv}}(\tau, K, N_\varepsilon(Q)) \leq r_{\mathrm{sep}}(\tau, \varepsilon, K, \Phi)$$

and therefore

$$h_{\mathrm{inv,out}}(K, Q) = \lim_{\varepsilon \searrow 0} h_{\mathrm{inv}}(K, N_\varepsilon(Q)) \leq h_{\mathrm{top},\varrho}(K, \Phi),$$

as claimed. This concludes the proof of the inequality "$\leq$" in (3.1).

Step 2. In this step, we prove the inequality "$\geq$" in (3.1) for the special case that all eigenvalues of $\varphi_{1,0}$ are greater than 1 in absolute value. If $\mathscr{S} \subset \mathscr{U}$ is a minimal (and hence finite) $(\tau, K, N_\varepsilon(Q))$-spanning set for some fixed $\varepsilon > 0$, we define

$$K_\omega := \{x \in K : \varphi(\tau, x, \omega) \in N_\varepsilon(Q)\}, \quad \omega \in \mathscr{S}.$$

One can write K_ω as the intersection of the compact set K and the open set $\varphi_{\tau,\omega}^{-1}(N_\varepsilon(Q))$, hence K_ω is a Borel set. Also the image of K_ω under $\varphi_{\tau,\omega}$ is a Borel set, since $\varphi_{\tau,\omega}$ is a diffeomorphism, which follows from the assumption about the eigenvalues of $\varphi_{1,0}$. By definition we have $K = \bigcup_{\omega \in \mathscr{S}} K_\omega$ and $\varphi_{\tau,\omega}(K_\omega) \subset N_\varepsilon(Q)$ for all $\omega \in \mathscr{S}$. Hence, we obtain

$$\lambda(N_\varepsilon(Q)) \geq \lambda\left(\varphi_{\tau,\omega}(K_\omega)\right) = |\det \varphi_{\tau,0}| \, \lambda(K_\omega).$$

It follows that

$$\lambda(K) \leq \sum_{\omega \in \mathscr{S}} \lambda(K_\omega) \leq \#\mathscr{S} \cdot \max_{\omega \in \mathscr{S}} \lambda(K_\omega) \leq \#\mathscr{S} \cdot \frac{\lambda(N_\varepsilon(Q))}{|\det \varphi_{\tau,0}|},$$

which implies

$$r_{\mathrm{inv}}(\tau, K, N_\varepsilon(Q)) \geq \underbrace{\frac{\lambda(K)}{\lambda(N_\varepsilon(Q))}}_{=:C \in (0, \infty)} |\det \varphi_{\tau,0}| \, .$$

Hence, we end up with the estimate

$$h_{\text{inv}}(K, N_\varepsilon(Q)) \geq \limsup_{\tau \to \infty} \frac{1}{\tau} \log \left(C \left| \det \varphi_{\tau,0} \right| \right).$$

Obviously, we can omit the constant C on the right-hand side. Each $\tau > 0$ can be written as $\tau = k_\tau + r_\tau$ with $k_\tau \in \mathbb{Z}_+$ and $r_\tau \in [0, 1)$. (In the discrete-time case, $r_\tau = 0$.) Then, using the semigroup property of the dynamical system Φ and elementary properties of the determinant, we obtain

$$\left| \det \varphi_{\tau,0} \right| = \left| \det \varphi_{r_\tau,0} \right| \left| \det \varphi_{1,0} \right|^{k_\tau},$$

which implies

$$h_{\text{inv}}(K, N_\varepsilon(Q)) \geq \limsup_{\tau \to \infty} \left(\frac{1}{\tau} \log \left| \det \varphi_{r_\tau,0} \right| + \frac{k_\tau}{\tau} \log \left| \det \varphi_{1,0} \right| \right)$$

$$= \log \left| \det \varphi_{1,0} \right| = \log \prod_{\lambda \in \sigma(\varphi_{1,0})} |\lambda|^{n_\lambda} = \sum_{\lambda \in \sigma(\varphi_{1,0})} n_\lambda \log |\lambda|.$$

Since this holds for every $\varepsilon > 0$, the second step is finished.

Step 3. If the absolute value of every eigenvalue of $\varphi_{1,0} : X \to X$ is less than or equal to 1, the assertion trivially holds, since $h_{\text{inv,out}}(K, Q) \geq 0$ anyway. Hence, we may assume that at least one eigenvalue is greater than 1 in absolute value. We write $\mathbb{E}^{cs}$ and $\mathbb{E}^u$ for the corresponding center-stable and unstable subspace of $\varphi_{1,0}$.[2] This furnishes the decomposition $X = \mathbb{E}^{cs} \oplus \mathbb{E}^u$. Both of these subspaces are invariant under each of the maps $\varphi_{t,0}, t \in \mathbb{T}_+$. In discrete time this is obvious, since $\varphi_{n,0} = \varphi_{1,0}^n$ for all $n \in \mathbb{Z}_+$. In continuous time, it follows from the fact that $\varphi_{t,0} = e^{At}$ with $A \in \mathscr{L}(X, X)$, since the generalized eigenspaces of A coincide with those of e^{At} for each $t > 0$. Let $\pi : X \to \mathbb{E}^u$ be the projection onto $\mathbb{E}^u$ along $\mathbb{E}^{cs}$. By Example 2.4 this establishes a topological semi-conjugacy $(\pi, \text{id}_{\mathscr{U}})$ from Σ onto the linear system $\Sigma' = (\mathbb{T}, \mathbb{E}^u, U, \mathscr{U}, \varphi^u)$ with $\varphi^u(t, x, \omega) = \pi \varphi(t, x, \omega)$. The system Σ' has the property that all eigenvalues of $\varphi_{1,0}^u$ are greater than 1 in absolute value. Hence, if the Lebesgue measure of $\pi K \subset \mathbb{E}^u$ is positive, we can apply the estimate of Step 2 to Σ' with the admissible pair $(\pi K, \pi Q)$. Then, by Proposition 2.13 we obtain

$$h_{\text{inv,out}}(K, Q; \Sigma) \geq h_{\text{inv,out}}(\pi K, \pi Q; \Sigma')$$

$$\geq \sum_{\lambda \in \sigma(\varphi_{1,0}^u)} n_\lambda \log |\lambda| = \sum_{\lambda \in \sigma(\varphi_{1,0})} \max\{0, n_\lambda \log |\lambda|\}.$$

---

[2]That is, $\mathbb{E}^{cs}$ and $\mathbb{E}^u$ are the sums of all generalized eigenspaces corresponding to eigenvalues λ with $|\lambda| \leq 1$ and $|\lambda| > 1$, respectively.

which completes the proof. To show that πK has positive Lebesgue measure, let $d_u = \dim \mathbb{E}^u$ and let λ^{d_u} denote the d_u-dimensional Lebesgue measure in $\mathbb{E}^u$. Assume to the contrary that $\lambda^{d_u}(\pi K) = 0$ and assume without loss of generality that the norm $|\cdot|$ comes from an inner product $\langle \cdot, \cdot \rangle$ such that $\mathbb{E}^u$ is the orthogonal complement of $\mathbb{E}^{cs}$. Using the theorem of Fubini, we obtain

$$\lambda(K) \le \lambda(\pi^{-1}\pi K) = \int_X \mathbb{1}_{\pi^{-1}\pi K}(x)\mathrm{d}\lambda(x)$$

$$= \int_{\mathbb{E}^u \oplus \mathbb{E}^{cs}} \mathbb{1}_{\pi^{-1}\pi K}(u + v)\mathrm{d}(u, v)$$

$$= \int_{\mathbb{E}^u} \int_{\mathbb{E}^{cs}} \mathbb{1}_{\pi^{-1}\pi K}(u + v)\mathrm{d}u\mathrm{d}v.$$

Since $\pi^{-1}\pi K = \pi K \oplus \mathbb{E}^{cs}$, we end up with the contradiction

$$\lambda(K) \le \int_{\mathbb{E}^u} \int_{\mathbb{E}^{cs}} \mathbb{1}_{\pi K \oplus \mathbb{E}^{cs}}(u, v)\mathrm{d}u\mathrm{d}v$$

$$= \int_{\mathbb{E}^u} \int_{\mathbb{E}^{cs}} \mathbb{1}_{\pi K}(u)\mathbb{1}_{\mathbb{E}^{cs}}(v)\mathrm{d}u\mathrm{d}v$$

$$= \int_{\mathbb{E}^{cs}} \mathbb{1}_{\mathbb{E}^{cs}}(v)\left(\int_{\mathbb{E}^u} \mathbb{1}_{\pi K}(u)\mathrm{d}u\right)\mathrm{d}v$$

$$= \int_{\mathbb{E}^{cs}} \mathbb{1}_{\mathbb{E}^{cs}}(v)\underbrace{\lambda^{d_u}(\pi K)}_{=0}\mathrm{d}v = 0.$$

This concludes the proof.   □

Remark 3.1.

- There are several possibilities to weaken the assumptions of the above theorem. First, it is not necessary to assume that Q is compact. Indeed, for the upper estimate it is sufficient that Q is closed. In the lower estimate we only used that the ε-neighborhoods of Q have finite Lebesgue measure. Also the assumption of K having positive Lebesgue measure can be weakened by assuming only that the projection of K to the unstable subspace $\mathbb{E}^u$ of $\varphi_{1,0}$ has positive Lebesgue measure.
- The preceding theorem generalizes [23, Theorem 5.1], the corresponding result for linear systems given by differential equations

$$\dot{x}(t) = Ax(t) + B\omega(t), \quad \omega \in \mathscr{U}, \tag{3.3}$$

with a compact control range $\Omega \subset \mathbb{R}^m$. Note that for such a system, formula (3.1) can be written as

$$h_{\text{inv,out}}(K, Q) = \sum_{\lambda \in \sigma(A)} \max\{0, n_\lambda \, \text{Re}(\lambda)\}.$$

In particular, we recover the one-dimensional result from Sect. 2.3. For the system defined by (3.3), the existence of a compact controlled invariant set Q with nonempty interior can be guaranteed if the matrix pair (A, B) is controllable, the matrix A is hyperbolic (that is, it has no eigenvalues on the imaginary axis), and the values of the admissible control functions are restricted to a compact and convex set with nonempty interior. Then there exists a unique control set D with nonempty interior and its closure $Q = \text{cl}\, D$ is compact (see Colonius and Spadini [26, Theorem 4.1]). By Proposition 1.11, Q is controlled invariant. Hence, in this case the existence of admissible pairs as required in Theorem 3.1 is guaranteed.

• In Hoock [59], the above result has been generalized to a certain class of infinite-dimensional linear systems on Hilbert spaces, namely systems described by abstract differential equations

$$\dot{x}(t) = \mathscr{A}x(t) + \mathscr{B}\omega(t), \quad \omega \in \mathscr{U},$$

where $\mathscr{A}$ is the infinitesimal generator of a strongly continuous semigroup and $\mathscr{B}$ is a bounded linear operator. Assuming the existence of a spectral decomposition with a finite-dimensional unstable subspace, the outer invariance entropy of such a system can be computed as in Theorem 3.1.

If the assumption of K (or $\pi(K) \subset \mathbb{E}^u$) having positive Lebesgue measure is not satisfied, it is hard to tell the exact value of $h_{\text{inv,out}}(K, Q)$. However, we can use the idea for obtaining the upper bound to prove the following general estimate.

Theorem 3.2. *Assume that (K, Q) is an admissible pair for Σ, where Q is closed (not necessarily compact). Let ϱ be a metric on X induced by a norm. Then*

$$h_{\text{inv,out}}(K, Q; \varrho) \le \sum_{j=1}^{r} \max\{0, \log r_j(\varphi_{1,0})\} \cdot \overline{\dim}_C(\pi_j K), \qquad (3.4)$$

where π_j denotes the projection from X onto the j-th Lyapunov space of $\varphi_{1,0}$ along the direct sum of the other Lyapunov spaces, and $r_j(\varphi_{1,0})$, $1 \le j \le r$, are the corresponding moduli of the eigenvalues of $\varphi_{1,0}$.

Proof. As in Step 1 of the proof of Theorem 3.1 one shows that

$$h_{\text{inv,out}}(K, Q; \varrho) \le h_{\text{top},\varrho}(K, \Phi).$$

Let $E_1, \ldots, E_r$ denote the different Lyapunov spaces of $\varphi_{1,0}$, that is, the sums of the generalized eigenspaces corresponding to eigenvalues of the same modulus. Associated with the direct sum decomposition $X = E_1 \oplus \cdots \oplus E_r$ we have r

projections $\pi_j : X \to E_j$, $j = 1, \ldots, r$. This gives the inclusion $K \subset \pi_1 K \oplus \cdots \oplus \pi_r K$ and the estimate

$$h_{\text{top},\varrho}(K, \Phi) \leq h_{\text{top},\varrho}(\pi_1 K \oplus \cdots \oplus \pi_r K, \Phi).$$

We have the inequality

$$h_{\text{top},\varrho}(\pi_1 K \oplus \cdots \oplus \pi_r K, \Phi) \leq \sum_{j=1}^{r} h_{\text{top},\varrho_j}(\pi_j K, \Phi|_{E_j}), \qquad (3.5)$$

where ϱ_j is a metric on E_j for $j = 1, \ldots, r$ such that $\varrho(x, y) = \max\{\varrho_1(\pi_1 x, \pi_1 y), \ldots, \varrho_r(\pi_r x, \pi_r y)\}$. Naturally, we assume that $\rho_1, \ldots, \rho_r$ are metrics induced by norms. (We can assume without loss of generality that ρ has this form.) In fact, this is proved as follows: If F_j $(\tau, \varepsilon, \Phi|_{F_j})$-spans the set $\pi_j K$, $j = 1, \ldots, r$, then $F_1 \oplus \cdots \oplus F_r$ $(\tau, \varepsilon, \Phi)$-spans $\pi_1 K \oplus \cdots \oplus \pi_r K$, which implies $r_{\text{span}}(\tau, \varepsilon, \pi_1 K \oplus \cdots \oplus \pi_r K, \Phi) \leq \prod_{j=1}^{r} r_{\text{span}}(\tau, \varepsilon, \pi_j K, \Phi|_{E_j})$ and hence (3.5). In order to estimate $h_{\text{top},\varrho_j}(\pi_j K, \Phi|_{E_j})$, we use Lemma B.4 and Proposition B.8.

$$h_{\text{top},\varrho_j}(\pi_j K, \Phi|_{E_j}) = \frac{1}{m} h_{\text{top},\varrho_j}(\pi_j K, \varphi_{m,0}|_{E_j})$$

$$\leq \max\left\{0, \log \left\| (\varphi_{1,0}|_{E_j})^m \right\|^{1/m}\right\} \cdot \overline{\dim}_C(\pi_j K).$$

It is well-known that $\|T^m\|^{1/m}$ converges to the spectral radius $r(T)$ for any finite-dimensional linear operator T. Hence, we conclude

$$h_{\text{inv,out}}(K, Q; \varrho) \leq h_{\text{top},\varrho}(K, \Phi)$$

$$\leq \sum_{j=1}^{r} h_{\text{top},\varrho_j}(\pi_j K, \Phi|_{E_j})$$

$$\leq \sum_{j=1}^{r} \max\left\{0, \lim_{m \to \infty} \log \left\| (\varphi_{1,0}|_{E_j})^m \right\|^{1/m}\right\} \cdot \overline{\dim}_C(\pi_j K)$$

$$= \sum_{j=1}^{r} \max\left\{0, \log r(\varphi_{1,0}|_{E_j})\right\} \cdot \overline{\dim}_C(\pi_j K),$$

which finishes the proof.   □

Remark 3.2. In the case where K has positive Lebesgue measure, one easily shows that the upper bound (3.4) reduces to the right-hand side in formula (3.1).

3.2 Inhomogeneous Bilinear Systems

In this section, we introduce inhomogeneous bilinear systems which are generalizations of linear systems. We first give a definition of homogeneous bilinear systems
which is motivated by smooth systems given by differential equations of the form

$$\dot{x}(t) = \left(A_0 + \sum_{i=1}^{m} \omega_i(t) A_i \right) x(t), \quad \omega \in \mathcal{U},$$

with a compact and convex control range $\Omega \subset \mathbb{R}^m$. Such a system is control-affine
and therefore the set $\mathcal{U}$ of admissible control functions endowed with the
weak*-topology of $L^\infty(\mathbb{R}, \mathbb{R}^m) = L^1(\mathbb{R}, \mathbb{R}^m)^*$ is a compact metrizable space,
and the associated control flow

$$\Phi_t(\omega, x) = (\Theta_t \omega, \varphi(t, x, \omega)),$$

which in this case is a linear flow on the trivial vector bundle $\pi : \mathcal{U} \times \mathbb{R}^d \to \mathcal{U}$,
$\pi(\omega, x) = \omega$, is continuous.

Definition 3.1. A time-invariant system $\Sigma = (\mathbb{T}, X, U, \mathcal{U}, \varphi)$ is called *(homogeneous) bilinear* if the following properties are satisfied:

(i) X is a d-dimensional vector space over $\mathbb{R}$;
(ii) $\mathcal{U}$ is endowed with a topology that makes it a compact metrizable space;
(iii) We can write

$$\varphi(t, x, \omega) = \varphi(t, \omega)x, \quad \varphi(t, \omega) \in \mathcal{L}(X, X),$$

such that the control semiflow

$$\mathbb{T}_+ \times (\mathcal{U} \times X) \to \mathcal{U} \times X, \quad (t, (\omega, x)) \mapsto (\Theta_t \omega, \varphi(t, \omega)x), \qquad (3.6)$$

is a linear semiflow on the trivial vector bundle $\pi : \mathcal{U} \times X \to \mathcal{U}$.

An inhomogeneous bilinear system is then defined as follows.

Definition 3.2. A time-invariant system $\Sigma = (\mathbb{T}, X, U, \mathcal{U}, \varphi)$ is called *inhomogeneous bilinear* or *affine* if there exists a homogeneous bilinear system of the form
$\Sigma' = (\mathbb{T}, X, U_1, \mathcal{U}_1, \alpha)$ such that the following properties are satisfied:

(i) The control-value space and the set of admissible control functions can be
written as Cartesian products of the form $U = U_1 \times U_2$ and $\mathcal{U} = \mathcal{U}_1 \times \mathcal{U}_2$,
respectively.
(ii) We can write

$$\varphi(t, x, (\omega_1, \omega_2)) = \alpha(t, \omega_1)x + \beta(t, \omega_1, \omega_2)$$

for all $t \in \mathbb{T}_+$, $x \in X$, and $(\omega_1, \omega_2) \in \mathcal{U} = \mathcal{U}_1 \times \mathcal{U}_2$, with a map $\beta :$ $\mathbb{T}_+ \times \mathcal{U}_1 \times \mathcal{U}_2 \to X$.

Example 3.1. A linear system $\Sigma = (\mathbb{T}, X, U, \mathcal{U}, \varphi)$ is a special inhomogeneous bilinear system. Indeed, the associated homogeneous bilinear system is the trivial linear system $\Sigma' = (\mathbb{T}, X, \{0\}, \{\mathbf{0}\}, \alpha)$ with $\alpha(t, x, \mathbf{0}) = \varphi(t, x, \mathbf{0})$.

Example 3.2. Consider a smooth system $\Sigma = (\mathbb{R}, \mathbb{R}^d, \mathbb{R}^m, \mathcal{U}, \varphi)$ given by differential equations of the form

$$\dot{x}(t) = \left[A_0 + \sum_{i=1}^{m_1} \omega_{1,i}(t) A_i \right] x(t) + g(\omega_1(t), \omega_2(t)),$$

$(\omega_1, \omega_2) \in \mathcal{U} = \mathcal{U}_1 \times \mathcal{U}_2.$

Here, $A_0, A_1, \ldots, A_{m_1} \subset \mathbb{R}^{d \times d}$ and $g : \mathbb{R}^{m_1} \times \mathbb{R}^{m_2} \to \mathbb{R}^d$ is a continuous map $(m = m_1 + m_2)$. The control range is of the form $\Omega_1 \times \Omega_2$ with $\Omega_i \subset \mathbb{R}^{m_i}$ such that Ω_1 is compact and convex. By linearity, the variation-of-constants formula gives the explicit expression

$$\varphi(t, x, (\omega_1, \omega_2)) = \Lambda_{\omega_1}(t, 0)x + \int_0^t \Lambda_{\omega_1}(t, s) g(\omega_1(s), \omega_2(s)) \mathrm{d}s$$

for the trajectories, where $\Lambda_{\omega}(t, s)$ are the corresponding evolution operators. By the assumptions on $\mathcal{U}_1$ it follows that $\mathcal{U}_1$ is a compact metrizable space with the weak*-topology of $L^\infty(\mathbb{R}, \mathbb{R}^{m_1}) = L^1(\mathbb{R}, \mathbb{R}^{m_1})^*$ and $(t, x, \omega_1) \mapsto \Lambda_{\omega_1}(t, 0)x$ is continuous. It follows that Σ satisfies Definition 3.2.

We cannot provide a formula for the (outer) invariance entropy of an inhomogeneous bilinear system as we did for linear systems. However, we can derive a lower bound in terms of volume growth rates which reduces to the right-hand side in formula (3.1) in the case that the given system is linear. The idea used here is essentially the same as that in the proof of Theorem 3.1. More precisely, we look at the volume growth on an invariant subbundle of the associated linear semiflow (3.6). In contrast to the linear case, here the invariant subspace is a fiber of the invariant subbundle and, in general, varies with time which causes some complications. We have to show that the volumes of the projected sets do not vary too much when the point in the base space is changed, which follows as a consequence of compactness of the base space. The next three lemmas provide the necessary tools to treat these rather technical problems.

The first problem is related to the fact that we have to look at the measures of images under projections. In general, linear maps do not map measurable sets onto measurable sets. The simplest counter-example is the projection $P : \mathbb{R}^2 \to \mathbb{R}$, $(x, y) \mapsto x$. If $A \subset \mathbb{R}$ is a set which is not measurable with respect to the one-dimensional Lebesgue measure in $\mathbb{R}$, then $A \times \{0\} \subset \mathbb{R}^2$ is measurable with two-dimensional Lebesgue measure zero and $P(A \times \{0\}) = A$. However, for a set which can be written as the intersection of a compact set and an F_σ-set, the image under a projection is again measurable.

Lemma 3.1. *Let X be a Euclidean space[3] and $P \in \mathscr{L}(X, X)$ a projection of rank k. Let $A = K \cap U \subset X$, where K is compact and U is an F_σ-set, that is, the countable union of closed sets. Then $P(A) \subset \mathrm{im}(P)$ is Lebesgue measurable with respect to the k-dimensional Lebesgue measure in $\mathrm{im}(P)$.*

Proof. By assumption, we can write $U = \bigcup_{n\in\mathbb{N}} C_n$ with closed sets C_n. Hence,

$$P(A) = P(K \cap U) = P\left(K \cap \bigcup_{n\in\mathbb{N}} C_n\right) = P\left(\bigcup_{n\in\mathbb{N}}(K \cap C_n)\right) = \bigcup_{n\in\mathbb{N}} P(K \cap C_n).$$

Since C_n is closed and K is compact, $K \cap C_n$ is compact and by continuity of P also $P(K \cap C_n)$. Therefore, $P(K \cap C_n)$ is closed in $\mathrm{im}(P)$ which implies that $P(A)$ is an F_σ-subset of $\mathrm{im}(P)$ and thus Lebesgue measurable. □

Remark 3.3. By Cohn [18, Proposition 1.1.5], every open set U is an F_σ-set. Of course, also compact sets are F_σ-sets.

Before we continue with the next lemma, we introduce some notation: For a compact set $K \subset X$ we denote the family of all sets $A \subset K$ which are intersections of K with F_σ-sets by $\mathscr{F}_\sigma(K)$ (these are the F_σ-subsets of K). If X is endowed with an inner product and $P \in \mathscr{L}(X, X)$ is a projection of rank k, we let λ_P^k denote the k-dimensional Lebesgue measure in $\mathrm{im}(P)$.

Lemma 3.2. *Let $(X, \langle \cdot, \cdot \rangle)$ be a d-dimensional Euclidean space and $\mathscr{P}_k(X)$ the set of all projections in $\mathscr{L}(X, X)$ of rank k, $1 \leq k \leq d$. Then for every $x \in X$ and $r > 0$, the map*

$$\mu : \mathscr{P}_k(X) \to \mathbb{R}_+, \quad P \mapsto \lambda_P^k(P(B(x, r))),$$

which assigns to a projection P the k-dimensional Lebesgue measure of the P-image of the ball $B(x, r)$, is continuous.

Proof. By Proposition A.1, every projection P maps the open ball $B(x, r)$ onto an open ellipsoid with semi-axes of lengths $r\sigma_1(P), \ldots, r\sigma_k(P)$, where $\sigma_1(P) \geq \cdots \geq \sigma_d(P)$ denote the singular values of P. Since the singular values depend continuously on P, also the k-dimensional volume of $P(B(x, r))$, which is proportional to the product of the nonzero singular values, depends continuously on P. This proves the assertion. □

Lemma 3.3. *Let $(X, \langle \cdot, \cdot \rangle)$ be a d-dimensional Euclidean space and $K \subset X$ a compact set with positive Lebesgue measure. Moreover, let $\mathscr{P}$ be a compact set of*

---

[3]Throughout the book, by a *Euclidean space* we understand a finite-dimensional vector space over $\mathbb{R}$ endowed with an inner product.

projections in X of constant rank $k \in \{1, \ldots, d\}$. Then there exists $\gamma > 0$ such that for every finite cover $\{K_1, \ldots, K_r\}$ of K with $K_j \in \mathscr{F}_\sigma(K)$ and for all $P_1, \ldots, P_r \in \mathscr{P}$ we have

$$\sum_{j=1}^{r} \lambda_{P_j}^{k}(P_j(K_j)) \geq \gamma.$$

Proof. Let λ^d denote the d-dimensional Lebesgue measure in X. Then it suffices to show that there exists a constant $C > 0$ such that

$$\lambda^d(A) \leq C\lambda_P^k(P(A)) \quad \text{for all } P \in \mathscr{P} \text{ and } A \in \mathscr{F}_\sigma(K), \tag{3.7}$$

since this implies

$$\sum_{j=1}^{r} \lambda_{P_j}^{k}(P_j(K_j)) \geq \frac{1}{C} \sum_{j=1}^{r} \lambda^d(K_j) \geq \frac{1}{C}\lambda^d(K) =: \gamma.$$

In order to prove (3.7), let $\hat{K}$ be a compact ball with $K \subset \hat{K}$. Then we obtain for every $P \in \mathscr{P}$ and for every $A \in \mathscr{F}_\sigma(K)$:

$$\lambda^d(A) \leq \lambda^d(P^{-1}(P(A)) \cap K) \leq \lambda^d(P^{-1}(P(A)) \cap \hat{K}).$$

Hence, it suffices to show the existence of $C > 0$ with

$$\lambda^d(P^{-1}(B) \cap \hat{K}) \leq C\lambda_P^k(B) \quad \text{for all } P \in \mathscr{P} \text{ and measurable } B \subset \operatorname{im}(P).$$

For every $P \in \mathscr{P}$ we define a new inner product on X by

$$\langle x, y \rangle_P := \langle Px, Py \rangle + \langle (I-P)x, (I-P)y \rangle.$$

This gives a norm $|\cdot|_P$ and a d-dimensional Lebesgue measure λ_P^d on X. For every $x \in X \backslash \{0\}$ we obtain

$$|x|^2 = |Px + (I-P)x|^2 = |Px|^2 + 2\langle Px, (I-P)x \rangle + |(I-P)x|^2$$

$$= \left(1 + 2\frac{\langle Px, (I-P)x \rangle}{|Px|^2 + |(I-P)x|^2}\right)\left(|Px|^2 + |(I-P)x|^2\right)$$

$$= \left(1 + 2\frac{\langle Px, (I-P)x \rangle}{|Px|^2 + |(I-P)x|^2}\right)|x|_P^2.$$

The function

$$f(P, x) := \frac{\langle Px, (I-P)x \rangle}{|Px|^2 + |(I-P)x|^2}, \quad f : \mathscr{P} \times (X \backslash \{0\}) \to \mathbb{R},$$

does not depend on the norm of the vector x, that is, $f(P, sx) = f(P, x)$ for every $s \in \mathbb{R}\backslash\{0\}$. Consequently, continuity of f and compactness of $\mathscr{P} \times S(X)$, $S(X) := \{x \in X : |x| = 1\}$, implies the existence of $M \geq 0$ with $f(P, x) \leq M$ for all $(P, x) \in \mathscr{P} \times (X \backslash \{0\})$, which yields

$$|x| \leq \sqrt{1 + 2M}\, |x|_P \quad \text{for all } P \in \mathscr{P} \text{ and } x \in X.$$

Hence, for the linear isomorphism $\Phi_P : (X, \langle \cdot, \cdot \rangle_P) \to (X, \langle \cdot, \cdot \rangle)$, $x \mapsto x$, we have

$$|\det \Phi_P| \leq (1 + 2M)^{d/2}.$$

Using this estimate, we obtain for every measurable set $A \subset \hat{K}$ that

$$\lambda^d(A) \leq (1 + 2M)^{d/2} \lambda_P^d(A).$$

In order to finish the proof, it suffices to show the existence of $\tilde{C} > 0$ with

$$\lambda_P^d\left(\Phi_P^{-1}(P^{-1}(B) \cap \hat{K})\right) \leq \tilde{C} \lambda_P^k(B)$$

for all $P \in \mathscr{P}$ and measurable $B \subset \mathrm{im}(P)$. Since $\ker(P)$ and $\mathrm{im}(P)$ are orthogonal with respect to the inner product $\langle \cdot, \cdot \rangle_P$, the theorem of Fubini yields

$$\lambda_P^d\left(\Phi_P^{-1}(P^{-1}(B) \cap \hat{K})\right) = \int_{\ker(P)} \int_{\mathrm{im}(P)} \mathbb{1}_{\Phi_P^{-1}(P^{-1}(B) \cap \hat{K})}(u, v)\, du dv.$$

Together with

$$P^{-1}(B) \cap \hat{K} \subset B \oplus \left(\ker(P) \cap (I - P)\hat{K}\right)$$

this leads to

$$
\begin{aligned}
\lambda_P^d\left(\Phi_P^{-1}(P^{-1}(B) \cap \hat{K})\right) &\leq \int_{\ker(P)} \int_{\mathrm{im}(P)} \mathbb{1}_{B \times (\ker(P) \cap (I-P)\hat{K})}(u, v)\, du dv \\
&= \int_{\ker(P)} \int_{\mathrm{im}(P)} \mathbb{1}_B(u) \cdot \mathbb{1}_{\ker(P) \cap (I-P)\hat{K}}(v)\, du dv \\
&= \int_{\ker(P)} \mathbb{1}_{\ker(P) \cap (I-P)\hat{K}}(v) \left(\int_{\mathrm{im}(P)} \mathbb{1}_B(u) du\right) dv \\
&= \lambda_P^k(B) \cdot \int_{\ker(P)} \mathbb{1}_{(I-P)\hat{K}}(v) dv \\
&= \lambda_P^k(B) \cdot \lambda_{I-P}^{d-k}((I - P)\hat{K}).
\end{aligned}
$$

Hence, it suffices to show that $\lambda_{I-P}^{d-k}((I-P)\hat{K})$ is bounded on $\mathscr{P}$. But this follows from Lemma 3.2. $\qquad\square$

Now we are in position to prove our main result.

Theorem 3.3. *Let (K, Q) be an admissible pair for the inhomogeneous bilinear system $\Sigma = (\mathbb{T}, X, U_1 \times U_2, \mathscr{U}_1 \times \mathscr{U}_2, \varphi)$, such that K has positive Lebesgue measure and Q is a bounded F_σ-set. Assume that for the linear semiflow associated with Σ (the control semiflow of the associated homogeneous system) there exists a decomposition*

$$\mathscr{U}_1 \times X = \mathscr{W}^+ \oplus \mathscr{W}^-$$

into invariant subbundles, that is, $\alpha(t, \omega)\mathscr{W}_\omega^{\pm} = \mathscr{W}_{\Theta_t \omega}^{\pm}$ for all $(t, \omega) \in \mathbb{T}_+ \times \mathscr{U}_1$. Moreover, suppose that the restrictions $\alpha(t, \omega)|_{\mathscr{W}_\omega^+} : \mathscr{W}_\omega^+ \to \mathscr{W}_{\Theta_t \omega}^+$ are invertible. Then the following estimate holds:

$$h_{\text{inv}}(K, Q) \geq \inf_{\omega \in \mathscr{U}_1} \limsup_{\tau \to \infty} \frac{1}{\tau} \log \left| \det \alpha(\tau, \omega)|_{\mathscr{W}_\omega^+} : \mathscr{W}_\omega^+ \to \mathscr{W}_{\Theta_\tau \omega}^+ \right|. \qquad (3.8)$$

Proof. On X we fix an inner product $\langle \cdot, \cdot \rangle$ with associated norm $|\cdot|$ and Lebesgue measure λ. For each $\omega \in \mathscr{U}_1$ we denote by $P(\omega) \in \mathscr{L}(X, X)$ the projection onto $\mathscr{W}_\omega^+$ along $\mathscr{W}_\omega^-$. By λ_ω^k we denote the k-dimensional Lebesgue measure in im $P(\omega)$, where k is the rank of $\mathscr{W}^+$.

We may assume that for all $\tau > 0$ finite (τ, K, Q)-spanning sets exist. Otherwise there would be some τ_0 such that for all $\tau \geq \tau_0$ every (τ, K, Q)-spanning set is infinite. In this case, $h_{\text{inv}}(K, Q) = \infty$ and the lower estimate becomes trivial. Now, for given $\tau \in \mathbb{T} \cap (0, \infty)$ let $\mathscr{S}$ be a minimal (and hence finite) (τ, K, Q)-spanning set. Define

$$K_\omega := \{x \in K : \varphi(\tau, x, \omega) \in Q\}, \qquad \omega = (\omega_1, \omega_2) \in \mathscr{S}.$$

Then $K = \bigcup_{\omega \in \mathscr{S}} K_\omega$. The set K_ω is an element of $\mathscr{F}_\sigma(K)$, since it is the intersection of K with $\varphi_{\tau, \omega}^{-1}(Q)$, which is itself an F_σ-set as the preimage of an F_σ-set under a continuous map. Hence, by Lemma 3.1, it follows that the k-dimensional Lebesgue measure $\lambda_{\omega_1}^k(P(\omega_1)K_\omega)$ is well-defined. From the definition of K_ω it immediately follows that

$$P(\Theta_\tau \omega_1)\varphi(\tau, K_\omega, \omega) = P(\Theta_\tau \omega_1)\alpha(\tau, \omega_1)K_\omega + P(\Theta_\tau \omega_1)\beta(\tau, \omega_1, \omega_2)$$

$$\subset P(\Theta_\tau \omega_1)Q,$$

or equivalently,

$$P(\Theta_\tau \omega_1)\alpha(\tau, \omega_1)K_\omega \subset P(\Theta_\tau \omega_1)Q - P(\Theta_\tau \omega_1)\beta(\tau, \omega_1, \omega_2).$$

Using that Lebesgue measure is invariant under translations, we find

$$\lambda^k_{\Theta_\tau \omega_1} \left(P(\Theta_\tau \omega_1) \alpha(\tau, \omega_1) K_\omega \right) \leq \lambda^k_{\Theta_\tau \omega_1} \left(P(\Theta_\tau \omega_1) Q \right). \tag{3.9}$$

By invariance of the subbundle $\mathscr{W}^+$ we get

$$P(\Theta_\tau \omega_1) \alpha(\tau, \omega_1) K_\omega = \alpha(\tau, \omega_1) P(\omega_1) K_\omega,$$

and therefore,

$$\lambda^k_{\Theta_\tau \omega_1} \left(P(\Theta_\tau \omega_1) \alpha(\tau, \omega_1) K_\omega \right) = \left| \det \alpha(\tau, \omega_1) \right|_{\mathscr{W}^+_{\omega_1}} \cdot \lambda^k_{\omega_1} \left(P(\omega_1) K_\omega \right). \tag{3.10}$$

Define

$$f(\tau) := \inf_{\omega \in \mathscr{U}_1} \left| \det \alpha(\tau, \omega) \right|_{\mathscr{W}^+_\omega} : \mathscr{W}^+_\omega \to \mathscr{W}^+_{\Theta_\tau \omega} \right|.$$

Both $\alpha(\tau, \omega)$ and the fiber $\mathscr{W}^+_\omega$ depend continuously on ω. Hence, by compactness of $\mathscr{U}_1$, the above infimum is a minimum which implies that $f(\tau) > 0$. We obtain

$$
\begin{aligned}
\sum_{\omega \in \mathscr{S}} \lambda^k_{\omega_1} \left(P(\omega_1) K_\omega \right) \ &\leq \ \# \mathscr{S} \cdot \max_{\omega \in \mathscr{S}} \lambda^k_{\omega_1} \left(P(\omega_1) K_\omega \right) \\[2mm]
&\overset{(3.10)}{\leq} \ \# \mathscr{S} \cdot f(\tau)^{-1} \max_{\omega \in \mathscr{S}} \lambda^k_{\Theta_\tau \omega_1} \left(P(\Theta_\tau \omega_1) \alpha(\tau, \omega_1) K_\omega \right) \\[2mm]
&\overset{(3.9)}{\leq} \ \# \mathscr{S} \cdot f(\tau)^{-1} \max_{\omega \in \mathscr{S}} \lambda^k_{\Theta_\tau \omega_1} \left(P(\Theta_\tau \omega_1) Q \right) \\[2mm]
&\leq \ \# \mathscr{S} \cdot f(\tau)^{-1} \sup_{\omega_1 \in \mathscr{U}_1} \lambda^k_{\Theta_\tau \omega_1} \left(P(\Theta_\tau \omega_1) Q \right).
\end{aligned}
$$

Since Q is bounded by assumption, we can find a ball $B \supset Q$. Then

$$
\begin{aligned}
\sum_{\omega \in \mathscr{S}} \lambda^k_{\omega_1} \left(P(\omega_1) K_\omega \right) &\leq \# \mathscr{S} \cdot f(\tau)^{-1} \sup_{\omega_1 \in \mathscr{U}_1} \lambda^k_{\Theta_\tau \omega_1} \left(P(\Theta_\tau \omega_1) B \right) \\[2mm]
&\leq \# \mathscr{S} \cdot f(\tau)^{-1} \sup_{\omega_1 \in \mathscr{U}_1} \lambda^k_{\omega_1} \left(P(\omega_1) B \right).
\end{aligned}
$$

By compactness of $\mathscr{U}_1$, continuity of the map $\omega_1 \mapsto P(\omega_1)$ (see Lemma B.1), and Lemma 3.2, we obtain an upper bound $\delta > 0$ for the supremum above. By Lemma 3.3 there exists a lower bound $\gamma > 0$ for the left-hand side. Hence,

$$\delta \leq \# \mathscr{S} \cdot f(\tau)^{-1} \gamma \quad \Rightarrow \quad \# \mathscr{S} \geq \frac{\delta}{\gamma} f(\tau).$$

Since $\#\mathscr{S} = r_{\mathrm{inv}}(\tau, K, Q)$, for $\tau \to \infty$ it follows that

$$h_{\mathrm{inv}}(K, Q) \geq \limsup_{\tau \to \infty} \frac{1}{\tau} \log \left(\frac{\delta}{\gamma} f(\tau) \right)$$

$$= \limsup_{\tau \to \infty} \inf_{\omega \in \mathscr{U}_1} \frac{1}{\tau} \log \left| \det \alpha(\tau, \omega) \right|_{\mathscr{W}_\omega^+} : \mathscr{W}_\omega^+ \to \mathscr{W}_{\Theta_\tau \omega}^+ \bigg|.$$

It remains to show that the order of the limit superior and the infimum can be reversed. To this end, we show that $a : \mathbb{T}_+ \times \mathscr{U}_1 \to \mathbb{R}$,

$$a(\tau, \omega) := \log \left| \det \alpha(\tau, \omega) \right|_{\mathscr{W}_\omega^+} : \mathscr{W}_\omega^+ \to \mathscr{W}_{\Theta_\tau \omega}^+ \bigg|,$$

is a continuous additive cocycle over the shift flow on $\mathscr{U}_1$. Continuity of a follows, since both $\alpha(\cdot, \cdot)$ and $\mathscr{W}_{(\cdot)}^+$ are continuous. More precisely, similarly as in the proof of Lemma B.1, we can choose around every $\omega^* \in \mathscr{U}_1$ a neighborhood W such that an orthonormal basis $\{e_1(\omega), \ldots, e_k(\omega)\}$, depending continuously on ω, exists for all $\omega \in W$. Then, writing $L(\tau, \omega) := \alpha(\tau, \omega) | \mathscr{W}_\omega^+$, we find

$$|\det L(\tau, \omega)|^2 = \det L(\tau, \omega)^* L(\tau, \omega)$$

$$= \det \left(\langle L(\tau, \omega)^* L(\tau, \omega) e_i(\omega), e_j(\omega) \rangle \right)_{i,j=1}^k$$

$$= \det \left(\langle \alpha(\tau, \omega) e_i(\omega), \alpha(\tau, \omega) e_j(\omega) \rangle \right)_{i,j=1}^k,$$

which clearly shows continuity of a. Moreover, a satisfies the cocycle property. Since (3.6) is a linear semiflow, we have

$$\alpha(t + s, \omega) = \alpha(t, \Theta_s \omega) \alpha(s, \omega) \quad \text{for all } t, s \in \mathbb{T}_+,\ \omega \in \mathscr{U}_1.$$

By invariance of the subbundle $\mathscr{W}^+$, this gives

$$\alpha(t + s, \omega)|_{\mathscr{W}_\omega^+} = \alpha(t, \Theta_s \omega)|_{\mathscr{W}_{\Theta_s \omega}^+}\, \alpha(s, \omega)|_{\mathscr{W}_\omega^+},$$

which implies that a is a cocycle. The assertion now follows with Theorem B.2. □

The following corollary is a simple consequence of the generalized Liouville formula (see Proposition A.5).

Corollary 3.1. *Assume that the inhomogeneous bilinear system is given by differential equations of the form*

$$\dot{x}(t) = \left(A_0 + \sum_{i=1}^{m_1} \omega_i(t) A_i \right) x(t) + B\mu(t), \quad (\omega, \mu) \in \mathscr{U}_1 \times \mathscr{U}_2.$$

Then, writing $A(\omega) := A_0 + \sum_{i=1}^{m_1} \omega_i A_i$, we have

$$h_{\mathrm{inv}}(K, Q) \geq \inf_{\omega \in \mathscr{U}_1} \limsup_{\tau \to \infty} \frac{1}{\tau} \int_0^{\tau} \mathrm{tr}\left[A(\omega(s)) \circ Q(\Theta_s \omega)\right] \mathrm{d}s,$$

where $Q(\omega) \in \mathscr{L}(\mathbb{R}^d, \mathbb{R}^d)$ denotes the orthogonal projection onto the fiber $\mathscr{W}_\omega^+$.

To obtain an optimal lower bound from Theorem 3.3, we must choose the decomposition of $\mathscr{U}_1 \times X$ such that the volume growth rate on $\mathscr{W}^+$ becomes maximal. At least for systems given by differential equations this can be done by using Selgrade's theorem B.1, which yields a unique decomposition of $\mathscr{U}_1 \times X$ into invariant subbundles

$$\mathscr{U}_1 \times \mathbb{R}^d = \mathscr{W}^1 \oplus \cdots \oplus \mathscr{W}^r, \tag{3.11}$$

where each $\mathscr{W}^i$ corresponds to a connected component of the chain recurrent set of the projective flow $\mathbb{P}\Phi$, where Φ is the control flow of the associated homogeneous system. Using this decomposition, one can choose $\mathscr{W}^+$ as the sum of all subbundles $\mathscr{W}^i$, where the linear flow is uniformly expanding distances. Then we can also derive a lower bound in terms of Lyapunov exponents as follows.

Corollary 3.2. *Consider the decomposition* (3.11) *and let d_i denote the rank of $\mathscr{W}^i$. Define for $i = 1, \ldots, r$ the minimal Lyapunov exponent on the subbundle $\mathscr{W}^i$ by*

$$\kappa_i := \inf_{\substack{(\omega, x) \in \mathscr{W}^i \\ x \neq 0}} \limsup_{\tau \to \infty} \frac{1}{\tau} \log |\alpha(\tau, \omega)x|.$$

Then the following estimate holds:

$$h_{\mathrm{inv}}(K, Q) \geq \sum_{i=1}^{r} \max\{0, d_i \kappa_i\}.$$

Proof. We may assume that $\kappa_i > 0$ for $i = 1, \ldots, l$, $0 \leq l \leq r$ and $\kappa_i \leq 0$ for $i = l+1, \ldots, r$. Then let $\mathscr{W}^+ := \mathscr{W}^1 \oplus \cdots \oplus \mathscr{W}^l$ and $\mathscr{W}^- := \mathscr{W}^{l+1} \oplus \cdots \oplus \mathscr{W}^r$. It is easy to check that

$$\inf_{\omega \in \mathscr{U}_1} \limsup_{\tau \to \infty} \frac{1}{\tau} \log \left| \det \alpha(\tau, \omega)|_{\mathscr{W}_\omega^+} \right|$$

$$= \limsup_{\tau \to \infty} \frac{1}{\tau} \inf_{\omega \in \mathscr{U}_1} \log \prod_{i=1}^{l} \left| \det \alpha(\tau, \omega)|_{\mathscr{W}_\omega^i} \right|$$

$$\geq \limsup_{\tau \to \infty} \sum_{i=1}^{l} \inf_{\substack{(\omega, x) \in \mathscr{W}^i \\ |x|=1}} \frac{1}{\tau} \log \left(|\alpha(\tau, \omega)x|^{d_i} \right)$$

$$\geq \sum_{i=1}^{l} d_i \liminf_{\tau \to \infty} \inf_{\substack{(\omega, x) \in \mathscr{W}^i \\ |x|=1}} \frac{1}{\tau} \log |\alpha(\tau, \omega)x|.$$

We can apply Theorem B.2 to reverse the order of the limit inferior and the infimum. Indeed, the dynamical system to be considered here is the control flow of the system on $(d-1)$-dimensional projective space $\mathbb{P}^{d-1}$ which is induced by the homogeneous bilinear system in $\mathbb{R}^d$ (see Sect. 7.4 for a precise description of this system), and the corresponding additive cocycle is given by $a_t(\omega, \mathbb{P}x) = \log|\alpha(t, \omega)x| - \log|x|$. Moreover, the compact invariant set consists of all pairs $(\omega, \mathbb{P}x) \in \mathcal{U} \times \mathbb{P}^{d-1}$ such that $(\omega, x) \in \mathcal{W}^i$. Hence, we obtain

$$h_{\mathrm{inv}}(K, Q) \geq \sum_{i=1}^{l} d_i \inf_{\substack{(\omega,x)\in\mathcal{W}^i \\ |x|=1}} \liminf_{\tau\to\infty} \frac{1}{\tau}\log|\alpha(\tau,\omega)x|$$

$$= \sum_{i=1}^{l} d_i \inf_{\substack{(\omega,x)\in\mathcal{W}^i \\ x\neq 0}} \limsup_{\tau\to\infty} \frac{1}{\tau}\log|\alpha(\tau,\omega)x|.$$

This concludes the proof.   □

3.3   Comments and Bibliographical Notes

The result about the outer invariance entropy of a linear system has been proved in Colonius and Kawan [23] for systems given by differential equations, and generalized in Hoock [59] to the infinite-dimensional case. Similar formulas for the minimal data rates to perform control tasks for (both deterministic and stochastic) linear systems can be found, for instance, in [7, 20, 57, 82–84, 107]. The result about inhomogeneous bilinear systems can be found in the thesis Kawan [62] for systems given by differential equations. This result of course leaves the question open if there is an analogous upper bound for $h_{\mathrm{inv}}(K, Q)$. A partial answer to this question is given in Sect. 7.3. Another unsolved problem concerns the value of $h_{\mathrm{inv}}(K, Q)$ for a linear system. A partial answer is given in Chap. 5, Corollary 5.3.

Chapter 4
General Estimates

In this chapter, we start to investigate nonlinear systems. Applying similar techniques as in the linear case, we can derive at least rough bounds for the invariance entropy. In particular, we can show that under appropriate compactness assumptions the outer invariance entropy of an admissible pair (K, Q) of a smooth system is finite, and an upper bound can be expressed in terms of a Lipschitz constant of the transition map and the upper capacitive dimension of K. Together with Example 2.3 this also shows that in general the quantities $h_{\text{inv}}(K, Q)$ and $h_{\text{inv,out}}(K, Q)$ are not identical. Lower bounds of $h_{\text{inv}}(K, Q)$ can be expressed in terms of volume growth rates. Some supplementary material about the differential calculus on manifolds, used in this chapter, can be found in Sects. A.2 and A.3 of Appendix A.

4.1 Upper Bounds of Ito-Type

In this section, we derive a general upper bound for the outer invariance entropy of a topological system, which then is adapted to smooth systems (in continuous and discrete time). The key idea behind the estimate comes from a classical result about topological entropy. Ito [60] proved that the topological entropy of a $\mathscr{C}^1$-diffeomorphism $f : M \to M$ on a compact Riemannian manifold is bounded by the maximum of the norm of the derivative $df : TM \to TM$ times the dimension of M. Even earlier, Kushnirenko [72] proved a similar estimate for the metric entropy of a diffeomorphism preserving a probability measure. There exist several generalizations and improvements of Ito's estimate, see for instance, [8, 10, 31, 61, 87, 91]. One of these can be found in Proposition B.8 of Appendix B. Here, $f : X \to X$ is assumed to be a globally Lipschitz continuous map on a metric space (X, ϱ) with Lipschitz constant $L(f)$. Then for every compact set $K \subset X$ of finite upper capacitive dimension the estimate

C. Kawan, *Invariance Entropy for Deterministic Control Systems*, Lecture Notes in Mathematics 2089, DOI 10.1007/978-3-319-01288-9_4, © Springer International Publishing Switzerland 2013

$$h_{\text{top},\varrho}(K, f) \leq \max\{0, \log L(f)\} \cdot \overline{\dim}_C(K)$$

holds. What we will do in the following is a simple adaptation of the proof given in Appendix B. To this end, we first introduce an auxiliary entropy-like quantity associated with an admissible pair (K, Q), which is closer to the definition of topological entropy defined via (n, ε)-spanning sets and turns out to be an upper bound for $h_{\text{inv,out}}(K, Q)$.

Definition 4.1. Let $\Sigma = (\mathbb{T}, X, U, \mathscr{U}, \varphi)$ be a topological time-invariant system and (K, Q) an admissible pair such that Q is closed. Define

$$\mathscr{K}_Q := \{(\omega, x) \in \mathscr{U} \times K : \varphi(\mathbb{T}_+, x, \omega) \subset Q\}.$$

Moreover, fix a metric ϱ on X. For $\tau \in \mathbb{T} \cap (0, \infty)$ and $\varepsilon > 0$, a set $\mathscr{S}^+ \subset \mathscr{K}_Q$ is called *strongly $(\tau, \varepsilon, K, Q)$-spanning* if

$$\forall x \in K \ \exists (\omega, y) \in \mathscr{S}^+ : \ \max_{t \in [0,\tau]} \varrho(\varphi(t, x, \omega), \varphi(t, y, \omega)) < \varepsilon.$$

We let $r_{\text{inv}}^+(\tau, \varepsilon, K, Q)$ denote the minimal cardinality of such a set and define

$$h_{\text{inv}}^+(\varepsilon, K, Q) := \limsup_{\tau \to \infty} \frac{1}{\tau} \log r_{\text{inv}}^+(\tau, \varepsilon, K, Q).$$

The *strong invariance entropy* of (K, Q) is then defined as

$$h_{\text{inv}}^+(K, Q) := \lim_{\varepsilon \searrow 0} h_{\text{inv}}^+(\varepsilon, K, Q) = \sup_{\varepsilon > 0} h_{\text{inv}}^+(\varepsilon, K, Q).$$

Basically with the same arguments as used for the outer invariance entropy it follows that $h_{\text{inv}}^+(K, Q)$ is a well-defined number in $[0, \infty]$, and that $r_{\text{inv}}^+(\tau, \varepsilon, K, Q) < \infty$.

Proposition 4.1. *The following assertions hold:*

(i) $h_{\text{inv}}(K, N_\varepsilon(Q)) \leq h_{\text{inv}}^+(\varepsilon, K, Q)$ and $h_{\text{inv,out}}(K, Q) \leq h_{\text{inv}}^+(K, Q)$;
(ii) If Q is compact, then $h_{\text{inv}}^+(K, Q)$ is independent of the metric ϱ.

Proof. (i) If $\mathscr{S}^+ \subset \mathscr{K}_Q$ is a strongly $(\tau, \varepsilon, K, Q)$-spanning set, then $\mathscr{S} := \{\omega : (\omega, y) \in \mathscr{S}^+\}$ is obviously $(\tau, K, N_\varepsilon(Q))$-spanning, which shows that $r_{\text{inv}}(\tau, K, N_\varepsilon(Q)) \leq r_{\text{inv}}^+(\tau, \varepsilon, K, Q)$, implying the assertion.
(ii) This follows with the same arguments as used in the proof of Proposition 2.5 (uniform equivalence of any two metrics on Q). $\square$

Our main result for topological systems reads as follows.

Theorem 4.1. *Consider a topological time-invariant system $\Sigma = (\mathbb{T}, X, U, \mathscr{U}, \varphi)$ with an admissible pair (K, Q) such that Q is closed. Let ϱ be a metric on the state* *space X. Further assume that for some $\varepsilon > 0$ there exists a constant $L_\varepsilon \in [1, \infty)$ such that*

$$\varrho(\varphi(s, x, \omega), \varphi(s, y, \omega)) \le L_\varepsilon \varrho(x, y)$$

whenever $s \in [0, 1] \cap \mathbb{T}$, $x \in Q$, and $\varrho(x, y) < \varepsilon$, and that $\overline{\dim}_C(K; \varrho) < \infty$. Then

$$h_{\mathrm{inv}}^+(\delta, K, Q) \le (\log L_\varepsilon) \cdot \overline{\dim}_C(K; \varrho)$$

for all $\delta \in (0, \varepsilon]$ and consequently, the same estimate holds for $h_{\mathrm{inv,out}}(K, Q)$.

Proof. The proof proceeds in two steps. It suffices to write down the proof for $\delta = \varepsilon$, since the assumptions are also fulfilled when ε is replaced by any smaller number.

Step 1. First assume that $L_\varepsilon > 1$. Fix $\tau \in \mathbb{T} \cap (0, \infty)$ and let

$$r(\varepsilon, \tau) := \varepsilon L_\varepsilon^{-(\lfloor \tau \rfloor + 1)}.$$

Let $(\omega, y) \in \mathscr{K}_Q$, $x \in B(y, r(\varepsilon, \tau))$, $t \in [0, \tau] \cap \mathbb{T}$, and set $s := t - \lfloor t \rfloor$. By the cocycle property the map $\varphi_{t,\omega}$ decomposes into $\lfloor t \rfloor + 1$ maps as follows:

$$\varphi_{t,\omega} = \varphi_{s,\Theta_{\lfloor t \rfloor}\omega} \circ \varphi_{1,\Theta_{\lfloor t \rfloor - 1}\omega} \circ \cdots \circ \varphi_{1,\Theta_1\omega} \circ \varphi_{1,\omega}.$$

We have

$$\varrho(\varphi_{1,\omega}(x), \varphi_{1,\omega}(y)) \le L_\varepsilon \varrho(x, y) < \varepsilon L_\varepsilon^{-\lfloor \tau \rfloor} \le \varepsilon,$$

where in the last inequality we used that $L_\varepsilon \ge 1$. Recursively, one proves that

$$\varrho\left(\varphi_{1,\Theta_l\omega} \circ \cdots \circ \varphi_{1,\omega}(x), \varphi_{1,\Theta_l\omega} \circ \cdots \circ \varphi_{1,\omega}(y)\right) \le L_\varepsilon^{-\lfloor \tau \rfloor + l} \varepsilon \le \varepsilon$$

for $l = 1, 2, \ldots, \lfloor t \rfloor - 1$, and finally

$$\varrho(\varphi_{t,\omega}(x), \varphi_{t,\omega}(y)) \le L_\varepsilon \varrho\left(\varphi_{\lfloor t \rfloor,\omega}(x), \varphi_{\lfloor t \rfloor,\omega}(y)\right)$$

$$< L_\varepsilon L_\varepsilon^{-\lfloor \tau \rfloor + \lfloor t \rfloor - 1} \varepsilon \le L_\varepsilon^{-\lfloor \tau \rfloor + \lfloor t \rfloor} \varepsilon = \varepsilon.$$

Hence, it holds that

$$x \in B(y, r(\varepsilon, \tau)) \quad \Rightarrow \quad \max_{t \in [0, \tau]} \varrho(\varphi(t, x, \omega), \varphi(t, y, \omega)) < \varepsilon,$$

which finishes the first step of the proof.

Step 2. Let $n := n(r(\varepsilon, \tau), K)$, that is, n is the minimal number of $r(\varepsilon, \tau)$-balls necessary to cover the set K (see Sect. B.3). By Proposition B.4 we may assume that these balls are centered at points $z_1, \ldots, z_n$ in K. Since (K, Q)

is admissible, we can assign to each z_j a control function $\omega_j \in \mathcal{U}$ such that $(\omega_j, z_j) \in \mathcal{K}_Q$. Hence, the set $\mathcal{S}^+ := \{(\omega_1, z_1), \ldots, (\omega_n, z_n)\}$ is strongly $(\tau, \varepsilon, K, Q)$-spanning, since for every $x \in K$ we have $x \in B(z_j, r(\varepsilon, \tau))$ for some z_j and in Step 1 we have shown that $\varrho(x, z_j) < r(\varepsilon, \tau)$ implies $\max_{t \in [0,\tau]} \varrho(\varphi(t, x, \omega_j), \varphi(t, z_j, \omega_j)) < \varepsilon$. Consequently,

$$r_{\mathrm{inv}}^+(\tau, \varepsilon, K, Q) \leq n(r(\varepsilon, \tau), K). \tag{4.1}$$

We have $\log r(\varepsilon, \tau) = \log(\varepsilon L_\varepsilon^{-(\lfloor \tau \rfloor + 1)}) = \log(\varepsilon) - (\lfloor \tau \rfloor + 1) \log(L_\varepsilon)$ and thus

$$\tau \geq \lfloor \tau \rfloor = \frac{\log \varepsilon - \log r(\varepsilon, \tau)}{\log L_\varepsilon} - 1 = -\frac{\log r(\varepsilon, \tau)}{\log L_\varepsilon} \left(1 + \frac{\log L_\varepsilon - \log \varepsilon}{\log r(\varepsilon, \tau)} \right). \tag{4.2}$$

Note that the last term in parentheses converges to 1 for $\tau \to \infty$. This yields

$$
\begin{aligned}
h_{\mathrm{inv}}^+(\varepsilon, K, Q) &= \limsup_{\tau \to \infty} \frac{\log r_{\mathrm{inv}}^+(\tau, \varepsilon, K, Q)}{\tau} \\[2mm]
&\overset{(4.1)}{\leq} \limsup_{\tau \to \infty} \frac{\log n(r(\varepsilon, \tau), K)}{\tau} \\[2mm]
&= \log L_\varepsilon \limsup_{\tau \to \infty} \frac{\log n(r(\varepsilon, \tau), K)}{\tau \log L_\varepsilon} \\[2mm]
&\overset{(4.2)}{\leq} \log L_\varepsilon \limsup_{\tau \to \infty} \frac{\log n(r(\varepsilon, \tau), K)}{-\log r(\varepsilon, \tau) \left(1 + \frac{\log L_\varepsilon - \log \varepsilon}{\log r(\varepsilon, \tau)} \right)} \\[2mm]
&= \log L_\varepsilon \limsup_{\tau \to \infty} \frac{\log n(r(\varepsilon, \tau), K)}{\log(r(\varepsilon, \tau)^{-1})} \leq (\log L_\varepsilon) \cdot \overline{\dim}_C(K).
\end{aligned}
$$

If $L_\varepsilon = 1$, we can prove the same estimate taking the Lipschitz constant $1 + \delta$ for an arbitrary $\delta > 0$ and hence for $\delta \searrow 0$ we obtain $h_{\mathrm{inv}}^+(\varepsilon, K, Q) = 0$. Therefore, the estimate for $h_{\mathrm{inv}}^+(\varepsilon, K, Q)$ holds. This finishes the proof. $\square$

Remark 4.1. Note that the preceding theorem implies that the outer invariance entropy vanishes if the maps $\varphi_{t,\omega}$ are weak contractions with respect to the metric ϱ, that is, if $\varrho(\varphi_{t,\omega}(x), \varphi_{t,\omega}(y)) \leq \varrho(x, y)$.

Example 4.1. Consider a one-dimensional linear system $\Sigma = (\mathbb{T}, \mathbb{R}, U, \mathcal{U}, \varphi)$ and let ϱ be the Euclidean metric on $\mathbb{R}$. From linearity it follows that

$$|\varphi(t, x, \omega) - \varphi(t, y, \omega)| = |\varphi_{t,0}| \, |x - y|$$

for all $x, y \in \mathbb{R}$, $t \in \mathbb{T}_+$, and $\omega \in \mathcal{U}$. Hence, if (K, Q) is an admissible pair for Σ such that Q is closed, for every $\varepsilon > 0$ the assumptions of Theorem 4.1 are satisfied with

$$L_\varepsilon := \max_{s \in [0,1]} |\varphi_{s,0}| = \max \{1, |\varphi_{1,0}|\}.$$

Therefore, we obtain

$$h_{\text{inv,out}}(K, Q) \leq h_{\text{inv}}^+(K, Q) \leq \max\{0, \log |\varphi_{1,0}|\} \cdot \overline{\dim}_C(K; \varrho). \qquad (4.3)$$

The same estimate can be obtained as a special case of Theorem 3.2.

Corollary 4.1. *Consider a smooth system* $\Sigma = (\mathbb{R}, M, \mathbb{R}^m, \mathcal{U}, \varphi)$ *given by differential equations*

$$\dot{x}(t) = F(x(t), \omega(t)), \quad \omega \in \mathcal{U},$$

where (M, g) *is a Riemannian* $\mathcal{C}^3$*-manifold and the control range* $\Omega \subset \mathbb{R}^m$ *is compact. Assume that* (K, Q) *is an admissible pair for* Σ *such that* Q *is compact. Then the estimate*

$$h_{\text{inv,out}}(K, Q) \leq \max \left\{0, \max_{(x,u) \in Q \times \Omega} \lambda_{\max}(S\nabla F_u(x))\right\} \cdot \overline{\dim}_C(K)$$

holds, where $\lambda_{\max}(\cdot)$ *denotes the maximal eigenvalue and* $S\nabla \cdot$ *the symmetrized covariant derivative of a vector field. In particular,* $h_{\text{inv,out}}(K, Q) < \infty$.

Proof. The proof is subdivided into two parts.

Step 1. Let $\varepsilon > 0$ be chosen small enough such that $\operatorname{cl} N_{2\varepsilon}(Q)$ is compact and for all $x \in Q$ the Riemannian exponential map $\exp_x$ is defined on the ball $B(0_x, \varepsilon) \subset T_x M$. By compactness of Q both is possible. For the first see Lemma A.3, and for the second Gallot et al. [48, Corollary 2.89].[1] By Proposition A.6 there exists a cut-off function $\theta : M \to [0, 1]$ of class $\mathcal{C}^1$ such that

$$\theta(x) \equiv 1 \text{ on } \operatorname{cl} N_\varepsilon(Q) \quad \text{and} \quad \theta(x) \equiv 0 \text{ on } M \setminus N_{2\varepsilon}(Q).$$

We define a map $\tilde{F} : M \times \mathbb{R}^m \to TM$ by

$$\tilde{F}(x, u) := \theta(x) F(x, u) \quad \text{for all } (x, u) \in M \times \mathbb{R}^m,$$

which is also an allowed right-hand side, that is, the associated system of differential equations

$$\dot{x}(t) = \tilde{F}(x(t), \omega(t)), \quad \omega \in \mathcal{U},$$

---

[1] This corollary asserts in particular that for every point x on a Riemannian manifold (M, g) there exists a neighborhood U of x and $\varepsilon > 0$ such that for all $y \in U$ the map $\exp_y$ is defined on $B(0_y, \varepsilon) \subset T_y M$.

defines another smooth system $\tilde{\Sigma} = (\mathbb{R}, M, \mathbb{R}^m, \mathcal{U}, \tilde{\varphi})$. From the definition of $\tilde{F}$ it follows that

$$\varphi(t, x, \omega) = \tilde{\varphi}(t, x, \omega) \quad \text{whenever } \varphi([0, t], x, \omega) \subset \text{cl } N_\varepsilon(Q).$$

In particular, this implies that (K, Q) is also admissible for $\tilde{\Sigma}$. Now we define for every $\tau > 0$ the set

$$\mathcal{D}(\tau) := [0, \tau] \times \text{cl } N_\varepsilon(Q) \times \mathcal{U}$$

and the number

$$L_\varepsilon(\tau) := \sup_{(t,x,\omega)\in\mathcal{D}(\tau)} \|d_x\tilde{\varphi}_{t,\omega}\|, \quad L_\varepsilon := L_\varepsilon(1),$$

where $\|\cdot\|$ denotes the operator norm induced by the Riemannian metric. Since $\tilde{\varphi}_{0,\omega}(x) \equiv x$ on $M \times \mathcal{U}$, we have

$$L_\varepsilon(\tau) \geq \sup_{(x,\omega)\in\text{cl } N_\varepsilon(Q)\times\mathcal{U}} \|d_x\tilde{\varphi}_{0,\omega}\| = \sup_{x\in\text{cl } N_\varepsilon(Q)} \|\text{id}_{T_x M}\| = 1.$$

Let $\lambda(t, x, \omega) := \lambda_{\max}(S\nabla\tilde{F}_{\omega(t)}(\tilde{\varphi}_{t,\omega}(x)))$ for all $(t, x, \omega) \in \mathbb{R}_+ \times M \times \mathcal{U}$. Then, by the Wazewski inequality (Proposition A.4), we obtain

$$L_\varepsilon(\tau) \leq \sup_{(t,x,\omega)\in\mathcal{D}(\tau)} \exp\left(\int_0^t \lambda(s, x, \omega)ds\right)$$

$$\leq \sup_{(t,x,\omega)\in\mathcal{D}(\tau)} \exp\left(\int_0^t \max\{0, \lambda(s, x, \omega)\}ds\right)$$

$$\leq \sup_{(x,\omega)\in\text{cl } N_\varepsilon(Q)\times\mathcal{U}} \exp\left(\int_0^\tau \max\{0, \lambda(s, x, \omega)\}ds\right)$$

$$\leq \sup_{(x,\omega)\in\text{cl } N_\varepsilon(Q)\times\mathcal{U}} \exp\left(\tau \operatorname*{ess\,sup}_{t\in[0,\tau]} \max\{0, \lambda(t, x, \omega)\}\right)$$

$$= \sup_{(x,\omega)\in\text{cl } N_\varepsilon(Q)\times\mathcal{U}} \exp\left(\tau \operatorname*{ess\,sup}_{t\in[0,\tau]} \max\{0, \lambda_{\max}(S\nabla\tilde{F}_{\omega(t)}(\tilde{\varphi}_{t,\omega}(x)))\}\right)$$

$$\leq \sup_{(z,v)\in\tilde{\varphi}(\mathcal{D}(\tau))\times\Omega} \exp\left(\tau \max\{0, \lambda_{\max}(S\nabla\tilde{F}_v(z))\}\right).$$

By definition of $\tilde{F}$ every trajectory of $\tilde{\Sigma}$ starting in $\text{cl } N_\varepsilon(Q)$ remains in $\text{cl } N_{2\varepsilon}(Q)$ for all positive times. Hence, $\tilde{\varphi}(\mathcal{D}(\tau)) \subset \text{cl } N_{2\varepsilon}(Q)$, which by continuity[2] of $(z, v) \mapsto \lambda_{\max}(S \nabla \tilde{F}_v(z))$ implies

$$L_\varepsilon(\tau) \leq \sup_{(z,v) \in \mathrm{cl}\, N_{2\varepsilon}(Q) \times \Omega} \exp\left(\tau \max\{0, \lambda_{\max}(S \nabla \tilde{F}_v(z))\}\right)$$

$$= \exp\left(\tau \max\left\{0, \max_{(z,v) \in \mathrm{cl}\, N_{2\varepsilon}(Q) \times \Omega} \lambda_{\max}(S \nabla \tilde{F}_v(z))\right\}\right) < \infty.$$

Therefore, $L_\varepsilon(\tau) \in [1, \infty)$ for all $\tau > 0$. We further obtain

$$\frac{1}{\tau} \log L_\varepsilon(\tau) \leq \sup_{(z,v) \in \tilde{\varphi}(\mathscr{D}(\tau)) \times \Omega} \max\{0, \lambda_{\max}(S \nabla \tilde{F}_v(z))\}. \qquad (4.4)$$

Now we apply Theorem 4.1 in order to obtain the estimate

$$h_{\mathrm{inv}}^+(\varepsilon, K, Q; \tilde{\Sigma}) \leq (\log L_\varepsilon) \cdot \dim_C(K). \qquad (4.5)$$

To verify the assumptions of Theorem 4.1, let $s \in [0, 1]$, $\omega \in \mathscr{U}$, $x \in Q$, and $y \in B(x, \varepsilon)$. By the choice of ε there exists a shortest geodesic $\gamma : [0, 1] \to M$ with $\gamma(0) = x$ and $\gamma(1) = y$. This implies

$$\varrho\big(\tilde{\varphi}(s, x, \omega), \tilde{\varphi}(s, y, \omega)\big) \leq \mathscr{L}(\tilde{\varphi}_{s,\omega} \circ \gamma)$$

$$= \int_0^1 \left|\frac{\mathrm{d}}{\mathrm{d}t} \tilde{\varphi}_{s,\omega}(\gamma(t))\right| \mathrm{d}t = \int_0^1 \left|\mathrm{d}_{\gamma(t)} \tilde{\varphi}_{s,\omega} \dot{\gamma}(t)\right| \mathrm{d}t$$

$$\leq \int_0^1 \left\|\mathrm{d}_{\gamma(t)} \tilde{\varphi}_{s,\omega}\right\| |\dot{\gamma}(t)|\, \mathrm{d}t \leq \sup_{(t,z,\omega) \in \mathscr{D}(1)} \left\|\mathrm{d}_z \tilde{\varphi}_{t,\omega}\right\| \varrho(x, y)$$

$$= L_\varepsilon \varrho(x, y).$$

Since $\overline{\dim}_C(K) \leq \dim M < \infty$ (cf. Proposition B.5), the estimate (4.5) follows.
Step 2. We complete the proof. To this end, consider for every $\tau > 0$ the system $\tilde{\Sigma}_\tau = (\mathbb{R}, M, \mathbb{R}^m, \mathscr{U}, \tilde{\varphi}^\tau)$ given by

$$\dot{x}(t) = \tau \cdot \tilde{F}(x(t), \omega(t)), \quad \omega \in \mathscr{U}.$$

Then, by Propositions 2.9 and 2.10, (K, Q) is admissible for each of these systems, and for every $\tau > 0$ we obtain the estimate

---

[2]Note that the eigenvalues depend continuously on the operator and continuity of $\nabla \tilde{F}_v(z)$ follows from the assumption that the derivative of F with respect to the first argument exists and is continuous as a function of (x, u).

$$h_{\mathrm{inv}}(K, N_\varepsilon(Q); \tilde{\Sigma}) = \frac{1}{\tau} h_{\mathrm{inv}}(K, N_\varepsilon(Q); \tilde{\Sigma}_\tau) \le \frac{1}{\tau} h^+_{\mathrm{inv}}(\varepsilon, K, Q; \tilde{\Sigma}_\tau). \quad (4.6)$$

Now we apply the estimate (4.5) to the system $\tilde{\Sigma}_\tau$. By Proposition 2.10, we have

$$\tilde{\varphi}^\tau \left(\frac{t}{\tau}, x, \tilde{\omega} \right) = \tilde{\varphi}(t, x, \omega) \quad \text{for all } (t, x, \omega) \in \mathbb{R} \times M \times \mathscr{U},$$

where $\tilde{\omega}(t) \equiv \omega(t\tau)$. Hence,

$$\sup_{(t,x,\omega) \in \mathscr{D}(1)} \left\| \mathrm{d}_x \tilde{\varphi}^\tau_{t,\omega} \right\| = \sup_{(t,x,\omega) \in \mathscr{D}(1)} \left\| \mathrm{d}_x \tilde{\varphi}_{t\tau,\omega} \right\|$$

$$= \sup_{(t,x,\omega) \in \mathscr{D}(\tau)} \left\| \mathrm{d}_x \tilde{\varphi}_{t,\omega} \right\| = L_\varepsilon(\tau).$$

Consequently, from (4.6) we obtain

$$h_{\mathrm{inv}}(K, N_\varepsilon(Q); \tilde{\Sigma}) \le \frac{1}{\tau} (\log L_\varepsilon(\tau)) \cdot \overline{\dim}_C(K)$$

$$\stackrel{(4.4)}{\le} \sup_{(z,v) \in \tilde{\varphi}(\mathscr{D}(\tau)) \times \Omega} \max\{0, \lambda_{\max}(S \nabla \tilde{F}_v(z))\} \cdot \overline{\dim}_C(K)$$

$$= \max \left\{ 0, \sup_{(z,v) \in \tilde{\varphi}(\mathscr{D}(\tau)) \times \Omega} \lambda_{\max}(S \nabla \tilde{F}_v(z)) \right\} \cdot \overline{\dim}_C(K).$$

Let $z \in \tilde{\varphi}(\mathscr{D}(\tau))$. Then $z = \tilde{\varphi}(t, x, \omega)$ for some $(t, x, \omega) \in [0, \tau] \times \mathrm{cl}\, N_\varepsilon(Q) \times \mathscr{U}$. If ω is a piecewise constant control function, then the corresponding solution $\tilde{\varphi}(\cdot, x, \omega)$ is piecewise continuously differentiable, and hence we can measure its length by taking the integral over $|(\mathrm{d}/\mathrm{d}t)\tilde{\varphi}_{x,\omega}(t)|$. This implies that for all $t \in [0, \tau]$ we have

$$\varrho(x, \tilde{\varphi}(t, x, \omega)) \le \int_0^t \left| \frac{\mathrm{d}}{\mathrm{d}t} \tilde{\varphi}_{x,\omega}(t) \right| \mathrm{d}t \le \int_0^\tau \left| \frac{\mathrm{d}}{\mathrm{d}t} \tilde{\varphi}_{x,\omega}(t) \right| \mathrm{d}t$$

$$= \int_0^\tau \left| \tilde{F}(\tilde{\varphi}(t, x, \omega), \omega(t)) \right| \mathrm{d}t$$

$$\le \underbrace{\max_{(z,v) \in \mathrm{cl}\, N_{2\varepsilon}(Q) \times \Omega} \left| \tilde{F}(z, v) \right|}_{=:C} \int_0^\tau \mathrm{d}t = C\tau.$$

The same inequality for arbitrary admissible control functions follows with Proposition 1.6. Hence,

$$\tilde{\varphi}(\mathscr{D}(\tau)) \subset \mathrm{cl}\, N_{\min\{2\varepsilon, \varepsilon + \tau C\}}(Q) \quad \text{for every } \tau > 0.$$

For $\tau > 0$ with $\varepsilon + \tau C < 2\varepsilon$ we obtain

$$h_{\mathrm{inv}}(K, N_\varepsilon(Q)) \leq \max \left\{ 0, \max_{(z,v) \in \mathrm{cl}\, N_{\varepsilon + \tau C}(Q) \times \Omega} \lambda_{\max}(S \nabla \tilde{F}_v(z)) \right\} \cdot \overline{\dim}_C (K).$$

Letting τ tend to zero and using standard compactness and continuity arguments, we get

$$h_{\mathrm{inv}}(K, N_\varepsilon(Q)) \leq \max \left\{ 0, \max_{(z,v) \in \mathrm{cl}\, N_\varepsilon(Q) \times \Omega} \lambda_{\max}(S \nabla \tilde{F}_v(z)) \right\} \cdot \overline{\dim}_C (K).$$

Since $\tilde{F}$ and F coincide on $\mathrm{cl}\, N_\varepsilon(Q) \times \Omega$, we can replace $\tilde{F}$ by F in the above estimate. Finally, sending ε to zero, we can conclude that

$$h_{\mathrm{inv,out}}(K, Q; \Sigma) = h_{\mathrm{inv,out}}(K, Q; \tilde{\Sigma}) - \lim_{\varepsilon \searrow 0} h_{\mathrm{inv}}(K, N_\varepsilon(Q); \tilde{\Sigma})$$

$$\leq \max \left\{ 0, \max_{(z,v) \in Q \times \Omega} \lambda_{\max}(S \nabla F_v(z)) \right\} \cdot \overline{\dim}_C (K),$$

which finishes the proof. $\qquad\square$

Remark 4.2. A formally analogous estimate for the topological entropy of a flow has been proved by A. Noack in her doctoral thesis [87]. Precisely, the result of Noack reads: If $\{\varphi^t\}$ is the flow induced by a differential equation $\dot{x}(t) = f(x(t))$, $f \in \mathscr{X}^1(M)$, on a Riemannian $\mathscr{C}^3$-manifold (M, g), and $K \subset M$ is a compact φ^t-invariant set, then

$$h_{\mathrm{top}}(\varphi^1, K) \leq \max \left\{ 0, \max_{x \in K} \lambda_{\max}(S \nabla f(x)) \right\} \cdot \underline{\dim}_C (K). \qquad (4.7)$$

The proof of Noack is based on a topological version of Ito's estimate, which involves the lower capacitive dimension $\underline{\dim}_C (K)$ of the invariant set K. This result can be found in Boichenko and Leonov [8, Theorem 3.2] or Boichenko et al. [9, Chap. IV, Theorem 6.1.2]. Furthermore, one finds the "Euclidean versions" of estimate (4.7) in both of these references.

The idea of the following corollary comes from Noack [87, Folgerung 2.11.1].

Corollary 4.2. *Under the assumptions of Corollary 4.1, let $W \subset M$ be an open neighborhood of Q and $\alpha : W \to \mathbb{R}$ a $\mathscr{C}^2$-function. Then*

$$h_{\mathrm{inv,out}}(K, Q) \leq \max \left\{ 0, \max_{(x,u) \in Q \times \Omega} (\lambda_{\max}(S \nabla F_u(x)) + (F_u \alpha)(x)) \right\} \cdot \overline{\dim}_C (K),$$

where $F_u \alpha$ stands for the $\mathscr{C}^1$-function obtained by application of the vector field F_u to α.

Proof. We define a new Riemannian metric $\tilde{g}$ of class $\mathscr{C}^2$ on W by

$$\tilde{g}(x) := e^{2\alpha(x)} g(x) \quad \text{for all } x \in W,$$

and we let $\tilde{\nabla}$ denote the Levi–Civita connection associated with $\tilde{g}$. Then, by (A.4), for every $f \in \mathscr{X}^1(M)$ the matrix representation of $S\tilde{\nabla} f$ with respect to a chart (ϕ, V) is given by

$$2\left[S\tilde{\nabla} f\right]_\nu^\mu = \frac{\partial f^\mu}{\partial \phi^\nu} + \frac{\partial f^\kappa}{\partial \phi^\theta} \tilde{g}^{\mu\theta} \tilde{g}_{\kappa\nu} + f^i \tilde{g}^{\mu l} \frac{\partial \tilde{g}_{\nu l}}{\partial \phi^i}$$

$$= \frac{\partial f^\mu}{\partial \phi^\nu} + \frac{\partial f^\kappa}{\partial \phi^\theta} g^{\mu\theta} g_{\kappa\nu} + f^i e^{-2\alpha} g^{\mu l} \frac{\partial (e^{2\alpha} g_{\nu l})}{\partial \phi^i}$$

$$= \frac{\partial f^\mu}{\partial \phi^\nu} + \frac{\partial f^\kappa}{\partial \phi^\theta} g^{\mu\theta} g_{\kappa\nu}$$
$$+ e^{-2\alpha} f^i g^{\mu l} \left[e^{2\alpha} \frac{\partial g_{\nu l}}{\partial \phi^i} + 2 e^{2\alpha} g_{\nu l} \frac{\partial \alpha}{\partial \phi^i} \right]$$

$$= \frac{\partial f^\mu}{\partial \phi^\nu} + \frac{\partial f^\kappa}{\partial \phi^\theta} g^{\mu\theta} g_{\kappa\nu}$$
$$+ f^i g^{\mu l} \frac{\partial g_{\nu l}}{\partial \phi^i} + 2 f^i g^{\mu l} g_{l\nu} \frac{\partial \alpha}{\partial \phi^i}.$$

Since $g^{\mu l} g_{l\nu} = \delta_\nu^\mu$, we obtain

$$\left[S\tilde{\nabla} f\right]_\nu^\mu = [S\nabla f]_\nu^\mu + \delta_\nu^\mu f^i \frac{\partial \alpha}{\partial \phi^i} = [S\nabla f]_\nu^\mu + (f\alpha)\delta_\nu^\mu.$$

This implies the assertion. $\square$

Finally, we prove the much simpler discrete-time version of Corollary 4.1.

Corollary 4.3. *Consider a smooth system* $\Sigma = (\mathbb{Z}, M, \mathbb{R}^m, \Omega^{\mathbb{Z}}, \varphi)$, *where* (M, g) *is a Riemannian* $\mathscr{C}^3$-*manifold and* $\Omega \subset \mathbb{R}^m$ *is compact. Assume that* (K, Q) *is an admissible pair for* Σ *such that* Q *is compact. Then the estimate*

$$h_{\mathrm{inv,out}}(K, Q) \leq \max\left\{0, \max_{(x,u)\in Q\times\Omega} \log \|\mathrm{d}_x \varphi_{1,u}\|\right\} \cdot \overline{\dim}_C(K)$$

holds. In particular, $h_{\mathrm{inv,out}}(K, Q) < \infty$.

Proof. Let $\varepsilon > 0$ be chosen small enough such that cl $N_\varepsilon(Q)$ is compact and for all $x \in Q$ the Riemannian exponential map $\exp_x$ is defined on the ball $B(0_x, \varepsilon) \subset T_x M$. Let

$$L_\varepsilon := \max\left\{1, \sup_{(x,\omega)\in \mathrm{cl}\, N_\varepsilon(Q)\times\mathscr{U}} \|\mathrm{d}_x\varphi_{1,\omega}\|\right\}.$$

Then $L_\varepsilon \in [1,\infty)$, since $L_\varepsilon < \infty$ by continuous differentiability of $\varphi(1,\cdot,\cdot)$ in the first argument and compactness of $\mathrm{cl}\, N_\varepsilon(Q) \times \Omega$. The estimate

$$h^+_{\mathrm{inv}}(\varepsilon, K, Q) \leq (\log L_\varepsilon) \cdot \overline{\dim}_C(K)$$

is proved exactly in the same manner as for continuous-time systems. Using that $h_{\mathrm{inv}}(K, N_\varepsilon(Q)) \leq h^+_{\mathrm{inv}}(\varepsilon, K, Q)$, one arrives at the desired estimate by letting ε go to zero. $\square$

4.2 Lower Bounds in Terms of Volume Growth Rates

In this section, we provide general lower bounds for the invariance entropy of an admissible pair by using the same volume argument as involved in the proof of Theorem 3.1.

Let $\Sigma = (\mathbb{T}, M, \mathbb{R}^m, \mathscr{U}, \varphi)$ be a smooth system, where (M, g) is a d-dimensional Riemannian $\mathscr{C}^2$-manifold and for each $t \in \mathbb{T}_+$ the map $\varphi_{t,\omega} : M \to M$ is a $\mathscr{C}^1$-diffeomorphism. The Riemannian volume on M is denoted by vol.

Theorem 4.2. *Let (K, Q) be an admissible pair for Σ such that Q is open or closed and both K and Q have finite and positive volume. Then the following estimate holds:*

$$h_{\mathrm{inv}}(K, Q) \geq \limsup_{\tau\to\infty} \frac{1}{\tau} \log \max\left\{1, \inf_{\substack{(x,\omega)\in K\times\mathscr{U} \\ \varphi([0,\tau],x,\omega)\subset Q}} |\det \mathrm{d}_x\varphi_{\tau,\omega}|\right\}. \tag{4.8}$$

Proof. Let $\mathscr{S} \subset \mathscr{U}$ be a minimal (τ, K, Q)-spanning set. With the same reasoning as in the proof of Theorem 3.3 we may assume that $\mathscr{S}$ is finite. We define

$$K_\omega := \{x \in K : \varphi([0,\tau], x, \omega) \subset Q\}, \quad \omega \in \mathscr{S}.$$

From the definition of spanning sets it follows that $K = \bigcup_{\omega\in\mathscr{S}} K_\omega$. In the discrete-time case it is clear that K_ω is measurable, since it can be written as the intersection of the compact set K and finitely many measurable sets $\varphi^{-1}_{t,\omega}(Q)$, $0 \leq t \leq \tau$. In the continuous-time case we distinguish two cases. If Q is closed, we can write

$$K_\omega = K \cap \bigcap_{t\in[0,\tau]\cap\mathbb{Q}} \varphi^{-1}_{t,\omega}(Q),$$

which shows that K_ω is the countable intersection of measurable sets. If Q is open, then continuity of φ in (t, x) implies that K_ω is relatively open in K, hence measurable. Since $\varphi_{\tau,\omega}$ is a diffeomorphism, also the image of K_ω under $\varphi_{\tau,\omega}$ is measurable. Now, using the transformation formula, we obtain

$$\operatorname{vol}(\varphi_{\tau,\omega}(K_\omega)) = \int_{\varphi_{\tau,\omega}(K_\omega)} \mathrm{dvol} = \int_{K_\omega} |\det \mathrm{d}\varphi_{\tau,\omega}|\, \mathrm{dvol}$$

$$\geq \operatorname{vol}(K_\omega) \cdot \underbrace{\inf_{\substack{(x,\omega)\in K\times \mathscr{U} \\ \varphi([0,\tau],x,\omega)\subset Q}} |\det \mathrm{d}_x \varphi_{\tau,\omega}|}_{=:v(\tau)}.$$

Since $\varphi_{\tau,\omega}(K_\omega) \subset Q$, we have $\operatorname{vol}(\varphi_{\tau,\omega}(K_\omega)) \leq \operatorname{vol}(Q)$ implying

$$\operatorname{vol}(K) \leq \sum_{\omega \in \mathscr{S}} \operatorname{vol}(K_\omega) \leq \#\mathscr{S} \cdot \max_{\omega \in \mathscr{S}} \operatorname{vol}(K_\omega) \leq \#\mathscr{S} \cdot \frac{\operatorname{vol}(Q)}{\max\{1, v(\tau)\}}.$$

Hence, we obtain

$$r_{\mathrm{inv}}(\tau, K, Q) \geq \underbrace{\frac{\operatorname{vol}(K)}{\operatorname{vol}(Q)}}_{=:C} \max\{1, v(\tau)\}.$$

From the assumptions it follows that $0 < C < \infty$, which implies (4.8). □

For smooth systems given by differential equations the Liouville formula can be applied to obtain the following corollary.

Corollary 4.4. *Let $\Sigma = (\mathbb{R}, M, \mathbb{R}^m, \mathscr{U}, \varphi)$ in Theorem 4.2 be a smooth system given by differential equations with right-hand side F and control range Ω. Then the following assertions hold:*

(i) *It holds that*

$$h_{\mathrm{inv}}(K, Q) \geq \limsup_{\tau \to \infty} \inf_{\substack{(x,\omega)\in K\times \mathscr{U} \\ \varphi([0,\tau],x,\omega)\subset Q}} \frac{1}{\tau} \max\left\{0, \int_0^\tau \operatorname{div} F_{\omega(s)}(\varphi(s, x, \omega))\mathrm{d}s\right\}.$$

$$(4.9)$$

(ii) *It holds that*

$$h_{\mathrm{inv}}(K, Q) \geq \max\left\{0, \inf_{(x,u)\in Q\times\Omega} \operatorname{div} F_u(x)\right\}. \qquad (4.10)$$

(iii) *If, additionally, Σ is control-affine and Q is compact and controlled invariant, then*

$$h_{\text{inv}}(K, Q) \geq \inf_{(\omega, x) \in \mathscr{Q}} \limsup_{\tau \to \infty} \frac{1}{\tau} \int_0^\tau \operatorname{div} F_{\omega(s)}(\varphi(s, x, \omega)) ds. \qquad (4.11)$$

(iv) If $\alpha : W \to \mathbb{R}$ is a $\mathscr{C}^1$-function, defined on an open neighborhood W of Q, then

$$h_{\text{inv}}(K, Q) \geq \max \left\{ 0, \inf_{(x, u) \in Q \times \Omega} (\operatorname{div} F_u(x) + (F_u \alpha)(x)) \right\}. \qquad (4.12)$$

Proof. The estimate (4.9) immediately follows from the Liouville formula (see Proposition A.5):

$$\log |\det d_x \varphi_{\tau, \omega}| = \int_0^\tau \operatorname{div} F_{\omega(s)}(\varphi(s, x, \omega)) ds.$$

The second estimate (4.10) is proved by replacing the integrand in (4.9) by the infimum of $\operatorname{div} F_u(x)$ over $(x, u) \in Q \times \Omega$. If Σ is control-affine, the set $\mathscr{U}$ of admissible control functions becomes a compact and metrizable space with the weak*-topology of $L^\infty(\mathbb{R}, \mathbb{R}^m) = L^1(\mathbb{R}, \mathbb{R}^m)^*$ and the control flow $(t, (\omega, x)) \mapsto (\Theta_t \omega, \varphi(t, x, \omega))$ is a continuous flow on $\mathscr{U} \times M$ (cf. Sect. 1.4). In this case, the function

$$(\tau, (\omega, x)) \mapsto \log |\det d_x \varphi_{\tau, \omega}|, \quad \mathbb{R} \times (\mathscr{U} \times M) \to \mathbb{R},$$

is a continuous additive cocycle over the control flow. Continuity follows from continuity of $d_x \varphi_{\tau, \omega}$ as a function of (τ, x, ω) (see Theorem 1.1) and the cocycle equation follows from the cocycle property of φ and the chain rule. Finally, by Proposition 1.10, the set $\mathscr{Q} = \{(\omega, x) \in \mathscr{U} \times M : \varphi(\mathbb{R}_+, x, \omega) \subset Q\}$ is a compact forward-invariant set for the control flow. Hence, by Theorem B.2, we can interchange the limit superior and the infimum and it follows that

$$h_{\text{inv}}(K, Q) \geq \limsup_{\tau \to \infty} \inf_{\substack{(x, \omega) \in K \times \mathscr{U} \\ \varphi([0, \tau], x, \omega) \subset Q}} \frac{1}{\tau} \max \left\{ 0, \int_0^\tau \operatorname{div} F_{\omega(s)}(\varphi(s, x, \omega)) ds \right\}$$

$$\geq \limsup_{\tau \to \infty} \inf_{(\omega, x) \in \mathscr{Q}} \frac{1}{\tau} \int_0^\tau \operatorname{div} F_{\omega(s)}(\varphi(s, x, \omega)) ds$$

$$= \inf_{(\omega, x) \in \mathscr{Q}} \limsup_{\tau \to \infty} \frac{1}{\tau} \int_0^\tau \operatorname{div} F_{\omega(s)}(\varphi(s, x, \omega)) ds.$$

It remains to prove (4.12). To this end, consider on W the Riemannian metric $\tilde{g} := \beta \cdot g$ with $\beta(x) \equiv e^{(2/d)\alpha(x)}$. Using a cut-off function we can extend $\tilde{g}$ to M without changing it on a smaller neighborhood $V \subset W$ of Q. Let $\tilde{\nabla}$ denote the associated Levi–Civita connection. For any $\mathscr{C}^1$-vector field f on M we have

$$\widetilde{\operatorname{div}} f(x) = \operatorname{tr} \tilde{\nabla} f(x) = \operatorname{tr} S \tilde{\nabla} f(x).$$

By the proof of Corollary 4.2 we thus obtain

$$\widetilde{\operatorname{div}} f(x) = \operatorname{tr} S \nabla f(x) + (f\alpha)(x) = \operatorname{div} f(x) + (f\alpha)(x),$$

which implies the assertion.   □

4.3   Comments and Bibliographical Notes

The main results of this chapter can be found in the paper Kawan [63] and in
the thesis Kawan [62]. Simpler versions of the upper and lower bound theorems
can also be found in Colonius and Kawan [23]. However, the topological version
of the upper bound (Theorem 4.1) has not appeared before in the literature. In
Liberzon and Hespanha [76], one finds a similar estimate for the minimal data
rate necessary for global asymptotic stabilization of a nonlinear system as the
one given by Theorem 4.1 for the outer invariance entropy. In [62, 63], the lower
bound is formulated in terms of the divergence with respect to a volume form. The
results formulated in Theorem 4.2 and Corollary 4.4, however, are more general,
since every volume form comes from a Riemannian metric and on the other hand,
only orientable manifolds admit volume forms, while Riemannian metrics exist on
arbitrary $\mathscr{C}^2$-manifolds.

Chapter 5
Controllability, Lyapunov Exponents, and Upper Bounds

In this chapter, we restrict our attention to smooth systems given by differential equations. Under additional controllability assumptions, we derive upper bounds for the invariance entropy in terms of Lyapunov exponents. These numbers measure the exponential rates of divergence for nearby trajectories, and hence are indicators for stability or instability of the system. In the entropy theory of classical dynamical systems, several relations between entropy and Lyapunov exponents are known. A classic result in this direction is *Pesin's formula* [90] which says that the metric entropy of a $\mathscr{C}^2$-diffeomorphism $f : M \to M$ on a compact Riemannian manifold M with respect to a smooth invariant probability measure μ is given by the μ-integral over the sum of the positive Lyapunov exponents which are defined almost everywhere.[1] Liu [77] generalized this result to the case of (not necessarily invertible) $\mathscr{C}^2$-maps. Ruelle [94] (and independently, Margulis) showed that without the assumption of μ being equivalent to the Riemannian volume and only assuming that f is a $\mathscr{C}^1$-map, the expression in Pesin's formula is still an upper bound for the entropy. The crowning achievement finally is a result by Ledrappier and Young [74] which provides a formula for the metric entropy of a $\mathscr{C}^2$-diffeomorphism which involves a weighted sum of positive Lyapunov exponents, where the weights are certain dimension-like characteristics of the conditional measures on unstable manifolds.

In Chap. 3, we have already seen relations between invariance entropy and Lyapunov exponents for (bi-)linear systems (cf. Theorems 3.1, 3.2, and Corollary 3.2). In this chapter, we use controllability assumptions to obtain further relations of this kind for nonlinear systems. The key idea stems from the paper of Nair et al. [85], who show that the infimal data rate for local uniform asymptotic stabilization of a discrete-time nonlinear system at an equilibrium pair (u_0, x_0) is given by the sum of the logarithms of the unstable eigenvalues associated with the linearization

---

[1] If the invariant measure is ergodic, the Lyapunov exponents are constant almost everywhere, and hence the integral in Pesin's formula can be replaced by the integrand, that is, the sum of those (almost everywhere constant) Lyapunov exponents which are positive. Moreover, the assumption of f being $\mathscr{C}^2$ can be weakened to $\mathscr{C}^{1+\alpha}$.

C. Kawan, *Invariance Entropy for Deterministic Control Systems*, Lecture Notes in Mathematics 2089, DOI 10.1007/978-3-319-01288-9_5,

at (u_0, x_0). An essential assumption needed for the proof of this result is that the linearization be controllable. This guarantees that appropriate coder–controllers can be constructed that achieve stabilization with data rates arbitrarily close to the sum of the unstable eigenvalues.

In this chapter, we are going to exploit this idea to obtain upper estimates for the invariance entropy in terms of Lyapunov exponents under appropriate infinitesimal and global controllability assumptions.

5.1   The Upper Bound Theorem for Control Sets

Controllable Topological Systems

Let $\Sigma = (\mathbb{T}, X, U, \mathscr{U}, \varphi)$ be a topological time-invariant system such that X has no isolated points. Recall from Sect. 1.4 that a set $Q \subset X$ has the *no-return property* if $x \in Q$, $\tau \in \mathbb{T}_+$ and $\omega \in \mathscr{U}$ with $\varphi(\tau, x, \omega) \in Q$ implies $\varphi([0, \tau], x, \omega) \subset Q$. That is, trajectories cannot leave the set Q and then return. In particular, all control sets with nonempty interior have this property (see Corollary 1.1). The following proposition contains the key observation which makes it possible to use the ideas of Nair et al. [85] to derive upper bounds for the invariance entropy.

Proposition 5.1. *Let $Q \subset X$ be a set with the no-return property. Assume that (K_1, Q) and (K_2, Q) are two admissible pairs for Σ such that K_2 has nonempty interior, and that for every $x \in K_1$ there exist $\omega_x \in \mathscr{U}$ and $\tau_x \in \mathbb{T}_+$ with $\varphi(\tau_x, x, \omega_x) \in \operatorname{int} K_2$. Then*

$$h_{\mathrm{inv}}(K_1, Q) \le h_{\mathrm{inv}}(K_2, Q).$$

Proof. If $r_{\mathrm{inv}}(\tau, K_2, Q) = \infty$ for all τ greater than some τ_0, we have $h_{\mathrm{inv}}(K_2, Q) = \infty$ and the assertion becomes trivial. If this is not the case, there exists a sequence $\tau_k \to \infty$ such that $r_{\mathrm{inv}}(\tau_k, K_2, Q)$ is finite for every k, which implies that $r_{\mathrm{inv}}(\tau, K_2, Q)$ is finite for all τ. In this case, for every $x \in K_1$ let $\omega_x \in \mathscr{U}$ and $\tau_x \in \mathbb{T}_+$ be as in the assumption. Since $\varphi(\tau_x, \cdot, \omega_x)$ is continuous, we find for every $x \in K_1$ an open neighborhood V_x of x such that $\varphi(\tau_x, V_x, \omega_x) \subset \operatorname{int} K_2$. By the no-return property we have $\varphi([0, \tau_x], y, \omega_x) \subset Q$ for all $y \in K_1 \cap V_x$. The family $\{V_x\}_{x \in K_1}$ is an open cover of K_1 and by compactness there exist $x_1, \ldots, x_n \in K_1$ with $K_1 \subset \bigcup_{i=1}^{n} V_{x_i}$. Now let $\mathscr{S} = \{\mu_1, \ldots, \mu_k\}$ be a minimal (τ, K_2, Q)-spanning set for some τ. For every index pair (i, j) with $1 \le i \le n$ and $1 \le j \le k$ such that there exists $x \in K_1$ with $y_x := \varphi(\tau_{x_i}, x, \omega_{x_i}) \in \operatorname{int} K_2$ and $\varphi([0, \tau], y_x, \mu_j) \subset Q$, we can define a control function $\nu_{ij} \in \mathscr{U}$ which satisfies

$$\nu_{ij}(t) = \begin{cases} \omega_{x_i}(t) & \text{for } t \in [0, \tau_{x_i}], \\ \mu_j(t - \tau_{x_i}) & \text{for } t > \tau_{x_i}. \end{cases}$$

The set $\tilde{\mathscr{S}}$ of all these control functions has cardinality $\leq nk$. Let $\tilde{\tau} := \tau + \min_{1 \leq i \leq n} \tau_{x_i}$. Then, by construction, $\tilde{\mathscr{S}}$ is a $(\tilde{\tau}, K_1, Q)$-spanning set and consequently

$$r_{\text{inv}}(\tau, K_1, Q) \leq r_{\text{inv}}(\tilde{\tau}, K_1, Q) \leq n \cdot r_{\text{inv}}(\tau, K_2, Q).$$

By sending τ to infinity, the assertion follows. $\qquad\square$

From the properties of control sets (namely, approximate controllability, controlled invariance, and the no-return property), the next corollary immediately follows.

Corollary 5.1. *Let $D \subset X$ be a control set of Σ. Further let $K_1, K_2 \subset D$ be two compact sets with nonempty interior. Then (K_1, D) and (K_2, D) are admissible and*

$$h_{\text{inv}}(K_1, D) = h_{\text{inv}}(K_2, D).$$

With similar arguments as above, the next result follows.

Proposition 5.2. *Let (K, D) be an admissible pair for Σ such that D is a control set. Assume that there exists a nonempty set $V \subset D$ which is open in X and $\mu \in \mathscr{U}$ such that for every $x \in V$ there is $y \in \text{int } D$ and a sequence $t_k \in \mathbb{T}_+$, $t_k \to \infty$, with $\varphi(t_k, x, \mu) \to y$. Then $h_{\text{inv}}(K, D) = 0$.*

Proof. By approximate controllability on D, for every $x \in K$ there exist $\omega_x \in \mathscr{U}$ and $t_x \geq 0$ with $\varphi(t_x, x, \omega_x) \in V$. By continuity of $\varphi(t_x, \cdot, \omega_x)$, there is a neighborhood W_x of x with $\varphi(t_x, W_x, \omega_x) \subset V$. Since K is compact, finitely many of these neighborhoods are sufficient to cover K, say $W_{x_1}, \dots, W_{x_n}$. We define n control functions by

$$\mu_i(t) := \begin{cases} \omega_{x_i}(t) & \text{for } t \in [0, t_{x_i}], \\ \mu(t - t_{x_i}) & \text{for } t > t_{x_i}. \end{cases}$$

Then for every $x \in K$ there exists $i \in \{1, \dots, n\}$ and a sequence $t_k \in \mathbb{T}_+$, $t_k \to \infty$, such that $\varphi(t_k, x, \mu_i) \in \text{int } D$ for all $k \in \mathbb{N}$. By the no-return property of control sets with nonempty interior, this implies $\varphi(\mathbb{T}_+, x, \mu_i) \subset D$. It follows that $r_{\text{inv}}(\tau, K, D) \leq n$ for all τ and hence $h_{\text{inv}}(K, D) = 0$. $\qquad\square$

The assumptions of the proposition are in particular satisfied if there exists a constant control function $\mu \in \mathscr{U}$ such that the classical dynamical system associated with μ, that is, the semigroup action $\mathbb{T}_+ \times X \to X$, $(t, x) \mapsto \varphi(t, x, \mu)$, has a compact attractor A in $\text{int } D$. Then V can be chosen as an open neighborhood of A such that A attracts all trajectories with initial values in V.

Controllable Continuous-Time Smooth Systems

Now we consider a smooth system $\Sigma = (\mathbb{R}, M, \mathbb{R}^m, \mathscr{U}, \varphi)$ given by differential equations

$$\dot{x}(t) = F(x(t), \omega(t)), \quad \omega \in \mathscr{U},$$

with compact control range $\Omega \subset \mathbb{R}^m$ satisfying int $\Omega \neq \emptyset$. Moreover, we assume that M is a $\mathscr{C}^3$-manifold and $F \in \mathscr{C}^1(M \times \mathbb{R}^m, TM)$.

First we show that under mild assumptions finiteness of $h_{\mathrm{inv}}(K, D)$ holds for a control set D.

Proposition 5.3. *If D is a control set of Σ with nonempty interior such that local accessibility holds on* int D, *then* $h_{\mathrm{inv}}(K, D) < \infty$ *for every compact set* $K \subset D$.

Proof. Any compact subset of D is contained in a compact subset with nonempty interior. Hence, by Proposition 2.1, we may assume that K has nonempty interior. Using local accessibility, we can construct a periodic controlled trajectory with period $\tau^* > 0$ in D corresponding to some $(x^*, \omega^*) \in$ int $D \times \mathscr{U}$, and by Proposition 1.23 (iv) it holds that $\varphi(\mathbb{R}_+, x^*, \omega^*) \subset$ int D. Since $\varphi(\mathbb{R}_+, x^*, \omega^*) = \varphi([0, \tau^*], x^*, \omega^*)$ is compact, we find a compact set $\tilde{K} \subset$ int D with nonempty interior and $\varphi(\mathbb{R}_+, x^*, \omega^*) \subset$ int $\tilde{K}$. By Corollary 5.1 we may assume that $K = \tilde{K}$. For every $x \in K \subset$ int D we can find a control function $\omega_x \in \mathscr{U}$ and a time $t_x \geq 0$ with $\varphi(t_x, x, \omega_x) = x^*$ by exact controllability in the interior of D (see Proposition 1.23 (iii)). By Proposition 1.23 (v) we may assume that $t_x \leq T_0$ for all $x \in K$ for some $T_0 > 0$. By switching to the control function ω^* after time t_x we can assume that

$$y_x := \varphi(T_0, x, \omega_x) \in \text{int } K \quad \text{for all } x \in K.$$

Let V_x be a neighborhood of y_x with $V_x \subset$ int K. By continuity there exists a neighborhood W_x of x with $\varphi(T_0, W_x, \omega_x) \subset V_x \subset$ int K. Since $\{W_x\}_{x \in K}$ covers the compact set K, we find $x_1, \ldots, x_n \in K$ with $K \subset \bigcup_{j=1}^n W_{x_j}$. Consequently, the set $\mathscr{S} := \{\omega_{x_1}, \ldots, \omega_{x_n}\}$ is (T_0, K, D)-spanning (by the no-return property). Obviously, one can construct (kT_0, K, D)-spanning sets $\mathscr{S}_k$ for all $k \in \mathbb{N}$ from $\mathscr{S}$ such that $\#\mathscr{S}_k \leq n^k$. This proves that $h_{\mathrm{inv}}(K, D) \leq (\log n)/T_0 < \infty$. $\qquad \square$

In the following, we provide a characterization of the interior of $\mathscr{U}$ as a subset of the Banach space $L^\infty(\mathbb{R}, \mathbb{R}^m)$. We denote the L^∞-norm by $\|\cdot\|_\infty$.

Lemma 5.1. *Let $\Omega \subset \mathbb{R}^m$ be a compact set, $(X, \mathscr{A})$ a measurable space, and $f : X \to \mathbb{R}^m$ a measurable function whose image is contained in Ω. Further assume that* dist$(f(x), \Omega^c) < \varepsilon/3$ *for all $x \in X$ and some $\varepsilon > 0$. Then there exists a measurable function $g : X \to \mathbb{R}^m$ such that $|f(x) - g(x)| < \varepsilon$ and $g(x) \in \Omega^c$ for all $x \in X$.*

Proof. By translation of the set Ω, we may assume that all coordinate functions $f_i : X \to \mathbb{R}$, $i = 1, \ldots, m$, are nonnegative measurable functions. It is well-known that such a function can be approximated by a (monotonically increasing) sequence of nonnegative simple functions. In particular, there are simple functions

$$s_i : X \to \mathbb{R}, \quad s_i(x) = \sum_{j=1}^{n_i} a_j^i \mathbb{1}_{A_j^i}(x), \quad i = 1, \ldots, m,$$

with $X = \bigcup_{j=1}^{n_i} A_j^i$ for each i, $A_j^i \subset X$ measurable, such that

$$|s_i(x) - f_i(x)| < \frac{\varepsilon}{3\sqrt{m}} \quad \text{for all } x \in X, \ i = 1, \ldots, m.$$

Here we used that f is a bounded function, and hence the sequences of simple functions can be chosen such that the convergence is uniform. By adding sets of measure zero, we may assume that the numbers n_i, $i = 1, \ldots, m$, are all equal to each other, say $n_i = n$. Now define the sets

$$A(j_1, \ldots, j_m) := A_{j_1}^1 \cap \ldots \cap A_{j_m}^m, \quad j_k \in \{1, \ldots, n\}.$$

These sets are obviously measurable and their union is equal to X. We define a measurable function

$$s(x) := \sum_{(j_1, \ldots, j_m)} (s_1(x), \ldots, s_m(x))^T \mathbb{1}_{A(j_1, \ldots, j_m)}(x), \quad s : X \to \mathbb{R}^m.$$

Taking the standard Euclidean norm $|\cdot|$ on $\mathbb{R}^m$, we find that

$$|f(x) - s(x)| < \frac{\varepsilon}{3} \quad \text{for all } x \in X.$$

The assumption that $\operatorname{dist}(f(x), \Omega^c) < \varepsilon/3$ implies

$$\operatorname{dist}(s(x), \Omega^c) = \inf_{u \in \Omega^c} |s(x) - u| \leq |s(x) - f(x)| + \operatorname{dist}(f(x), \Omega^c) < \frac{2\varepsilon}{3}$$

for all $x \in X$. By construction, the values of s are the vectors $a(j_1, \ldots, j_m) := (a_{j_1}^1, \ldots, a_{j_m}^m)^T$. Therefore, for each $(j_1, \ldots, j_m)$, there exists $b(j_1, \ldots, j_m) := (b_{j_1}^1, \ldots, b_{j_m}^m)^T \in \Omega^c$ with $|a(j_1, \ldots, j_m) - b(j_1, \ldots, j_m)| < (2\varepsilon)/3$. Define the desired function g as

$$g(x) := \sum_{(j_1, \ldots, j_m)} b(j_1, \ldots, j_m) \mathbb{1}_{A(j_1, \ldots, j_m)}(x), \quad g : X \to \mathbb{R}^m.$$

This gives

$$|f(x) - g(x)| \leq |f(x) - s(x)| + |s(x) - g(x)| < \frac{\varepsilon}{3} + \frac{2\varepsilon}{3} = \varepsilon,$$

which concludes the proof.   □

Proposition 5.4. *For a function $\omega \in L^\infty(\mathbb{R}, \mathbb{R}^m)$ it holds that $\omega \in \mathrm{int}\,\mathscr{U}$ if and only if there exists a compact set $K \subset \mathrm{int}\,\Omega$ with $\omega(t) \in K$ for almost all $t \in \mathbb{R}$.*

Proof. We start with the easier direction: Assume that $\omega(t) \in K$ for almost all $t \in \mathbb{R}$ and a compact set $K \subset \mathrm{int}\,\Omega$. Then, by compactness, we find $\varepsilon > 0$ such that the ε-neighborhood of K is contained in Ω. Hence, if $\|\mu - \omega\|_\infty < \varepsilon$ for some $\mu \in L^\infty(\mathbb{R}, \mathbb{R}^m)$, then $\mu(t) \in \Omega$ almost everywhere, that is, $\mu \in \mathscr{U}$. This shows that $\omega \in \mathrm{int}\,\mathscr{U}$.

Now, conversely, assume that $\omega \in \mathrm{int}\,\mathscr{U}$. Then there exists $\varepsilon > 0$ such that $\|\omega - \mu\|_\infty < \varepsilon$ with $\mu \in L^\infty(\mathbb{R}, \mathbb{R}^m)$ implies $\mu \in \mathscr{U}$, that is, if $|\omega(t) - \mu(t)| < \varepsilon$ for almost all $t \in \mathbb{R}$, then $\mu(t) \in \Omega$ for almost all $t \in \mathbb{R}$.

By a general fact in real analysis, $\mathrm{int}\,\Omega$ can be written as the countable union of the elements of an increasing sequence of compact sets, that is, $\mathrm{int}\,\Omega = \bigcup_{n \geq 1} K_n$, K_n compact with $K_n \subset K_{n+1}$. Indeed, such a sequence can be constructed as follows: Let $\{u_k\}$ be a countable dense subset of $\mathrm{int}\,\Omega$ and consider for each u_k all compact balls centered at u_k of rational radius which are contained in $\mathrm{int}\,\Omega$. The family of all these balls is countable and its union is easily seen to be $\mathrm{int}\,\Omega$. Enumerate the members of this family and define K_n to be the union of the first n members. This gives the desired increasing sequence of compact sets. Moreover, from this construction it can easily be seen that every $u \in \mathrm{int}\,\Omega$ is contained in the interior of one of the sets K_n.

This construction also implies that there is $n_0 \geq 1$ such that

$$u \in \mathrm{int}\,\Omega \backslash K_{n_0} \quad \Rightarrow \quad \mathrm{dist}(u, \mathbb{R}^m \backslash \Omega) < \frac{\varepsilon}{3}. \tag{5.1}$$

We prove this by contradiction: Assume that such n_0 does not exist. Then for every $n \geq 1$ there is $v_n \in \mathrm{int}\,\Omega \backslash K_n$ with $\mathrm{dist}(v_n, \mathbb{R}^m \backslash \Omega) \geq \varepsilon/3$, that is, $|v_n - w| \geq \varepsilon/3$ for all $w \notin \Omega$. By compactness of Ω we may assume that $v_n \to v \in \Omega$. The limit v on the one hand satisfies $|v - w| \geq \varepsilon/3$ for all $w \notin \Omega$. On the other hand, $v \in \partial\Omega$, since $v \in \mathrm{int}\,\Omega$ implies $v \in \mathrm{int}\,K_{n_1}$ for some n_1 which gives $v_n \in K_{n_1}$ for all sufficiently large n, contradicting the definition of the sequence v_n.

Now consider the compact set $K := K_{n_0} \subset \mathrm{int}\,\Omega$ which satisfies (5.1). We claim that $\omega(t) \in K$ for almost all $t \in \mathbb{R}$. Indeed, if this was not true, there would be a set $I \subset \mathbb{R}$ of positive measure with $|\omega(t) - w| < \varepsilon/3$ for all $t \in I$ and all $w \notin \Omega$. By Lemma 5.1 there exists a measurable function $\mu : I \to \mathbb{R}^m \backslash \Omega$ with $|\mu(t) - \omega(t)| < \varepsilon$ for all $t \in I$. We can extend this function to a measurable function $\mu : \mathbb{R} \to \mathbb{R}^m$ by putting $\mu(t) := \omega(t)$ for all $t \in \mathbb{R} \backslash I$. This gives $\|\omega - \mu\|_\infty < \varepsilon$ which is a contradiction to the choice of ε.   □

Given a Riemannian metric g on M, to every trajectory $\varphi(\cdot, x, \omega)$ of the smooth system Σ we can associate a finite set of Lyapunov exponents. For the control function ω, the Lyapunov exponent at x in direction $v \in T_x M$, $v \neq 0_x$, is given by

$$\lambda(v) = \lambda(v; x, \omega) := \limsup_{t \to \infty} \frac{1}{t} \log |d_x \varphi_{t, \omega}(v)| \in \mathbb{R} \cup \{-\infty, +\infty\}.$$

We also call these numbers the *Lyapunov exponents at* (ω, x). Some basic and well-known properties are summarized in the following proposition (see also Arnold [4, Sect. 3.2.1]).[2]

Proposition 5.5. *The following assertions hold:*

(i) $\lambda(\alpha v) = \lambda(v)$ *for all nonzero* $v \in T_x M$ *and* $\alpha \in \mathbb{R} \backslash \{0\}$.

(ii) $\lambda(v + w) \leq \max\{\lambda(v), \lambda(w)\}$ *for all nonzero* $v, w \in T_x M$ *with* $w \neq -v$, *with equality if* $\lambda(v) \neq \lambda(w)$.

(iii) *The number of different Lyapunov exponents* $\lambda(v; x, \omega)$, $v \in T_x M \backslash \{0_x\}$, *is bounded by* $d = \dim M$.

(iv) *If* (u, x) *is an equilibrium pair, the Lyapunov exponents* $\lambda(v; x, u)$ *are the real parts of the eigenvalues of* $\nabla F_u(x) : T_x M \to T_x M$.

(v) *If there is a compact set* $K \subset M$ *with* $\varphi(\mathbb{R}_+, x, \omega) \subset K$, *then the Lyapunov exponents* $\lambda(v; x, \omega)$ *are all* $< \infty$.

(vi) *If two Riemannian metrics are equivalent on the image of a trajectory* $\varphi(\cdot, x, \omega)$, *then the Lyapunov exponents with respect to these two metrics are the same. In particular, if* M *is compact, the Lyapunov exponents of a trajectory are independent of the metric.*

(vii) *For a periodic trajectory, the Lyapunov exponents are independent of the metric.*

Remark 5.1. From the statements of Proposition 5.5 we mainly use the fourth and the seventh. The proof of statement (vii) is contained in the proof of the next theorem. Statement (iv) is an easy consequence of the Riemannian variational equation (see Proposition A.3). Indeed, for an equilibrium pair (ω, x) the variational equation becomes an autonomous linear equation on $T_x M$ whose solutions have the form $z(t) = \exp(t \nabla F_\omega(x)) v$, $v \in T_x M$, which immediately implies the assertion.

Each Lyapunov exponent has a *multiplicity* which can be defined as follows. For every (ω, x) let $\lambda_1(\omega, x) < \lambda_2(\omega, x) < \cdots < \lambda_{s(\omega, x)}(\omega, x)$ be the associated Lyapunov exponents. Then there exists a filtration

$$\{0_x\} = V_0(\omega, x) \subsetneq V_1(\omega, x) \subsetneq \cdots \subsetneq V_{s(\omega, x)}(\omega, x) = T_x M,$$

---

[2]In the dynamical systems literature, usually the notion of *Lyapunov exponents* refers to the Lyapunov exponents associated with an invariant measure. Sometimes, the Lyapunov exponents as we define them are called *upper Lyapunov exponents* because of the upper limit in their definition.

such that

$$V_i(\omega, x) = \{0_x\} \cup \{v \in T_x M \setminus \{0_x\} \ : \ \lambda(v; x, \omega) \le \lambda_i(\omega, x)\}.$$

The multiplicity of the Lyapunov exponent $\lambda_i(\omega, x)$ is defined as the natural number $\dim V_i(\omega, x) - \dim V_{i-1}(\omega, x)$.

Before we state the main result of this section, let us recall the fundamental lemma of *Floquet theory*. A proof can be found, for instance, in Chicone [17, Theorem 2.47].

Lemma 5.2 (Fundamental Lemma of Floquet Theory). *Let C be a nonsingular real $n \times n$-matrix. Then there exists a (possibly complex) $n \times n$-matrix A with $\exp(A) = C$. Moreover, there exists a real $n \times n$-matrix B with $\exp(B) = C^2$.*

In the formulation of our theorem we already use the knowledge that the Lyapunov exponents of a periodic trajectory are metric-independent, as asserted in statement (vii) of Proposition 5.5. This fact also becomes clear in the first step of the proof.

Theorem 5.1. *Let $D \subset M$ be a control set with nonempty interior and compact closure. Let $(\varphi(\cdot, x_0, \omega_0), \omega_0(\cdot))$ be a τ_0-periodic controlled trajectory which is regular on $[0, \tau_0]$ such that $(x_0, \omega_0) \in \operatorname{int} D \times \operatorname{int} \mathcal{U}$. Moreover, let $\rho_1, \ldots, \rho_r$ be the different Lyapunov exponents at (ω_0, x_0) with corresponding multiplicities $d_1, \ldots, d_r$. Then for every compact subset $K \subset D$ and every superset $Q \supset D$ the pair (K, Q) is admissible and*

$$h_{\mathrm{inv}}(K, Q) \le \sum_{j=1}^{r} \max\{0, d_j \rho_j\}. \tag{5.2}$$

The basic idea of the proof of Theorem 5.1 is to steer close to the point x_0 on the periodic trajectory and then use local controllability along the trajectory to stay in a neighborhood of the periodic orbit for arbitrary future times, that is, to stabilize the system at the periodic trajectory. This can be done by using a collection of control functions whose cardinality is arbitrarily close to the sum of the positive Lyapunov exponents (up to log and dividing by the time), which can be regarded as a measure for how fast one is driven away from the periodic trajectory on average without applying controls. The actual proof is quite lengthy and technical, so we give a short overview of the main ideas involved before we start: We proceed in three steps. In the first step, we use the fundamental lemma of Floquet theory in order to write the solutions of the linearization along the controlled trajectory $(\varphi(\cdot, x_0, \omega_0), \omega_0(\cdot))$ in terms of the matrix exponential of an endomorphism R of $T_{x_0} M$. Then we construct an adapted Riemannian metric, which yields an orthonormal Jordan basis for R. In the second step, we define several constants. In particular, a (large) time step $\tau \in \tau_0 \mathbb{N}$ and a (small) radius $b_0 > 0$ are defined such that the controllability of the linearization can be used in order to steer the system from the ball $B(x_0, b_0)$ to itself in time τ, using a finite number of control functions that is related to the eigenvalues of R and hence to the Lyapunov exponents $\rho_1, \ldots, \rho_r$. This is done in Step 3 by subdividing a cube of side length $2b_0$ centered at the origin of $T_{x_0}M$ into an appropriate number of subcuboids whose midpoints are steered to $0_{x_0} \in T_{x_0}M$ in time τ via the linearization. Using the Riemannian exponential map at x_0, it is shown that the corresponding control functions also work for the nonlinear system in order to get back to $B(x_0, b_0)$ in time τ. This process can be repeated and thus yields $(k\tau, B(x_0, b_0), Q)$-spanning sets for all $k \in \mathbb{N}$. By choosing τ big enough and b_0 small enough, the corresponding cardinality growth rate of these sets comes arbitrarily close to $\sum_j \max\{0, d_j \rho_j\}$. Since $h_{\mathrm{inv}}(K, Q)$ does not depend on the set K as long as it has a nonempty interior, this proves the assertion.

Proof (of Theorem 5.1). By controlled invariance of D, it is clear that every pair (K, Q) with $K \subset D$ and $Q \supset D$ is admissible. For brevity in notation, the map ϕ^{x_0, ω_0} associated with the linearization along $(\varphi(\cdot, x_0, \omega_0), \omega_0(\cdot))$ is simply denoted by ϕ (cf. Sect. 1.5). The proof of estimate (5.2) now proceeds in three steps.

Step 1. Let M be endowed with an arbitrary Riemannian metric and consider the automorphism

$$A := \mathrm{D}\varphi_{2\tau_0}(x_0, \omega_0)(\cdot, \mathbf{0}) \overset{(1.6)}{=} \phi(2\tau_0, \cdot, \mathbf{0}) : T_{x_0}M \to T_{x_0}M.$$

From Proposition 1.26 (iv) it follows that $A = \phi(\tau_0, \cdot, \mathbf{0})^2$, and hence from Lemma 5.2 it follows that there exists $R \in \mathscr{L}(T_{x_0}M, T_{x_0}M)$ with

$$A = \exp(2\tau_0 R).$$

From Proposition 1.26 (iv) we get

$$\phi(2\tau_0 k, \lambda, \mathbf{0}) = A^k \lambda = \exp(2\tau_0 k R)\lambda \quad \text{for all } \lambda \in T_{x_0}M, \ k \in \mathbb{Z}_+. \qquad (5.3)$$

We claim that the real parts of the eigenvalues of R coincide with the Lyapunov exponents at (ω_0, x_0). To show this, we write every $t > 0$ as $t = 2\tau_0 k + s$ with $k \in \mathbb{Z}_+$ and $s \in [0, 2\tau_0)$. Then for all $\lambda \in T_{x_0}M$ we obtain

$$\phi(t, \lambda, \mathbf{0}) = \phi(s, \phi(k(2\tau_0), \lambda, \mathbf{0}), \mathbf{0}) \overset{(5.3)}{=} \phi(s, \cdot, \mathbf{0})\exp(2\tau_0 k R)\lambda.$$

Hence, it follows that

$$l_1 |\exp(2k\tau_0 R)\lambda| \le |\phi(t, \lambda, \mathbf{0})| \le l_2 |\exp(2k\tau_0 R)\lambda|$$

with the positive constants

$$l_1 := \min_{s \in [0, 2\tau_0]} \left\| \phi(s, \cdot, \mathbf{0})^{-1} \right\|^{-1}, \quad l_2 := \max_{s \in [0, 2\tau_0]} \left\| \phi(s, \cdot, \mathbf{0}) \right\|.$$

By Proposition 1.26 (ii) we have

$$d_{x_0}\varphi_{t,\omega_0}(\lambda) = \phi(t,\lambda,\mathbf{0}),$$

and hence the exponential growth rate of $|d_{x_0}\varphi_{t,\omega_0}(\lambda)|$ for $t \to \infty$ equals the growth rate of $|\exp(2\tau_0\lfloor t/(2\tau_0)\rfloor R)\lambda|$ for all nonzero $\lambda \in T_{x_0}M$, which implies the claim.

Now choose a basis B_{x_0} of $T_{x_0}M$ adapted to the real Jordan structure of R and let $L_1(R),\ldots,L_r(R)$ be the different Lyapunov spaces of R, that is, the sums of the generalized eigenspaces corresponding to eigenvalues with the same real part. Then we have the decomposition

$$T_{x_0}M = L_1(R) \oplus \cdots \oplus L_r(R).$$

Let $d_j = \dim L_j(R)$ and denote by $\lambda^{(j)} \in L_j(R)$ the j-th component of a vector $\lambda \in T_{x_0}M$ with respect to this decomposition. Moreover, denote by ρ_j the common real part of the eigenvalues corresponding to $L_j(R)$. The restriction of R to $L_j(R)$ is denoted by R_j. Now let g be a Riemannian metric on M of class $\mathscr{C}^2$ such that the basis B_{x_0} is orthonormal with respect to g_{x_0}, and let ϱ denote the Riemannian distance induced by g. In order to obtain a metric with this property, one can start with an arbitrary $\mathscr{C}^2$-metric $\tilde{g}$ on M. Then one takes a chart (ψ, V) around x_0 and an inner product $(\cdot,\cdot)$ on $\mathbb{R}^d$ such that B_{x_0} is orthonormal with respect to the induced inner product $(d_{x_0}\psi(\cdot), d_{x_0}\psi(\cdot))$ on $T_{x_0}M$. On V consider the pullback $\hat{g}$ of $(\cdot,\cdot)$ by ψ, that is,

$$\hat{g}(x)(v,w) := (d_x\psi(v), d_x\psi(w)) \quad \text{for all } x \in V, \ v,w \in T_xM.$$

Let $\theta : M \to [0,1]$ be a cut-off function of class $\mathscr{C}^2$ such that $\mathrm{supp}\,\theta \subset V$ and $\theta(x) \equiv 1$ on a compact neighborhood W of x_0 (see Proposition A.6). Define g by

$$g(x) := \begin{cases} \theta(x)\hat{g}(x) + (1 - \theta(x))\tilde{g}(x) & \text{for all } x \in V, \\ \tilde{g}(x) & \text{for all } x \in M\setminus V. \end{cases}$$

It can easily be seen that g is a Riemannian metric on M with g_{x_0} having the desired property.

Step 2. We fix some constants: Let S_0 be a real number which satisfies

$$S_0 > \sum_{j=1}^r \max\{0, d_j\rho_j\}.$$

Choose $\xi = \xi(S_0) > 0$ such that

$$0 < d\xi < S_0 - \sum_{j=1}^r \max\{0, d_j\rho_j\}. \tag{5.4}$$

Let $\delta \in (0, \xi)$ be chosen small enough such that $\rho_j < 0$ implies $\rho_j + \delta < 0$ for all $j \in \{1, \ldots, r\}$. From Lemma B.2 it follows that there exists a constant $c = c(\delta) \geq 1$ such that

$$\forall j \in \{1, \ldots, r\} \ \forall k \in \mathbb{Z}_+ : \ \left\| \exp(k \tau_0 R_j) \right\| \leq c e^{(\rho_j + \delta) k \tau_0}, \tag{5.5}$$

where $\|\cdot\|$ denotes the operator norm on $\mathscr{L}(T_{x_0} M, T_{x_0} M)$ induced by g_{x_0}. For every $t > 0$ we define positive integers

$$M_j(t) := \begin{cases} \lfloor e^{(\rho_j + \xi)t} \rfloor + 1 & \text{if } \rho_j \geq 0 \\ 1 & \text{if } \rho_j < 0 \end{cases}, \quad j = 1, \ldots, r. \tag{5.6}$$

Moreover, we define a function $\beta : (0, \infty) \to (0, \infty)$ by

$$\beta(t) := c \sqrt{r} \ \max_{1 \leq j \leq r} \left[e^{(\rho_j + \delta)t} \frac{\sqrt{d_j}}{M_j(t)} \right]. \tag{5.7}$$

If $\rho_j < 0$, then (by definition) $\rho_j + \delta < 0$ and $M_j(t) \equiv 1$. This implies that $e^{(\rho_j + \delta)t}(\sqrt{d_j}/M_j(t))$ converges to zero for $t \to \infty$. If $\rho_j \geq 0$, we have $M_j(t) \geq e^{(\rho_j + \xi)t}$ by (5.6) and hence

$$e^{(\rho_j + \delta)t} \frac{\sqrt{d_j}}{M_j(t)} \leq e^{(\rho_j + \delta)t} \frac{\sqrt{d_j}}{e^{(\rho_j + \xi)t}} = \sqrt{d_j} e^{(\delta - \xi)t}.$$

Since $\delta \in (0, \xi)$, we have $\delta - \xi < 0$ and hence the term above converges to zero for $t \to \infty$. Thus, also $\beta(t) \to 0$ for $t \to \infty$. This implies that for given $\varepsilon > 0$ we can choose a number $\tau = 2k\tau_0$ with $k \in \mathbb{N}$ big enough such that

$$\beta(\tau) < 1 \quad \text{and} \quad \frac{d}{\tau} \log(2) < \varepsilon. \tag{5.8}$$

Since we assume regularity of $(\varphi(\cdot, x_0, \omega_0), \omega_0(\cdot))$ on $[0, \tau_0]$, by Proposition 1.30 there exists a constant $C > 0$ with the following property (note that regularity on $[0, \tau_0]$ implies regularity on $[0, \tau]$):

$$\forall \lambda \in T_{x_0} M \ \exists \mu \in L^\infty([0, \tau], \mathbb{R}^m) : \begin{cases} \phi(\tau, \lambda, \mu) = 0_{x_0} \\ \text{and} \\ \|\mu\|_{[0, \tau]} \leq C |\lambda|. \end{cases} \tag{5.9}$$

Let $W_1 \subset T_{x_0} M$ and $W_2 \subset M$ be open neighborhoods of 0_{x_0} and x_0, respectively, such that $\exp_{x_0} : W_1 \to W_2$ is a $\mathscr{C}^1$-diffeomorphism. The inverse of $\exp_{x_0}|_{W_1}$ is simply denoted by $\exp_{x_0}^{-1}$. Now choose $b_0 > 0$ small enough such that the following conditions are satisfied:

$$\begin{cases} \operatorname{cl} B(0_{x_0}, b_0) \subset W_1, \\ \operatorname{cl} B(x_0, b_0) \subset D, \\ \operatorname{cl} B(\omega_0(t), C\sqrt{d}\,b_0) \subset \Omega \quad \text{for almost all } t \in [0, \tau_0], \\ \varphi(\tau, \operatorname{cl} B(x_0, b_0), \omega) \subset W_2 \quad \text{if } \|\omega - \omega_0\|_{[0,\tau]} \leq C\sqrt{d}\,b_0. \end{cases} \tag{5.10}$$

The second and third inclusion are possible, since $x_0 \in \operatorname{int} D$ and, by Proposition 5.4, $\omega_0(t)$ is contained in a compact subset of $\operatorname{int} \Omega$ for almost all $t \in [0, \tau_0]$. The last one is possible by continuity of $(x, \omega) \mapsto \varphi(\tau, x, \omega)$. By Proposition 1.29 there exists a function $\zeta = \zeta_{\tau, \sqrt{d}C} : [0, \alpha) \to \mathbb{R}_+$ ($\alpha > 0$) with

$$\left| \exp_{x_0}^{-1}(\varphi(\tau, x, \omega)) - \phi(\tau, \exp_{x_0}^{-1}(x), \omega - \omega_0) \right| \leq \zeta(b)b \tag{5.11}$$

for all $(x, \omega) \in M \times \mathcal{U}$ with $\varrho(x, x_0) \leq b \leq b_0$ and $\|\omega - \omega_0\|_{[0,\tau]} \leq C\sqrt{d}\,b$, and $\zeta(b) \to 0$ for $b \to 0$. We can assume that $b_0 < \alpha$ and hence $\zeta(b_0)$ is defined. Because of the strict inequality $\beta(\tau) < 1$ we can also assume that b_0 is chosen small enough such that

$$\sqrt{r}\zeta(b_0) + \beta(\tau) \leq 1. \tag{5.12}$$

Step 3. By Corollary 5.1 and (5.10) we can assume that $K = \operatorname{cl} B(x_0, b_0)$. Consider a d-dimensional compact cube $\mathscr{C}$ in $T_{x_0}M$ centered at the origin with sides of length $2b_0$ parallel to the vectors of the basis B_{x_0}. Then $\exp_{x_0}^{-1}(K) = \operatorname{cl} B(0_{x_0}, b_0) \subset T_{x_0}M$, since $\exp_{x_0}$ is a radial isometry, and hence $\exp_{x_0}^{-1}(K) \subset \mathscr{C}$. Partition $\mathscr{C}$ by dividing each coordinate axis corresponding to a component of the j-th Lyapunov space of R into $M_j(\tau)$ intervals of equal length. The total number of subcuboids in this partition is $\prod_{j=1}^{r} M_j(\tau)^{d_j}$. Now pick an arbitrary $x \in \operatorname{cl} B(x_0, b_0)$. Let $\gamma_0 : [0, 1] \to M$ be a shortest geodesic from x_0 to x and let $\lambda_x \in \mathscr{C}$ be the center of a subcuboid which contains $\exp_{x_0}^{-1}(x) = \dot{\gamma}_0(0)$. (Note that $|\dot{\gamma}_0(0)| = \mathscr{L}(\gamma_0) = \varrho(x_0, x) \leq b_0$.) Then the following estimate holds, where the additional superscripts denote components of vectors within the corresponding Lyapunov spaces of R:

$$\left| \dot{\gamma}_0(0)^{(j)} - \lambda_x^{(j)} \right| = \left[\sum_{l=1}^{d_j} \left(\dot{\gamma}_0(0)^{(j,l)} - \lambda_x^{(j,l)} \right)^2 \right]^{1/2}$$

$$\leq \left[\sum_{l=1}^{d_j} \left(\frac{b_0}{M_j(\tau)} \right)^2 \right]^{1/2} = \frac{\sqrt{d_j}}{M_j(\tau)} b_0. \tag{5.13}$$

By (5.9) there exists $\omega_x \in L^\infty([0, \tau], \mathbb{R}^m)$ such that $\phi(\tau, \lambda_x, \omega_x - \omega_0) = 0_{x_0}$ or equivalently,

$$\phi(\tau, \lambda_x, \omega_x) = \phi(\tau, 0_{x_0}, \omega_0) \tag{5.14}$$

and

$$\|\omega_x - \omega_0\|_{[0,\tau]} \le C\,|\lambda_x| \le C\left[\sum_{j=1}^{r}\sum_{l=1}^{d_j}|\lambda_x^{(j,l)}|^2\right]^{1/2} \le C\sqrt{d}\,b_0,$$

since $\lambda_x \in \mathscr{C}$ implies $|\lambda_x^{(j,l)}| \le b_0$ for each component. By (5.10) it holds that $\omega_x \in \mathscr{U}$ and

$$\varphi(\tau, x, \omega_x) \in W_2.$$

Let $\gamma_1 : [0, 1] \to M$ be a shortest geodesic from x_0 to $\varphi(\tau, x, \omega_x)$. Then

$$\varrho\left(\varphi(\tau, x, \omega_x), x_0\right) = \mathscr{L}(\gamma_1) = \int_0^1 \underbrace{|\dot{\gamma}_1(t)|}_{=\text{ constant}} \, \mathrm{d}t = |\dot{\gamma}_1(0)|.$$

By the triangle inequality we have

$$\left|\dot{\gamma}_1(0)^{(j)}\right| \le \left|\dot{\gamma}_1(0)^{(j)} - \phi\left(\tau, \dot{\gamma}_0(0), \omega_x - \omega_0\right)^{(j)}\right|$$
$$+ \left|\phi(\tau, \dot{\gamma}_0(0), \omega_x - \omega_0)^{(j)}\right|.$$

Since g is chosen such that the Lyapunov spaces of R are orthogonal, for the first term we obtain

$$\left|\dot{\gamma}_1(0)^{(j)} - \phi(\tau, \dot{\gamma}_0(0), \omega_x - \omega_0)^{(j)}\right|$$
$$= \left|\left[\dot{\gamma}_1(0) - \phi(\tau, \dot{\gamma}_0(0), \omega_x - \omega_0)\right]^{(j)}\right|$$
$$\le \left|\dot{\gamma}_1(0) - \phi(\tau, \dot{\gamma}_0(0), \omega_x - \omega_0)\right|$$
$$= \left|\exp_{x_0}^{-1}(\varphi(\tau, x, \omega_x)) - \phi(\tau, \exp_{x_0}^{-1}(x), \omega_x - \omega_0)\right|$$
$$\overset{(5.11)}{\le} \zeta(b_0)b_0.$$

By linearity of $\phi(\tau, \cdot, \cdot)$, for the second term we obtain

$$\left|\phi(\tau, \dot{\gamma}_0(0), \omega_x - \omega_0)^{(j)}\right| = \left|\phi(\tau, \dot{\gamma}_0(0), \omega_x)^{(j)} - \phi(\tau, 0_{x_0}, \omega_0)^{(j)}\right|$$
$$\overset{(5.14)}{=} \left|\phi(\tau, \dot{\gamma}_0(0), \omega_x)^{(j)} - \phi(\tau, \lambda_x, \omega_x)^{(j)}\right|$$
$$= \left|\phi(\tau, \dot{\gamma}_0(0) - \lambda_x, \mathbf{0})^{(j)}\right|$$
$$\overset{(5.3)}{=} \left|[\exp(2k\tau_0 R)(\dot{\gamma}_0(0) - \lambda_x)]^{(j)}\right|$$
$$= \left|[\exp(\tau R)(\dot{\gamma}_0(0) - \lambda_x)]^{(j)}\right|.$$

By invariance of the Lyapunov spaces of R under $\exp(\tau R)$, we get

$$
\begin{aligned}
\left|\phi(\tau, \dot{\gamma}_0(0), \omega_x - \omega_0)^{(j)}\right| &= \left|\exp(\tau R)(\dot{\gamma}_0(0) - \lambda_x)^{(j)}\right| \\
&\le \left\|\exp(\tau R_j)\right\| \left|(\dot{\gamma}_0(0) - \lambda_x)^{(j)}\right| \\
&\overset{(5.5)}{\le} ce^{(\rho_j + \delta)\tau} \left|(\dot{\gamma}_0(0) - \lambda_x)^{(j)}\right|.
\end{aligned}
$$

Altogether, we have

$$
\begin{aligned}
\left|\dot{\gamma}_1(0)^{(j)}\right| &\le \zeta(b_0)b_0 + ce^{(\rho_j + \delta)\tau} \left|(\dot{\gamma}_0(0) - \lambda_x)^{(j)}\right| \\
&\overset{(5.13)}{\le} \zeta(b_0)b_0 + ce^{(\rho_j + \delta)\tau} \frac{\sqrt{d_j}}{M_j(\tau)} b_0.
\end{aligned}
$$

By orthogonality of the Lyapunov spaces of R, it follows that

$$
\begin{aligned}
\varrho\left(\varphi(\tau, x, \omega_x), x_0\right) = |\dot{\gamma}_1(0)| &= \left(\sum_{j=1}^{r} \left|\dot{\gamma}_1(0)^{(j)}\right|^2\right)^{1/2} \\
&\le \left(\sum_{j=1}^{r} \left(\zeta(b_0)b_0 + ce^{(\rho_j + \delta)\tau} \frac{\sqrt{d_j}}{M_j(\tau)} b_0\right)^2\right)^{1/2} \\
&\overset{(\Delta)}{\le} \sqrt{r}\zeta(b_0)b_0 + \left(\sum_{j=1}^{r} \left(ce^{(\rho_j + \delta)\tau} \frac{\sqrt{d_j}}{M_j(\tau)} b_0\right)^2\right)^{1/2} \\
&\le \sqrt{r}\zeta(b_0)b_0 + c\sqrt{r} \max_{1 \le j \le r} \left[e^{(\rho_j + \delta)\tau} \frac{\sqrt{d_j}}{M_j(\tau)}\right] b_0 \\
&\overset{(5.7)}{=} \left[\sqrt{r}\zeta(b_0) + \beta(\tau)\right] b_0 \overset{(5.12)}{\le} b_0.
\end{aligned}
$$

The estimate (Δ) follows from the triangle inequality in $\mathbb{R}^r$. Hence, we have proved that $\prod_{j=1}^{r} M_j(\tau)^{d_j}$ admissible control functions are sufficient to steer the system from all states in K back to K in time τ. By the no-return property of control sets it follows that the trajectories do not leave D within the time interval $(0, \tau)$. By iterated concatenation of these control functions we can construct an $(n\tau, K, D)$-spanning set for each $n \in \mathbb{N}$ with $(\prod_{j=1}^{r} M_j(\tau)^{d_j})^n$ elements and hence we obtain

$$
r_{\mathrm{inv}}(n\tau, K, D) \le \left(\prod_{j=1}^{r} M_j(\tau)^{d_j}\right)^n = \left(\prod_{j: \rho_j \ge 0} \left(\lfloor e^{(\rho_j + \xi)\tau}\rfloor + 1\right)^{d_j}\right)^n,
$$

which implies

$$
\begin{aligned}
h_{\mathrm{inv}}(K, Q) \;\le\; h_{\mathrm{inv}}(K, D) &= \limsup_{n\to\infty} \frac{1}{n\tau} \log r_{\mathrm{inv}}(n\tau, K, D) \\[2mm]
&\le \frac{1}{\tau} \sum_{j:\,\rho_j \ge 0} \log\left(\lfloor e^{(\rho_j+\xi)\tau}\rfloor + 1\right)^{d_j} \\[2mm]
&= \sum_{j:\,\rho_j \ge 0} d_j \frac{1}{\tau} \log\left(\lfloor e^{(\rho_j+\xi)\tau}\rfloor + 1\right) \\[2mm]
&\le \sum_{j:\,\rho_j \ge 0} d_j \frac{1}{\tau} \log\left(2 e^{(\rho_j+\xi)\tau}\right) \\[2mm]
&= \sum_{j:\,\rho_j \ge 0} d_j \left(\frac{\log(2)}{\tau} + (\rho_j + \xi)\right) \\[2mm]
&\le \frac{d}{\tau} \log(2) + d\xi + \sum_{j=1}^{r} \max\{0, d_j \rho_j\} \\[2mm]
&\overset{(5.4)}{<} \frac{d}{\tau} \log(2) + S_0 \overset{(5.8)}{<} S_0 + \varepsilon.
\end{aligned}
$$

The first equality follows from Proposition 2.6. Since ε can be chosen arbitrarily small and S_0 arbitrarily close to $\sum_{j=1}^{r} \max\{0, d_j \rho_j\}$, the assertion of the theorem follows.   □

Remark 5.2. It is clear that the above theorem implies the estimate

$$
h_{\mathrm{inv}}(K, Q) \le \inf_{(\omega, x)} \sum_{j=1}^{r(\omega, x)} \max\{0, d_j(\omega, x)\rho_j(\omega, x)\}, \tag{5.15}
$$

where the infimum is taken over all $(\omega, x) \in \mathscr{U} \times M$ such that the controlled trajectory $(\varphi(\cdot, x, \omega), \omega(\cdot))$ is periodic and regular with $x \in \mathrm{int}\, D$ and $\omega \in \mathrm{int}\,\mathscr{U}$. In general, it is not clear if any such trajectory exists. However, in many cases we can guarantee their existence. A quite general approach in this direction is worked out in Sect. 5.2.

Remark 5.3. Estimates for the topological entropy of diffeomorphisms, which are formally similar to (5.15), can be found in the work of Catalan and Tahzibi [16]. However, these results are of generic nature and use the variational principle.

Since an equilibrium pair is a τ-periodic controlled trajectory for every $\tau > 0$, the following result immediately follows (using Proposition 5.5 (iv)).

Corollary 5.2. *Let* $D \subset M$ *be a control set with nonempty interior and compact closure. Let* $(\omega_0, x_0) \in \text{int} \, \Omega \times \text{int} \, D$ *be a regular equilibrium pair. Then for every compact set* $K \subset D$ *and every superset* $Q \supset D$ *we have*

$$h_{\text{inv}}(K, Q) \leq \sum_{\lambda \in \sigma(\nabla F_{\omega_0}(x_0))} \max \{0, n_\lambda \, \text{Re}(\lambda)\} .$$

Corollary 5.3. *Consider a linear system* $\Sigma = (\mathbb{R}, \mathbb{R}^d, \mathbb{R}^m, \mathcal{U}, \varphi)$ *given by differential equations associated with a controllable matrix pair* (A, B) *such that* A *is hyperbolic (that is, A has no eigenvalues on the imaginary axis). Further assume that the control range* Ω *is a compact and convex set with* $0 \in \text{int} \, \Omega$. *Let* $D \subset \mathbb{R}^d$ *be the unique control set of* Σ *with nonempty interior. Then for every compact set* $K \subset D$ *it holds that*

$$h_{\text{inv}}(K, D) \leq \sum_{\lambda \in \sigma(A)} \max\{0, n_\lambda \, \text{Re}(\lambda)\}. \tag{5.16}$$

If, additionally, K has positive Lebesgue measure and $Q = \text{cl} \, D$, then

$$h_{\text{inv}}(K, Q) = h_{\text{inv,out}}(K, Q) = \sum_{\lambda \in \sigma(A)} \max\{0, n_\lambda \, \text{Re}(\lambda)\}. \tag{5.17}$$

Proof. As noted in Remark 3.1, the assumptions about the matrix pair (A, B) and the control range Ω guarantee the existence of a unique control set $D = \text{cl} \, \mathcal{O}^+(0) \cap \mathcal{O}^-(0)$ with nonempty interior and compact closure. In particular, $0 \in \text{int} \, D$. Then the pair $(0, 0) \in \mathbb{R}^m \times \mathbb{R}^d$ is an equilibrium pair which is regular by the controllability assumption. Hence, Corollary 5.2 implies (5.16). Formula (5.17) follows from the combination of Theorem 3.1 with (5.16) and the fact that $h_{\text{inv,out}}(K, Q) \leq h_{\text{inv}}(K, Q)$. $\quad\square$

Recall the definition of inner control sets (Definition 2.6). For such sets, the estimate of Theorem 5.1 holds for the outer invariance entropy without the assumption that the periodic trajectory is contained in the interior.

Corollary 5.4. *Let* D *be an inner control set of* Σ *with closure* $Q = \text{cl} \, D$. *Let* $(\varphi(\cdot, x_0, \omega_0), \omega_0(\cdot))$ *be a regular* τ_0-*periodic controlled trajectory with* $x_0 \in Q$ *and* $\omega_0 \in \mathcal{U}_1$. *Then*

$$h_{\text{inv,out}}(Q) \leq \sum_{j=1}^{r} \max\{0, d_j \rho_j\}$$

holds, where $\lambda_1, \ldots, \lambda_r$ *are the different Lyapunov exponents at* (ω_0, x_0) *with corresponding multiplicities* $d_1, \ldots, d_r$.

Proof. Note that the definition of inner control sets implies that Q is compact. From Theorem 5.1 it follows that

$$h_{\text{inv}}(Q, \text{cl}\, D_\rho; \Sigma_\rho) \leq \sum_{j=1}^{r} \max\{0, d_j \lambda_j\} \quad \text{for all } \rho \in [0, 1).$$

Now for given $\varepsilon > 0$ choose $\rho \in [0, 1)$ such that $\text{cl}\, D_\rho \subset N_\varepsilon(Q)$. Then

$$h_{\text{inv}}(Q, N_\varepsilon(Q); \Sigma_0) \leq h_{\text{inv}}(Q, N_\varepsilon(Q); \Sigma_\rho)$$

$$\leq h_{\text{inv}}(Q, \text{cl}\, D_\rho; \Sigma_\rho) \leq \sum_{j=1}^{r} \max\{0, d_j \lambda_j\}.$$

The first two inequalities follow from $\mathcal{U}_\rho \subset \mathcal{U}_0$ and Proposition 2.1. Since $h_{\text{inv,out}}(Q) = \lim_{\varepsilon \searrow 0} h_{\text{inv}}(Q, N_\varepsilon(Q); \Sigma_0)$, the assertion follows. $\qquad\square$

Remark 5.4. For discrete-time smooth systems given by difference equations $x_{k+1} = F(x_k, u_k)$ it is no problem to prove the analog of Theorem 5.1. In fact, the proof of Theorem 5.1 has been developed using a discrete-time blueprint which can be found in Nair et al. [85, Theorem 3]. As mentioned in the beginning of this chapter, this result of Nair et al. asserts that the infimal data rate for local uniform asymptotic stabilization of a discrete-time nonlinear system at a regular equilibrium pair (u_0, x_0) is given by the sum of the logarithms of the unstable eigenvalues associated with the linearization at (u_0, x_0). These numbers are identical with the positive Lyapunov exponents at (u_0, x_0). Essentially, all the arguments needed for a discrete-time version of Theorem 5.1 are contained in the proof of [85, Theorem 3].

5.2 Approximation Results for Lyapunov Exponents

The main result of the preceding section, Theorem 5.1, naturally leads to the following questions:

1. Are there easy-to-verify conditions which guarantee that a regular periodic controlled trajectory as required exists?
2. Can the assumptions of regularity and periodicity be weakened?

In this section, we show that there are indeed conditions which imply the existence of plenty of regular periodic trajectories in the interior of a control set, and which in many cases are relatively easy to check. Under a weak hyperbolicity assumption these trajectories then can be used to weaken the assumptions of regularity and periodicity in the upper estimate of Theorem 5.1. To this end, we first have to introduce the notion of *strong accessibility*. A well-known result of Sontag asserts that real-analytic systems with this property possess so-called *universally regular control functions*. These can be used to construct regular periodic trajectories as required.

Strong Accessibility

Assume that $\Sigma = (\mathbb{R}, M, \mathbb{R}^m, \mathscr{U}, \varphi)$ is a smooth system given by differential equations

$$\dot{x}(t) = F(x(t), \omega(t)), \quad \omega \in \mathscr{U},$$

where M is a real-analytic manifold of dimension d and $F : M \times \mathbb{R}^m \to TM$ is a real-analytic map. Moreover, assume that the control range $\Omega \subset \mathbb{R}^m$ is a compact, locally path-connected[3] set with nonempty and connected interior such that $\Omega = \mathrm{cl\,int}\,\Omega$. We also consider the associated system $\Sigma^0 = (\mathbb{R}, M, \mathbb{R}^m, \mathscr{U}^0, \varphi^0)$ with control range $\Omega^0 := \mathrm{int}\,\Omega$ and the same right-hand side F. Then $\varphi^0(t, x, \omega) = \varphi(t, x, \omega)$ for all $(t, x, \omega) \in \mathbb{R} \times M \times \mathscr{U}^0$.

Definition 5.1. A topological time-invariant system is called *strongly accessible* if for each $x \in M$ there is some $\tau > 0$ such that $\mathrm{int}\,\mathscr{O}_\tau(x) \neq \emptyset$.

Recall from Sect. 1.5 that we call a control function ω regular for a state x on a time interval $[0, \tau]$ if the linearization along $(\varphi(\cdot, x, \omega), \omega(\cdot))$ is controllable on $[0, \tau]$.

Definition 5.2. A control function $\omega \in \mathscr{U}$ is said to be *universally regular* if it is regular for every $x \in M$ on some time interval $[0, \tau]$, $\tau = \tau(x) > 0$.

The following proposition summarizes some well-known results about strong accessibility.

Proposition 5.6. *The following assertions hold:*

(i) *Let $\mathscr{L}$ denote the Lie subalgebra of vector fields on M generated by the vector fields F_u, $u \in \mathrm{int}\,\Omega$. Then Σ^0 is strongly accessible if and only if the ideal $\mathscr{L}_0$ in $\mathscr{L}$ generated by the vector fields*

$$F_{u,v} := F_u - F_v, \quad u, v \in \mathrm{int}\,\Omega,$$

satisfies $\dim \mathscr{L}_0(x) = d$ for all $x \in M$, where $\mathscr{L}_0(x) := \{f(x)\}_{f \in \mathscr{L}_0}$. (See Sussmann and Jurdjevic [106, Corollary 4.7].)

(ii) *System Σ^0 is strongly accessible if and only if for every $x \in M$ there is some $\omega \in \mathscr{U}^0$ which is regular for x on some time interval $[0, \tau]$, $\tau > 0$. (See Sontag [100] and [101, Sect. 1].)*

(iii) *If $\omega \in \mathscr{U}^0$ is an analytic control function, then ω is regular for $x \in M$ on some time interval $[0, \tau]$, $\tau > 0$, if and only if it is regular for x on every interval of this form. (See Sontag [101, Sect. 1].)*

---

[3]Recall that a topological space X is called *locally path-connected* if every neighborhood of a point $x \in X$ contains a path-connected neighborhood of x.

(iv) *Assume that Σ^0 is strongly accessible. Then there exists an analytic universally regular control function $\omega \in \mathcal{U}^0$. (See Sontag [101, Theorem 1].)*[4]

(v) *If the universal covering space of M is compact, then strong accessibility of Σ^0 is equivalent to local accessibility. (See Sussmann and Jurdjevic [106, Theorem 4.9].)*

(vi) *If Σ is control-affine with right-hand side $F(x,u) = f_0 + \sum_{i=1}^{m} u_i f_i$, then Σ is strongly accessible if and only if Σ^0 is strongly accessible if and only if the ideal $\mathcal{L}_0$ generated by the vector fields $f_1, \ldots, f_m$ satisfies $\dim \mathcal{L}_0(x) = d$ for all $x \in M$.*

Remark 5.5. Statement (iv) is proved in Sontag [101] for systems whose state space is an open subset of $\mathbb{R}^d$, but can easily be generalized to systems on arbitrary real-analytic manifolds as noted in [101, Remark 2.3]. Its proof is based on Sussmann's theorem about the existence of universally distinguishing control functions (cf. Sussmann [105, Theorem 2.1]).

Lemma 5.3. *Let $D \subset M$ be a control set of Σ with nonempty interior. If Σ^0 is strongly accessible, then for every $x \in \operatorname{int} D$ there exist $\tau > 0$ and $\omega \in \operatorname{int} \mathcal{U}$ such that $(\varphi(\cdot, x, \omega), \omega(\cdot))$ is τ-periodic and regular on $[0, \tau]$.*

Proof. By Proposition 5.6 (iii) and (iv) we can apply a universally regular control function $\omega_* \in \mathcal{U}^0$ to x and obtain a trajectory $\varphi(\cdot, x, \omega_*)$ which is regular on every nontrivial interval of the form $[0, \tau_1]$. For τ_1 chosen sufficiently small we have $\varphi([0, \tau_1], x, \omega_*) \subset \operatorname{int} D$. Let $y := \varphi(\tau_1, x, \omega_*)$. Since $\omega_*(t) \in \operatorname{int} \Omega$ and ω_* is continuous, $\omega_*([0, \tau_1])$ is a compact subset of $\operatorname{int} \Omega$. Hence, by Proposition 5.4, we can assume that $\omega_* \in \operatorname{int} \mathcal{U}$. Strong accessibility implies local accessibility and the latter implies exact controllability on $\operatorname{int} D$ by Proposition 1.23 (iii). Hence, we find an admissible control function $\mu \in \mathcal{U}$ and a time $\tau_2 \geq 0$ with $\varphi(\tau_2, y, \mu) = x$. This gives the desired periodic trajectory with corresponding period $\tau := \tau_1 + \tau_2$ and control function $\omega := \omega_*|_{[0, \tau_1]} \mu^{\tau_1}$. By Proposition 1.28 this periodic trajectory is regular on $[0, \tau]$. To conclude the proof, we have to show that μ can be chosen such that $\mu \in \operatorname{int} \mathcal{U}$. In fact, we can assume that μ is piecewise constant with values in $\operatorname{int} \Omega$ which by Proposition 5.4 guarantees that $\mu \in \operatorname{int} \mathcal{U}$. This easily follows from the fact that local accessibility and approximate controllability on D also hold for the class of piecewise constant control functions with values in $\operatorname{int} \Omega$. □

The First Approximation Result

The aim of this subsection is to prove an approximation result, which shows that the sum of positive Lyapunov exponents of an arbitrary periodic trajectory in the

---

[4]Sontag also proves a stronger result which asserts that the set of smooth universally regular control functions is generic in $\mathscr{C}^\infty([0, T], \operatorname{int} \Omega)$ for all $T > 0$.

interior of a control set can be approximated by the corresponding sums for regular periodic trajectories. Let the following assumptions be satisfied:

(a) There is a control set D of Σ with nonempty interior and compact closure;
(b) System Σ^0 is strongly accessible.

Furthermore, let g be an arbitrary $\mathscr{C}^\infty$-Riemannian metric on M.

In the following, we speak of subadditive cocycles over the control flow Φ : $\mathbb{R} \times (\mathscr{U} \times M) \to \mathscr{U} \times M$ of Σ. However, note that we do not impose any continuity assumptions here (neither on the control flow nor on the cocycles). In particular, we do not assume that Σ is control-affine.

Proposition 5.7. *Let $(\varphi(\cdot, x, \omega), \omega(\cdot))$ be a τ-periodic controlled trajectory with $(x, \omega) \in \operatorname{int} D \times \operatorname{int} \mathscr{U}$. Moreover, let $a : \mathbb{R} \times (\mathscr{U} \times M) \to \mathbb{R}$, $(t, (\omega, x)) \mapsto a_t(\omega, x)$, be a subadditive cocycle over the control flow which satisfies the following two assumptions:*

(a) $a_\tau(\omega, x) \geq 0$;
(b) For all $T > 0$, $y \in M$, and $\omega_1, \omega_2 \in \mathscr{U}$ it holds that

$$\omega_1(t) = \omega_2(t) \text{ a.e. on } [0, T] \quad \Rightarrow \quad a_T(\omega_1, y) = a_T(\omega_2, y). \tag{5.18}$$

Then for every $\varepsilon > 0$ there exists a regular periodic controlled trajectory $(\varphi(\cdot, x, \omega_*), \omega_*(\cdot))$ with $\omega_* \in \operatorname{int} \mathscr{U}$ and period $\tau_* > 0$ such that

$$\frac{1}{\tau_*} a_{\tau_*}(\omega_*, x) \leq \frac{1}{\tau} a_\tau(\omega, x) + \varepsilon.$$

Proof. For the given periodic trajectory $\varphi(\cdot, x, \omega)$ we construct a family of approximating trajectories as follows. By Lemma 5.3 there exists a regular periodic trajectory $\varphi(t, x, \mu)$, $t \in [0, \rho]$. For every $N \in \mathbb{N}$ we define

$$\omega_N(t) := \begin{cases} \omega(t) & \text{for } t \in [0, N\tau) \\ \mu(t - N\tau) & \text{for } t \in [N\tau, N\tau + \rho] \end{cases},$$

and we extend ω_N $(N\tau + \rho)$-periodically. By construction and Proposition 5.4, ω_N is an admissible control function in $\operatorname{int} \mathscr{U}$. Moreover, from Proposition 1.28 it follows that ω_N is regular for x on $[0, N\tau + \rho]$. Using subadditivity of a, we obtain

$$a_{N\tau+\rho}(\omega_N, x) \leq a_\rho(\Theta_{N\tau}\omega_N, \varphi_{N\tau, \omega_N}(x)) + \sum_{i=0}^{N-1} a_\tau(\Theta_{i\tau}\omega_N, \varphi_{i\tau, \omega_N}(x)).$$

By construction we have $\varphi_{i\tau, \omega_N}(x) = x$ for $i = 0, 1, \ldots, N$. Moreover, we have $\Theta_{i\tau}\omega_N(t) = \omega(t)$ for all $t \in [0, \tau]$ and $i = 0, \ldots, N-1$. By assumption (5.18) this implies

$$a_{N\tau+\rho}(\omega_N, x) \le a_\rho(\mu, x) + Na_\tau(\omega, x).$$

Hence, for given $\varepsilon > 0$ we can choose N sufficiently large so that

$$\frac{1}{N\tau + \rho} a_{N\tau+\rho}(\omega_N, x) \le \frac{N}{N\tau + \rho} a_\tau(\omega, x) + \frac{1}{N\tau + \rho} a_\rho(\mu, x)$$

$$\le \frac{1}{\tau + \frac{\rho}{N}} a_\tau(\omega, x) + \varepsilon \le \frac{1}{\tau} a_\tau(\omega, x) + \varepsilon.$$

In the last inequality we used that $a_\tau(\omega, x) \ge 0$. Consequently, the desired estimate follows with $\omega_* = \omega_N$ and $\tau_* = N\tau + \rho$. $\quad\square$

Next we introduce some notation. For given $(t, x, \omega) \in \mathbb{R} \times M \times \mathscr{U}$, the derivative

$$d_x \varphi_{t,\omega} : T_x M \to T_{\varphi(t,x,\omega)} M$$

is a linear isomorphism between d-dimensional Euclidean spaces, and hence has well-defined (positive) singular values, which we denote by

$$\sigma_1(t, x, \omega) \ge \cdots \ge \sigma_d(t, x, \omega) > 0.$$

For $0 \le k \le d$, the singular value function of order k of $d_x \varphi_{t,\omega}$ is denoted by

$$\alpha_k(t, x, \omega) = \begin{cases} \sigma_1(t, x, \omega)\sigma_2(t, x, \omega) \cdots \sigma_k(t, x, \omega) & \text{for } k > 0, \\ 1 & \text{for } k = 0. \end{cases}$$

Proposition 5.8. *For every $k \in \{0, 1, \ldots, d\}$ the function*

$$a_t^k(\omega, x) := \log \alpha_k(t, x, \omega), \quad a^k : \mathbb{R} \times (\mathscr{U} \times M) \to \mathbb{R},$$

is a subadditive cocycle over the control flow which satisfies assumption (5.18).

Proof. To prove subadditivity, let $t, s \in \mathbb{R}_+$. Then, using Horn's inequality (cf. Sect. A.1), we find

$$a_{t+s}^k(\omega, x) = \log \alpha_k(d_x \varphi_{t+s,\omega})$$

$$= \log \alpha_k \left(d_{\varphi(t,x,\omega)} \varphi_{s,\Theta_t\omega} \circ d_x \varphi_{t,\omega}\right)$$

$$\le \log \alpha_k \left(d_{\varphi(t,x,\omega)} \varphi_{s,\Theta_t\omega}\right) + \log \alpha_k \left(d_x \varphi_{t,\omega}\right)$$

$$= a_t^k(\omega, x) + a_s^k(\Phi_t(\omega, x)).$$

Finally, assumption (5.18) is satisfied. Indeed, $\omega_1(t) = \omega_2(t)$ almost everywhere on $[0, \tau]$ implies $\varphi(t, x, \omega_1) = \varphi(t, x, \omega_2)$ for all $t \in [0, \tau]$ and $x \in M$. In particular, $\varphi_{\tau,\omega_1} = \varphi_{\tau,\omega_2}$ and hence $d_x \varphi_{\tau,\omega_1} \equiv d_x \varphi_{\tau,\omega_2}$. $\quad\square$

Lemma 5.4. *For every* $k \in \{1, \ldots, d\}$ *and all* $t \geq 0$, $(\omega, x) \in \mathcal{U} \times M$, *the following estimate holds:*

$$a_t^k(\omega, x) \leq k \int_0^t \lambda_{\max} \left(S \nabla F_{\omega(s)}(\varphi(s, x, \omega)) \right) \mathrm{d}s.$$

Therefore, if $\varphi(t, x, \omega)$ *is contained in a compact set for all* $t \geq 0$, *there is a constant* $C \geq 0$ *(which does not depend on* (ω, x)*) with*

$$a_t^k(\omega, x) \leq Ct \quad \text{for all } t \geq 0. \tag{5.19}$$

Proof. First note that $\sigma_1(t, x, \omega)$ equals the operator norm of $\mathrm{d}_x \varphi_{t,\omega}$. Hence,

$$\alpha_k(t, x, \omega) = \sigma_1(t, x, \omega) \cdots \sigma_k(t, x, \omega) \leq \sigma_1(t, x, \omega)^k = \|\mathrm{d}_x \varphi_{t,\omega}\|^k.$$

Using the Wazewski inequality (Proposition A.4) gives

$$a_t^k(\omega, x) \leq k \log \|\mathrm{d}_x \varphi_{t,\omega}\| \leq k \int_0^t \lambda_{\max} \left(S \nabla F_{\omega(s)}(\varphi(s, x, \omega)) \right) \mathrm{d}s.$$

If $\varphi(t, x, \omega)$ is contained in a compact set K, then $C := k \max_{(z,u) \in K \times \Omega} \lambda_{\max}$ $(S \nabla F_u(z))$ gives $a_t^k(\omega, x) \leq Ct$ for all $t \geq 0$. $\square$

We introduce the *local Lyapunov exponents* at (ω, x),[5] defined recursively by

$$\nu_1(\omega, x) + \cdots + \nu_k(\omega, x) := \limsup_{t \to \infty} \frac{1}{t} a_t^k(\omega, x), \quad k = 1, 2, \ldots, d.$$

Then we obtain the first improvement over Theorem 5.1 which shows that under the assumption that all periodic trajectories have the same number of positive Lyapunov exponents, the condition of regularity is no longer necessary.

Lemma 5.5. *If the controlled trajectory* $(\varphi(\cdot, x, \omega), \omega(\cdot))$ *in* $\mathcal{Q}$ *(the forward lift of* $Q = \mathrm{cl}\, D$*) is periodic, then for every* $k \in \{1, \ldots, d\}$ *the identities*

$$\nu_1(\omega, x) + \cdots + \nu_k(\omega, x) = \lim_{t \to \infty} \frac{1}{t} a_t^k(\omega, x)$$

$$= \lambda_1(\omega, x) + \cdots + \lambda_k(\omega, x)$$

hold, where $\lambda_1(\omega, x) \geq \cdots \geq \lambda_k(\omega, x)$ *denote the k largest Lyapunov exponents at* (ω, x). *In particular,* $\nu_i(\omega, x) = \lambda_i(\omega, x)$ *for* $i = 1, \ldots, d$.

---

[5] See, for instance, Boichenko et al. [9, Chap. IV, Sect. 8.1].

Proof. Let $\tau > 0$ be the period of $(\varphi(\cdot, x, \omega), \omega(\cdot))$ and fix $k \in \{1, \ldots, d\}$. From the first step of the proof of Theorem 5.1 we know that there exists a linear operator $R : T_x M \to T_x M$ such that

$$d_x \varphi_{2\tau n, \omega} = e^{2\tau n R} \quad \text{for all } n \in \mathbb{Z},$$

and that the Lyapunov exponents are the real parts of the eigenvalues of R. Using subadditivity of a^k and writing each $t \geq 0$ as $t = 2\tau n(t) + r(t)$ with $n(t) \in \mathbb{Z}_+$ and $r(t) \in [0, 2\tau)$, we find

$$a_t^k(\omega, x) \leq a_{2\tau n(t)}^k(\omega, x) + a_{r(t)}^k(\omega, x).$$

Since $a_{(\cdot)}^k(\omega, x)$ is bounded on the compact set $[0, 2\tau]$ by Lemma 5.4, we thus obtain

$$\limsup_{t \to \infty} \frac{1}{t} a_t^k(\omega, x) \leq \limsup_{t \to \infty} \frac{1}{t} a_{2\tau n(t)}^k(\omega, x) = \frac{1}{2\tau} \limsup_{\mathbb{N} \ni n \to \infty} \frac{1}{n} a_{2\tau n}^k(\omega, x).$$

On the other hand, for each $t \geq 0$ we find $n(t) \in \mathbb{Z}_+$ and $r(t) \in [0, 2\tau)$ such that $t + r(t) = 2\tau n(t)$. Subadditivity gives $a_{2\tau n(t)}^k(\omega, x) \leq a_t^k(\omega, x) + a_{r(t)}^k(\Phi_t(\omega, x))$. Using that $\varphi(t, x, \omega)$ is contained in the compact set Q for all t, Lemma 5.4 implies boundedness of $a_{r(t)}^k(\Phi_t(\omega, x))$. Hence,

$$\frac{1}{2\tau} \liminf_{n \to \infty} \frac{1}{n} a_{2\tau n}^k(\omega, x) = \liminf_{t \to \infty} \frac{1}{2\tau n(t)} a_{2\tau n(t)}^k(\omega, x) \leq \liminf_{t \to \infty} \frac{1}{t} a_t^k(\omega, x).$$

We have the relations $(e^{2\tau n R})^{\wedge k} = e^{2\tau n R_k} = (e^{2\tau R_k})^n$, where R_k denotes the k-th derivation operator of R. This gives

$$\frac{1}{n} a_{2\tau n}^k(\omega, x) = \frac{1}{n} \log \prod_{i=1}^{k} \sigma_i(e^{2\tau n R}) = \frac{1}{n} \log \left\| (e^{2\tau n R})^{\wedge k} \right\| = \frac{1}{n} \log \left\| (e^{2\tau R_k})^n \right\|.$$

We know that the limit for $n \to \infty$ of the last expression exists and is equal to the logarithm of the spectral radius of $e^{2\tau R_k}$. The eigenvalues of R_k are the sums $\lambda_{i_1} + \cdots + \lambda_{i_k}$, where $\{\lambda_{i_1}, \ldots, \lambda_{i_k}\}$ is any subset of the spectrum of R consisting of k elements. Since the real parts of these eigenvalues are the Lyapunov exponents $\lambda_1(\omega, x) \geq \cdots \geq \lambda_d(\omega, x)$, it follows that

$$\frac{1}{2\tau} \lim_{n \to \infty} \frac{1}{n} a_{2\tau n}^k(\omega, x) = \lambda_1(\omega, x) + \cdots + \lambda_k(\omega, x).$$

Putting everything together, the proof is finished.   □

Proposition 5.9. *Assume that every periodic trajectory corresponding to some* $(x, \omega) \in \text{int } D \times \text{int } \mathscr{U}$ *has exactly* k *positive Lyapunov exponents (counted* *with multiplicities), where $k \in \{0, 1, \ldots, d\}$. Then for every periodic controlled trajectory $(\varphi(\cdot, x, \omega), \omega(\cdot))$ with $(x, \omega) \in \operatorname{int} D \times \operatorname{int} \mathcal{U}$ and every compact set $K \subset D$ it holds that*

$$h_{\mathrm{inv}}(K, D) \leq \sum_{j=1}^{r} \max \{0, d_j \lambda_j\},$$

where $\lambda_1, \ldots, \lambda_r$ are the different Lyapunov exponents at (ω, x) with corresponding multiplicities $d_1, \ldots, d_r$.

Proof. The case $k = 0$ is trivial, since here anyway $h_{\mathrm{inv}}(K, D) = 0$ (by Lemma 5.3 combined with Theorem 5.1). Hence, we may assume that $1 \leq k \leq d$. Given a τ_0-periodic controlled trajectory $(\varphi(\cdot, x, \omega), \omega(\cdot))$ with $(x, \omega) \in \operatorname{int} D \times \operatorname{int} \mathcal{U}$, we write $\lambda_1(\omega, x) \geq \cdots \geq \lambda_d(\omega, x)$ for the Lyapunov exponents at (ω, x) (here every Lyapunov exponent can appear several times according to its multiplicity). By assumption, the first k of these Lyapunov exponents are positive. From Lemma 5.5 it follows that

$$\lambda_1(\omega, x) + \cdots + \lambda_k(\omega, x) = \nu_1(\omega, x) + \cdots + \nu_k(\omega, x) = \lim_{t \to \infty} \frac{1}{t} a_t^k(\omega, x).$$

Now fix some $\varepsilon > 0$ and choose $n_0 \in \mathbb{N}$ sufficiently large such that

$$\left| \frac{1}{n_0 \tau_0} a_{n_0 \tau_0}^k(\omega, x) - \lim_{t \to \infty} \frac{1}{t} a_t^k(\omega, x) \right| \leq \frac{\varepsilon}{2}. \tag{5.20}$$

The limit $\lim_{t \to \infty}(1/t) a_t^k(\omega, x)$ is positive. Hence, we can choose n_0 large enough that also $a_{n_0 \tau_0}^k(\omega, x) > 0$. Applying Proposition 5.7, we obtain a regular periodic trajectory $(\varphi(\cdot, x, \omega_*), \omega_*(\cdot))$ with $\omega_* \in \operatorname{int} \mathcal{U}$ of some period $\tau_* > 0$ such that

$$\frac{1}{\tau_*} a_{\tau_*}^k(\omega_*, x) \leq \frac{1}{n_0 \tau_0} a_{n_0 \tau_0}^k(\omega, x) + \frac{\varepsilon}{2}. \tag{5.21}$$

Now Theorem 5.1 gives

$$h_{\mathrm{inv}}(K, D) \leq \lambda_1(\omega_*, x) + \cdots + \lambda_k(\omega_*, x).$$

The sequence $n \mapsto a_{n \tau_*}^k(\omega_*, x)$ is easily seen to be subadditive and hence, the subadditivity Lemma B.3 implies

$$\lim_{n \to \infty} \frac{1}{n \tau_*} a_{n \tau_*}^k(\omega_*, x) = \inf_{n \in \mathbb{N}} \frac{1}{n \tau_*} a_{n \tau_*}^k(\omega_*, x) \leq \frac{1}{\tau_*} a_{\tau_*}^k(\omega_*, x)$$

$$\overset{(5.21)}{\leq} \frac{1}{n_0 \tau_0} a_{n_0 \tau_0}^k(\omega, x) + \frac{\varepsilon}{2}.$$

Using Lemma 5.5 again, we find

$$
\lambda_1(\omega_*, x) + \cdots + \lambda_k(\omega_*, x) = \lim_{t \to \infty} \frac{1}{t} a_t^k(\omega_*, x)
$$

$$
= \lim_{n \to \infty} \frac{1}{n\tau_*} a_{n\tau_*}^k(\omega_*, x) \le \frac{1}{n_0\tau_0} a_{n_0\tau_0}^k(\omega, x) + \frac{\varepsilon}{2}.
$$

Altogether, we obtain

$$
h_{\mathrm{inv}}(K, D) \;\le\; \frac{1}{n_0\tau_0} a_{n_0\tau_0}^k(\omega, x) + \frac{\varepsilon}{2}
$$

$$
\overset{(5.20)}{\le} \lim_{t \to \infty} \frac{1}{t} a_t^k(\omega, x) + \frac{\varepsilon}{2} + \frac{\varepsilon}{2}
$$

$$
= \lambda_1(\omega, x) + \cdots + \lambda_k(\omega, x) + \varepsilon.
$$

Since ε can be chosen arbitrarily small, this finishes the proof. $\square$

The Second Approximation Result

Proposition 5.9 shows that under appropriate assumptions we can do without regularity of the periodic trajectory in Theorem 5.1. Let us impose the same assumptions on the system Σ as before (real-analytic, strongly accessible, compact control range). By using a second approximation result for subadditive cocycles, we can also weaken the periodicity assumption.

Proposition 5.10. *Let $a : \mathbb{R} \times (\mathcal{U} \times M) \to \mathbb{R}$ be a subadditive cocycle over the control flow satisfying assumption (5.18) and the boundedness property (5.19) of a^k. Furthermore, let $(x, \omega) \in \mathrm{int}\, D \times \mathrm{int}\, \mathcal{U}$ such that $\varphi(t, x, \omega)$ is contained in a compact set $K \subset \mathrm{int}\, D$ for all $t \ge 0$, and suppose that there exists $t_0 \ge 0$ with $a_t(\omega, x) \ge 0$ for all $t \ge t_0$. Then for every $\varepsilon > 0$ there exists a periodic trajectory with initial state x corresponding to a periodic control function $\omega_* \in \mathrm{int}\, \mathcal{U}$ of the same period $\tau_* > 0$ such that*

$$
\frac{1}{\tau_*} a_{\tau_*}(\omega_*, x) \le \limsup_{t \to \infty} \frac{1}{t} a_t(\omega, x) + \varepsilon.
$$

Proof. Let $(t_n)_{n \in \mathbb{N}}$ be a sequence of positive times with $t_n \to \infty$ such that

$$
\sigma := \limsup_{t \to \infty} \frac{1}{t} a_t(\omega, x) = \lim_{n \to \infty} \frac{1}{t_n} a_{t_n}(\omega, x).
$$

Now define the first hitting time

$$\tau := \inf \left\{ t \geq 0 \ : \ x \in \mathcal{O}_{\leq t}^+(z) \text{ for all } z \in K \right\}.$$

By Proposition 1.23 (v), local accessibility (which follows from strong accessibility) guarantees that $\tau < \infty$. There is $n_1 \in \mathbb{N}$ such that for all $n \geq n_1$ and all $T \in [0, \tau]$ it holds that

$$\frac{1}{t_n + T} \sup_{\substack{(t,z,v) \in [0,\tau] \times K \times \mathcal{U} \\ \varphi([0,\tau],z,v) \subset Q}} |a_t(v, z)| \leq \frac{\varepsilon}{2}. \tag{5.22}$$

Finiteness of the above supremum follows from the boundedness assumption imposed on a. Finally, there is $N \geq n_1$ such that (by assumption)

$$a_{t_N}(\omega, x) \geq 0 \tag{5.23}$$

and such that

$$\left| \frac{1}{t_N} a_{t_N}(\omega, x) - \sigma \right| \leq \frac{\varepsilon}{2}. \tag{5.24}$$

By definition of τ we can choose a control function $v \in \mathcal{U}[0, T)$ with $T \leq \tau$ and $\varphi(T, \varphi(t_N, x, \omega), v) = x$, and we may assume that v is piecewise constant taking values in int Ω. Define the control function ω_* on $[0, t_N + T]$ as

$$\omega_*(t) := \begin{cases} \omega(t) & \text{for } t \in [0, t_N] \\ v(t - t_N) & \text{for } t \in (t_N, t_N + T] \end{cases},$$

and extend ω_* $(t_N + T)$-periodically. This yields a $(t_N + T)$-periodic trajectory in int D, and $\omega_* \in$ int $\mathcal{U}$. Then, with $\tau_* := t_N + T$, we have

$$
\begin{aligned}
\frac{1}{\tau_*} a_{\tau_*}(\omega_*, x) &\leq \frac{1}{t_N + T} \left(a_{t_N}(\omega_*, x) + a_T(\Theta_{t_N} \omega_*, \varphi(t_N, x, \omega_*)) \right) \\
&= \frac{1}{t_N + T} \left(a_{t_N}(\omega, x) + a_T(v, \varphi(t_N, x, \omega)) \right) \\
&\overset{(5.22)}{\leq} \frac{1}{t_N + T} a_{t_N}(\omega, x) + \frac{\varepsilon}{2} \\
&\overset{(5.23)}{\leq} \frac{1}{t_N} a_{t_N}(\omega, x) + \frac{\varepsilon}{2} \overset{(5.24)}{\leq} \sigma + \varepsilon.
\end{aligned}
$$

This finishes the proof. $\square$

Proposition 5.11. *Let $(x, \omega) \in \text{int } D \times \text{int } \mathscr{U}$ such that $\varphi(t, x, \omega)$ is contained in a compact subset of $\text{int } D$ for all $t \geq 0$. Furthermore, assume that there exists $k \in \{0, 1, \ldots, d\}$ such that the following assumptions are satisfied:*

(i) Every periodic trajectory corresponding to some $(y, \mu) \in \text{int } D \times \text{int } \mathscr{U}$ has exactly k positive Lyapunov exponents (counted with multiplicities);
(ii) There exists $t_0 \geq 0$ such that $a_t^k(\omega, x) \geq 0$ for all $t \geq t_0$.

Then for every compact set $K \subset D$ it holds that

$$h_{\text{inv}}(K, D) \leq \nu_1(\omega, x) + \cdots + \nu_k(\omega, x).$$

Proof. Note that the assumptions of Proposition 5.10 are satisfied for the subadditive cocycle a^k. Hence, for given $\varepsilon > 0$ we find a periodic controlled trajectory of the form $(\varphi(\cdot, x, \omega_*), \omega_*(\cdot))$ with $\omega_* \in \text{int } \mathscr{U}$ of some period $\tau_* > 0$ such that

$$\frac{1}{\tau_*} a_{\tau_*}^k(\omega_*, x) \leq \limsup_{t \to \infty} \frac{1}{t} a_t^k(\omega, x) + \varepsilon$$

$$= (\nu_1(\omega, x) + \cdots + \nu_k(\omega, x)) + \varepsilon. \qquad (5.25)$$

By Proposition 5.9 we have

$$h_{\text{inv}}(K, D) \leq \lambda_1(\omega_*, x) + \cdots + \lambda_k(\omega_*, x)$$

$$= \lim_{t \to \infty} \frac{1}{t} a_t^k(\omega_*, x)$$

$$= \lim_{m \to \infty} \frac{1}{m\tau_*} a_{m\tau_*}^k(\omega_*, x)$$

$$= \inf_{m \in \mathbb{N}} \frac{1}{m\tau_*} a_{m\tau_*}^k(\omega_*, x) \leq \frac{1}{\tau_*} a_{\tau_*}^k(\omega_*, x).$$

Here we used that the sequence $m \mapsto a_{m\tau_*}^k(\omega_*, x)$ is subadditive. Combining this inequality with (5.25) gives the desired result. $\qquad \square$

Remark 5.6. Notice that the assumption that Σ is real-analytic and strongly accessible has only been used to guarantee that for every point in the interior of the given control set there exists a regular periodic trajectory going through this point. To have that (together with local accessibility) it is sufficient and necessary that there are two points in the interior of the control set which can be joined by a regular trajectory. At first sight, this seems to be a much weaker condition than strong accessibility, but a result of Sontag [100, Proposition 4.2] shows that (under mild assumptions) for real-analytic systems this is equivalent to strong accessibility. However, for control-affine systems there is an easy trick which can be used to show that the assumption of strong accessibility can be weakened to local accessibility. Moreover, using a result of Coron [30, Theorem 1.3 and Corollary 1.8] it can be shown that analyticity can be weakened to smoothness. (In fact, this works not only for control-affine systems, but we do not go into the quite technical details involved here.)

Proposition 5.12. *Assume that Σ is control-affine, $F(x,u) = f_0(x) + \sum_{i=1}^{m} u_i f_i(x)$ with a (compact and convex) control range with nonempty interior. Then the assertions of Propositions 5.9 and 5.11 also hold if the vector fields $f_0, f_1, \ldots, f_m$ are of class $\mathscr{C}^\infty$ and the Lie algebra rank condition holds on D.*

Proof. The proof proceeds in four steps.

Step 1. We show that if Σ satisfies the particular assumptions of Propositions 5.9 and 5.11, then they are also satisfied for each of the time-transformed systems $\Sigma^\alpha = (\mathbb{R}, M, \mathbb{R}^{m+1}, \mathscr{U}^\alpha, \varphi^\alpha), \alpha > 1$, given by the differential equations

$$\dot{x}(t) = \gamma(t) \cdot F(x(t), \omega(t)), \qquad (\gamma, \omega) \in \mathscr{U}^\alpha = \mathscr{V}^\alpha \times \mathscr{U},$$

where $\mathscr{V}^\alpha = \{\gamma \in L^\infty(\mathbb{R}, \mathbb{R}) : \gamma(t) \in [1/\alpha, \alpha]\}$. First we prove that the trajectories of Σ^α are just time reparametrizations of the trajectories of Σ. To this end, for every $\gamma \in \mathscr{V}^\alpha$ define

$$\sigma(t) := \int_0^t \gamma(s)\mathrm{d}s, \qquad t \geq 0.$$

It is clear that $\sigma : \mathbb{R}_+ \to \mathbb{R}_+$ is absolutely continuous with $\sigma(0) = 0$. It is bijective, since $\gamma \geq 1/\alpha$ implies that σ is strictly increasing and $\sigma(t) \to \infty$ for $t \to \infty$. We claim that

$$\varphi(\sigma(t), x, \omega) = \varphi^\alpha(t, x, (\gamma, \omega \circ \sigma)) \qquad (5.26)$$

for all $x \in M, \omega \in \mathscr{U}$, and $t \geq 0$. Indeed, for almost all $t \geq 0$ we have

$$\frac{\mathrm{d}}{\mathrm{d}t}\varphi(\sigma(t), x, \omega) = \dot{\sigma}(t) \cdot F(\varphi(\sigma(t), x, \omega), \omega(\sigma(t)))$$

$$= \gamma(t) \cdot F(\varphi(\sigma(t), x, \omega), \omega \circ \sigma(t)).$$

By uniqueness of solutions, the identity (5.26) follows. From this identity it can easily be seen that if D is a control set of Σ, then D is a control set of Σ^α. Now assume that every periodic trajectory of Σ corresponding to some $(x, \omega) \in \mathrm{int}\, D \times \mathrm{int}\, \mathscr{U}$ has exactly k positive Lyapunov exponents as required in Proposition 5.9. Then the analogous statement for Σ^α is true (with $(x, (\gamma, \omega)) \in \mathrm{int}\, D \times \mathrm{int}(\mathscr{V}^\alpha \times \mathscr{U})$). Indeed, let $(\varphi^\alpha(\cdot, x, (\gamma, \omega)), (\gamma, \omega))$ be a τ-periodic controlled trajectory with $x \in \mathrm{int}\, D$ and $(\gamma, \omega) \in \mathrm{int}(\mathscr{V}^\alpha \times \mathscr{U}) = \mathrm{int}\, \mathscr{V}^\alpha \times \mathrm{int}\, \mathscr{U}$. The number of positive Lyapunov exponents of the given trajectory is given by the number of eigenvalues of $\mathrm{d}_x \varphi^\alpha_{\tau, (\gamma, \omega)} : T_x M \to T_x M$ of absolute value bigger than one. From (5.26) it follows that $\mathrm{d}_x \varphi^\alpha_{\tau, (\gamma, \omega)} = \mathrm{d}_x \varphi_{\sigma(\tau), \omega \circ \sigma^{-1}}$. From τ-periodicity of γ it follows that

$t + \tau = \sigma^{-1}(\sigma(t) + \sigma(\tau))$ for all $t \geq 0$. This implies $(\omega \circ \sigma^{-1})(t + \sigma(\tau)) = \omega(\sigma^{-1}(t) + \tau) = \omega(\sigma^{-1}(t))$. Hence, $\omega \circ \sigma^{-1}$ is $\sigma(\tau)$-periodic. Thus, $(\varphi(\cdot, x, \omega \circ \sigma^{-1}), \omega \circ \sigma^{-1})$ is a $\sigma(\tau)$-periodic controlled trajectory of Σ with $(x, \omega \circ \sigma^{-1}) \in$ int $D \times$ int $\mathcal{U}$ and hence has exactly k positive Lyapunov exponents. This implies the assertion. Analogously, one shows that assumption (ii) in Proposition 5.11 carries over from Σ to Σ^α.

Step 2. We show that the invariance entropies of (K, D) with respect to Σ and Σ^α, respectively, are related by

$$h_{\mathrm{inv}}(K, D; \Sigma) \leq \alpha \cdot h_{\mathrm{inv}}(K, D; \Sigma^\alpha). \tag{5.27}$$

To this end, let $\mathscr{S} \subset \mathcal{V}^\alpha \times \mathcal{U}$ be a (τ, K, D)-spanning set for Σ^α. We claim that

$$\mathscr{S}' := \{\omega \circ \sigma^{-1} \mid \exists \gamma \in \mathcal{V}^\alpha : (\gamma, \omega) \in \mathscr{S}\}$$

is a $(\tau/\alpha, K, D)$-spanning set for Σ. Indeed, let $x \in K$. Then there is $(\gamma, \omega) \in \mathscr{S}$ with

$$\varphi\left(\sigma(t), x, \omega \circ \sigma^{-1}\right) = \varphi^\alpha(t, x, (\gamma, \omega)) \in D \quad \text{for all } t \in [0, \tau],$$

which implies $\varphi(t, x, \omega \circ \sigma^{-1}) \in D$ for all $t \in [0, \tau/\alpha]$, since $\sigma(\tau) \geq \int_0^\tau 1/\alpha \mathrm{d}s = \tau/\alpha$. It follows that $r_{\mathrm{inv}}(\tau/\alpha, K, D; \Sigma) \leq r_{\mathrm{inv}}(\tau, K, D; \Sigma^\alpha)$ and hence

$$h_{\mathrm{inv}}(K, D; \Sigma) = \limsup_{\tau \to \infty} \frac{\alpha}{\tau} \log r_{\mathrm{inv}}(\tau/\alpha, K, D; \Sigma)$$

$$\leq \limsup_{\tau \to \infty} \frac{\alpha}{\tau} \log r_{\mathrm{inv}}(\tau, K, D; \Sigma^\alpha) = \alpha \cdot h_{\mathrm{inv}}(K, D; \Sigma^\alpha),$$

which finishes Step 2.

Step 3. We prove the assertion for the case that $f_0, f_1, \ldots, f_m$ are analytic vector fields. Since we assume that the Lie algebra rank condition holds for Σ, the smallest Lie algebra spanned by the vector fields $f_0, f_1, \ldots, f_m$ has full rank at every point (see Proposition 1.8). Note that the strong accessibility algebra of Σ^α, that is, the ideal generated by the differences $v[f_0 + \sum_{i=1}^m u_i f_i] - v'[f_0 + \sum_{i=1}^m u'_i f_i]$, contains the vector fields $f_1, \ldots, f_m$ as well as the vector field f_0 (put $(v, u) := (\alpha, 0) \in \mathbb{R} \times \mathbb{R}^m$ and $(v', u') := (1, 0) \in \mathbb{R} \times \mathbb{R}^m$, then $v[f_0 + \sum_{i=1}^m u_i f_i] - v'[f_0 + \sum_{i=1}^m u'_i f_i] = (\alpha - 1) f_0)$. By Proposition 5.6 (vii) this implies that Σ^α is strongly accessible. Hence, we find that the Propositions 5.9 and 5.11 can be applied to the systems Σ^α, if $f_0, f_1, \ldots, f_m$ are analytic. The inequality (5.27) shows that the corresponding estimates for the invariance entropy $h_{\mathrm{inv}}(K, D; \Sigma^\alpha)$ carry over to $h_{\mathrm{inv}}(K, D; \Sigma)$ by letting $\alpha \to 1$.

Step 4. We show that the assumption of analyticity can be weakened to smoothness. Observe that analyticity (in combination with strong accessibility) was only used in the proof of Lemma 5.3 to show the existence of arbitrarily short regular trajectories in the interior of D. However, this also follows as a consequence of Coron [30, Corollary 1.8] if the right-hand side of the system is of class $\mathscr{C}^\infty$ and polynomial with respect to the control variable, and if the strong accessibility algebra has full rank at every point. Since these assumptions are satisfied for the time-transformed systems Σ^α, if the given system is smooth and satisfies the Lie algebra rank condition, we are done. $\square$

Remark 5.7.

- Of course, one would like to have a third approximation result to get rid of the assumptions that $\varphi(t, x, \omega)$ be contained in a compact subset of int D and $\omega \in$ int $\mathscr{U}$. As can be seen in Sect. 7.1, for one-dimensional systems things are easier than in the general case, since here only equilibria instead of arbitrary trajectories have to be considered. The same holds for particular control sets of projective systems, as we show in Sect. 7.4.
- The existence of universally regular control functions and regular periodic trajectories inside of control sets for discrete-time systems has been studied in Wirth [110–112] and Sontag and Wirth [103]. Hence, it should be an easy task to adapt the results of this section to the discrete-time setting.

5.3   Comments and Bibliographical Notes

The main theorem of this chapter, Theorem 5.1, has appeared before in Kawan [62, 64]. All results about the invariance entropy in Sect. 5.2 are new and have not been published before. The methods used in the proofs of the approximation results for subadditive cocycles are basically taken from Colonius and Kliemann [25, Theorem 6.2.17], a result which relates the Lyapunov and Floquet spectra of certain control systems on vector bundles to each other. Further note that the estimate for a^k given in Lemma 5.4 can be improved (see Boichenko et al. [9, Chap. I, Corollary 4.2.1]). Of course, the results of this chapter leave many questions open. For instance, what can be said about the value of $h_{\mathrm{inv}}(Q)$ when Q is the closure of a relatively compact control set D? Is it the same as $h_{\mathrm{inv}}(K, Q)$ for $K \subset D$ or can it be strictly greater? Another question concerns the existence of regular periodic trajectories without the regularity assumptions of Sect. 5.2. One could ask, for instance, if they exist generically. Finally, notice that in this chapter we have seen a second example for the equality $h_{\mathrm{inv,out}}(K, Q) = h_{\mathrm{inv}}(K, Q)$, namely Corollary 5.3.

Chapter 6
Escape Rates and Lower Bounds

In this chapter, we derive lower estimates for the invariance entropy using an approach which is based on the observation that $h_{\mathrm{inv}}(K, Q)$ is bounded from below by a quantity which can be regarded as a uniform escape rate from the set Q. Section 6.1 explains our basic approach which stems from Young [115]. In Sects. 6.2 and 6.3, two similar strategies how to derive explicit lower bounds from this approach are presented. These strategies are based on two different results about the estimation of the volumes of Bowen-balls, the first one by Franz [44] and Gelfert [49, 50], and the second by Bowen and Ruelle [13, 14] and in its nonautonomous formulation by Liu [78]. Throughout the whole chapter, we concentrate on smooth systems given by differential equations.

6.1 Escape Rates and Invariance Entropy

Consider a topological time-invariant system $\Sigma = (\mathbb{T}, X, U, \mathcal{U}, \varphi)$ with an admissible pair (K, Q) such that $r_{\mathrm{inv}}(\tau, K, Q) < \infty$ for all $\tau \in \mathbb{T} \cap (0, \infty)$. Then every minimal (τ, K, Q)-spanning set $\mathcal{S}$ yields a finite cover of the set K consisting of the subsets

$$K(\omega, \tau) := \{x \in K \ : \ \varphi([0, \tau], x, \omega) \subset Q\}, \quad \omega \in \mathcal{S}.$$

Let μ denote an outer measure on X satisfying $0 < \mu(K) < \infty$. Then

$$0 < \mu(K) \leq \sum_{\omega \in \mathcal{S}} \mu(K(\omega, \tau)) \leq r_{\mathrm{inv}}(\tau, K, Q) \cdot \sup_{\omega \in \mathcal{U}(K,Q)} \mu(K(\omega, \tau)),$$

where

$$\mathcal{U}(K, Q) := \{\omega \in \mathcal{U} \mid \exists x \in K \ : \ \varphi([0, \tau], x, \omega) \subset Q\}.$$

C. Kawan, *Invariance Entropy for Deterministic Control Systems*, Lecture Notes in Mathematics 2089, DOI 10.1007/978-3-319-01288-9_6,
© Springer International Publishing Switzerland 2013

This implies

$$\log r_{\mathrm{inv}}(\tau, K, Q) \geq \log \mu(K) - \log \sup_{\omega \in \mathscr{U}(K,Q)} \mu(K(\omega, \tau)),$$

which gives the estimate

$$h_{\mathrm{inv}}(K, Q) \geq -\liminf_{\tau \to \infty} \sup_{\omega \in \mathscr{U}(K,Q)} \frac{1}{\tau} \log \mu(K(\omega, \tau)). \qquad (6.1)$$

Our aim is to relate this lower bound to quantities which are better accessible to computation, in particular to Lyapunov exponents and quantities similar to topological entropy. To explain how this can be accomplished we have to take a short digression into the theory of *escape rates* for classical dynamical systems.

Consider a continuous map $f : X \to X$ on a compact metric space (X, ϱ). Let m denote a reference (Borel-)measure on X. For a closed set $Q \subset X$ define

$$Q_n := \{x \in Q \,:\, f^i(x) \in Q \text{ for } i = 0, 1, \ldots, n-1\}, \quad n \in \mathbb{Z}_+.$$

This gives a decreasing sequence of compact sets

$$Q = Q_0 \supset Q_1 \supset Q_2 \supset \cdots$$

We define the *lower escape rate* from Q by

$$\underline{\lambda} = \underline{\lambda}(f, Q, m) := \liminf_{n \to \infty} \frac{1}{n} \log m(Q_n),$$

where $\log 0 := -\infty$ by convention. The *upper escape rate* $\overline{\lambda}$ is defined analogously, replacing the lower limit by an upper limit. Let us denote the Bowen-ball of order n with radius ε centered at $x \in X$ by $B^n(x, \varepsilon)$. Then, taking a maximal (n, ε, f)-separated subset E_n of Q_n (for a fixed $\varepsilon > 0$), we obtain

$$m(Q_n) \leq \sum_{x \in E_n} m(B^n(x, \varepsilon)) \leq r_{\mathrm{sep}}(n, \varepsilon, Q_n, f) \cdot \sup_{x \in Q_n} m(B^n(x, \varepsilon)),$$

since E_n also (n, ε, f)-spans Q_n (cf. Sect. B.3). Now one can obtain bounds for $m(Q_n)$ by estimating $r_{\mathrm{sep}}(n, \varepsilon, Q_n, f)$ and $m(B^n(x, \varepsilon))$ separately. Let us assume that each of the sets Q_n is nonempty and that there exists a continuous function $\varphi : X \to \mathbb{R}$ satisfying an estimate of the form

$$m(B^n(x, \varepsilon)) \leq C \exp\left(-\sum_{i=0}^{n-1} \varphi(f^i(x))\right)$$

for all $x \in X$ and $n \in \mathbb{Z}_+$. Then, a similar idea as described above together with arguments from the proof of the variational principle for pressure (as can be found, for instance, in Katok and Hasselblatt [61, Lemma 20.2.3]) yields the existence of an f-invariant measure μ supported on Q such that

$$\overline{\lambda} \le h_\mu(f) - \int \varphi \mathrm{d}\mu.$$

This construction can be found in Young [115]. In the smooth setting, where f is a diffeomorphism on a compact manifold, one can express both $h_\mu(f)$ and $\int \varphi \mathrm{d}\mu$ in terms of the positive Lyapunov exponents of μ and other dimension-like characteristics (cf. [115]) in case that the diffeomorphism satisfies some hyperbolicity conditions.

Returning to the problem of estimating $h_{\mathrm{inv}}(K, Q)$, we see that the lower bound (6.1) can be regarded as a uniform escape rate. So there is some hope that in the smooth setting we can use similar techniques to find relations between this expression and the Lyapunov exponents of the given system. In the following sections, we describe two similar approaches to obtain such relations.

6.2   The First Lower Bound Theorem

Let $\Sigma = (\mathbb{R}, M, \mathbb{R}^m, \mathscr{U}, \varphi)$ be a smooth system given by differential equations on a d-dimensional Riemannian $\mathscr{C}^3$-manifold (M, g) with right-hand side F and a compact control range Ω. By ϱ we denote the Riemannian distance on M.

Let (K, Q) be an admissible pair for Σ such that Q is compact and controlled invariant. Furthermore, assume that the d-dimensional Hausdorff measure[1] $\mu_H(K, d) = \mu_H(K, d; \varrho)$ is positive (which is equivalent to $\mathrm{vol}(K) > 0$, see Federer [42, Theorem 2.10.35]), and that $h_{\mathrm{inv}}(Q) < \infty$ (which is equivalent to $r_{\mathrm{inv}}(\tau, Q) < \infty$ for all $\tau > 0$ by Proposition 2.3).

With the set $\mathscr{Q}$ we can associate a vector bundle of rank d:

$$\pi_{\mathscr{Q}} : \bigcup_{(\omega, x) \in \mathscr{Q}} \{\omega\} \times T_x M \to \mathscr{Q}, \quad \pi_{\mathscr{Q}}(\omega, v) = (\omega, \pi_{TM}(v)), \qquad (6.2)$$

where $\pi_{TM} : TM \to M$ is the map sending a tangent vector $v \in T_x M$ to its base point x. On $\mathscr{U}$ we may consider the relative topology of $L^\infty(\mathbb{R}, \mathbb{R}^m)$ which turns $\mathscr{Q} \subset \mathscr{U} \times M$ into a metrizable topological space. The vector space structure on the fibers $\{\omega\} \times T_x M = \pi_{\mathscr{Q}}^{-1}(\omega, x)$ is the natural one induced by the vector space structure of $T_x M$. We call (6.2) the *extended tangent bundle over* $\mathscr{Q}$.

---

[1]For the notion of Hausdorff measure and related notions used in the following see Sect. B.3.

By $\pi_{\mathscr{U}} : \mathscr{U} \times M \to \mathscr{U}$ we denote the projection onto the first factor, $\pi_{\mathscr{U}}(\omega, x) = \omega$. We define

$$\mathscr{K}_Q := \{(\omega, x) \in \mathscr{Q} : x \in K\}.$$

Moreover, for each $\omega \in \pi_{\mathscr{U}} \mathscr{K}_Q$ we introduce the nonempty compact sets

$$K(\omega, \tau) := \{x \in K : \varphi(t, x, \omega) \in Q \text{ for all } t \in [0, \tau]\}, \quad \tau > 0.$$

For each $\omega \in \mathscr{U}$ and $\tau > 0$ we define the *Bowen-metric*

$$\varrho_{\omega, \tau}(x, y) := \max_{t \in [0, \tau]} \varrho\left(\varphi(t, x, \omega), \varphi(t, y, \omega)\right).$$

It is easy to see that $\varrho_{\omega, \tau}$ indeed is a metric on M which is topologically equivalent to ϱ.[2] For each $(\omega, x) \in \mathscr{U} \times M$, $\tau > 0$, and $\varepsilon > 0$, the *Bowen-ball of order τ and radius ε* centered at $x \in M$, is denoted by

$$B_{\omega}^{\tau}(x, \varepsilon) = \{y \in M : \varrho_{\omega, \tau}(x, y) < \varepsilon\}.$$

A set $S \subset M$ is called $(\omega, \tau, \varepsilon)$-*separated* if for all $x_1, x_2 \in S$ with $x_1 \neq x_2$ one has $\varrho_{\omega, \tau}(x_1, x_2) \geq \varepsilon$. By $r_{\text{sep}}(\omega, \tau, \varepsilon, K, Q)$ we denote the maximal cardinality of an $(\omega, \tau, \varepsilon)$-separated subset of $K(\omega, \tau)$. We say that a set $D \subset M$ $(\omega, \tau, \varepsilon)$-*spans* another set $E \subset M$ if for every $x \in E$ there is $y \in D$ such that $\varrho_{\omega, \tau}(x, y) < \varepsilon$. By $r_{\text{span}}(\omega, \tau, \varepsilon, K, Q)$ we denote the minimal cardinality of a set which $(\omega, \tau, \varepsilon)$-spans $K(\omega, \tau)$. It is easy to see that a maximal $(\omega, \tau, \varepsilon)$-separated subset S of $K(\omega, \tau)$ also $(\omega, \tau, \varepsilon)$-spans $K(\omega, \tau)$ (cf. the proof of Proposition 6.1 (i)) and hence

$$K(\omega, \tau) \subset \bigcup_{x \in S} B_{\omega}^{\tau}(x, \varepsilon).$$

In addition, we call a set $S \subset M$ ε-*separated* if $\varrho(x_1, x_2) \geq \varepsilon$ holds for each pair of distinct points $x_1, x_2 \in S$.

Definition 6.1. The *escape entropy of* (K, Q) is defined as follows:

$$\overline{r}_{\text{sep}}(\tau, \varepsilon, K, Q) := \sup_{\omega \in \pi_{\mathscr{U}} \mathscr{K}_Q} \varepsilon^d r_{\text{sep}}(\omega, \tau, \varepsilon, K, Q),$$

$$\overline{r}_{\text{sep}}(\tau, K, Q) := \limsup_{\varepsilon \searrow 0} \overline{r}_{\text{sep}}(\tau, \varepsilon, K, Q),$$

$$h_{\text{esc}}(K, Q) = h_{\text{esc}}(K, Q; g) := \limsup_{\tau \to \infty} \frac{1}{\tau} \log \overline{r}_{\text{sep}}(\tau, K, Q).$$

---

[2]See also Kolyada and Snoha [70], where topological entropy is defined for nonautonomous dynamical systems.

By definition, $h_{esc}(K, Q)$ is an element of the extended real line $\mathbb{R} \cup \{-\infty, \infty\}$ and it might depend on the Riemannian metric g. As for the topological entropy, one obtains an alternative definition of $h_{esc}(K, Q)$ by replacing maximal $(\omega, \tau, \varepsilon)$-separated subsets of $K(\omega, \tau)$ by minimal $(\omega, \tau, \varepsilon)$-spanning sets (cf. Proposition 6.1 (ii)). We define

$$\overline{r}_{span}(\tau, \varepsilon, K, Q) := \sup_{\omega \in \pi_{\mathscr{U}} \mathscr{K}_Q} \varepsilon^d r_{span}(\omega, \tau, \varepsilon, K, Q),$$

$$\overline{r}_{span}(\tau, K, Q) := \limsup_{\varepsilon \searrow 0} \overline{r}_{span}(\tau, \varepsilon, K, Q).$$

Proposition 6.1. *The following assertions hold:*

(i) *For all $\tau, \varepsilon > 0$ and $\omega \in \mathscr{U}$ it holds that*

$$r_{span}(\omega, \tau, \varepsilon, K, Q) \leq r_{sep}(\omega, \tau, \varepsilon, K, Q) \leq r_{span}\left(\omega, \tau, \frac{\varepsilon}{2}, K, Q\right) < \infty.$$

(ii) *The escape entropy can be expressed in terms of the cardinalities of minimal $(\omega, \tau, \varepsilon)$-spanning sets as*

$$h_{esc}(K, Q) = \limsup_{\tau \to \infty} \frac{1}{\tau} \log \overline{r}_{span}(\tau, K, Q).$$

(iii) *If the distance functions induced by two Riemannian metrics g and $\tilde{g}$ on M are equivalent on the set Q, then*

$$h_{esc}(K, Q; g) = h_{esc}(K, Q; \tilde{g}).$$

In particular, this is satisfied if g and $\tilde{g}$ both are complete Riemannian metrics.

(iv) $|h_{esc}(K, Q)| < \infty$. *In particular, $h_{esc}(K, Q) \geq -h_{inv}(K, Q)$.*

Proof. (i) Let $S \subset K(\omega, \tau)$ be an $(\omega, \tau, \varepsilon)$-separated set of maximal cardinality. Assume to the contrary that there is $y \in K(\omega, \tau)$ with $\varrho_{\omega,\tau}(x, y) \geq \varepsilon$ for all $x \in S$. Then also $S \cup \{y\}$ is $(\omega, \tau, \varepsilon)$-separated in contradiction to the maximality of S. Hence, S also $(\omega, \tau, \varepsilon)$-spans $K(\omega, \tau)$ implying that $r_{span}(\omega, \tau, \varepsilon, K, Q) \leq r_{sep}(\omega, \tau, \varepsilon, K, Q)$. Now let $S \subset K(\omega, \tau)$ be any $(\omega, \tau, \varepsilon)$-separated set and $E \subset M$ any set which $(\omega, \tau, \varepsilon/2)$-spans $K(\omega, \tau)$. Define a map $\alpha : S \to E$ by assigning to each $x \in S$ one $\alpha(x) \in E$ such that $\varrho_{\omega,\tau}(x, \alpha(x)) < \varepsilon/2$. Assume that $\alpha(x_1) = \alpha(x_2)$ for some $x_1, x_2 \in S$. Then

$$\varrho_{\omega,\tau}(x_1, x_2) \leq \varrho_{\omega,\tau}(x_1, \alpha(x_1)) + \varrho_{\omega,\tau}(\alpha(x_2), x_2) < \varepsilon.$$

Hence, $x_1 = x_2$, which shows that α is injective. Therefore, $\#S \leq \#E$ implying $r_{sep}(\omega, \tau, \varepsilon, K, Q) \leq r_{span}(\omega, \tau, \varepsilon/2, K, Q)$. By compactness of $K(\omega, \tau)$ it is clear that minimal $(\omega, \tau, \varepsilon)$-spanning sets are finite.

(ii) From statement (i) it follows that

$$\varepsilon^d r_{\text{span}}(\omega, \tau, \varepsilon, K, Q) \leq \varepsilon^d r_{\text{sep}}(\omega, \tau, \varepsilon, K, Q)$$

$$\leq 2^d \left(\frac{\varepsilon}{2}\right)^d r_{\text{span}}\left(\omega, \tau, \frac{\varepsilon}{2}, K, Q\right),$$

which implies the assertion.

(iii) We first show that complete Riemannian metrics g and $\tilde{g}$ induce distance functions ϱ and $\tilde{\varrho}$ which are equivalent on Q. Subsequently, we prove that this equivalence implies $h_{\text{esc}}(K, Q; g) = h_{\text{esc}}(K, Q; \tilde{g})$. To start the argument, note that for every $x \in M$ the norms induced by g_x and $\tilde{g}_x$ on $T_x M$ are equivalent, so in particular there is $L(x) > 0$ such that $g_x(v, v)^{1/2} \leq L(x)\tilde{g}_x(v, v)^{1/2}$ for all $v \in T_x M$. Since g_x and $\tilde{g}_x$ depend continuously on x, we can assume the same for $L(x)$. Now let $x, y \in Q$ and let $\gamma : [0, 1] \to M$ be a shortest geodesic from x to y with respect to $\tilde{g}$, which exists by completeness. Then

$$\varrho(x, y) \leq \int_0^1 g_{\gamma(s)}\left(\dot{\gamma}(s), \dot{\gamma}(s)\right)^{1/2} ds$$

$$\leq \int_0^1 L(\gamma(s))\tilde{g}_{\gamma(s)}\left(\dot{\gamma}(s), \dot{\gamma}(s)\right)^{1/2} ds$$

$$\leq \max_{s \in [0,1]} L(\gamma(s)) \int_0^1 \tilde{g}_{\gamma(s)}\left(\dot{\gamma}(s), \dot{\gamma}(s)\right)^{1/2} ds$$

$$= \left(\max_{s \in [0,1]} L(\gamma(s))\right) \tilde{\varrho}(x, y).$$

Let $A \subset M$ be the set defined as the union of the images of all shortest geodesics with respect to $\tilde{g}$ joining two points in Q. This set is obviously bounded, and hence, by the theorem of Hopf–Rinow, its closure is compact. Consequently, for all $x, y \in Q$ we have

$$\varrho(x, y) \leq L\tilde{\varrho}(x, y) \quad \text{with } L := \max_{x \in \text{cl } A} L(x).$$

Changing the roles of g and $\tilde{g}$ yields the equivalence of the metrics restricted to Q. Now let $S \subset K(\omega, \tau)$ be a maximal $(\omega, \tau, \varepsilon)$-separated set with respect to the metric ϱ. Then for all $x \neq y$ in S we have

$$\varepsilon \leq \varrho_{\omega, \tau}(x, y) = \max_{t \in [0, \tau]} \varrho(\varphi(t, x, \omega), \varphi(t, y, \omega))$$

$$\leq L \max_{t \in [0, \tau]} \tilde{\varrho}(\varphi(t, x, \omega), \varphi(t, y, \omega)) = L\tilde{\varrho}_{\omega, \tau}(x, y).$$

Hence, S is $(\omega, \tau, \varepsilon/L)$-separated with respect to $\tilde{\varrho}$ implying that

$$L^d \left(\frac{\varepsilon}{L}\right)^d r_{\text{sep}}\left(\omega, \tau, \frac{\varepsilon}{L}, K, Q; \tilde{g}\right) \geq \varepsilon^d r_{\text{sep}}(\omega, \tau, \varepsilon, K, Q; g)$$

which yields $h_{\text{esc}}(K, Q; \tilde{g}) \geq h_{\text{esc}}(K, Q; g)$. By changing the roles of g and $\tilde{g}$ the converse inequality follows.

(iv) We first show that $h_{\text{esc}}(K, Q) < \infty$. To this end, let $c, a > 0$ be constants such that

$$\varrho(\varphi(t, x, \omega), \varphi(t, y, \omega)) \leq c e^{at} \varrho(x, y)$$

holds on a compact neighborhood of Q, for all $x, y \in Q$ with $\varrho(x, y) < \varepsilon$ for some sufficiently small $\varepsilon > 0$, for all $t \geq 0$, and $\omega \in \mathcal{U}$ (existence of such constants can be shown with similar arguments as in the proof of Corollary 4.1). Then, for each $t \in [0, \tau]$, $\varrho(x, y) < e^{-a\tau} \varepsilon$ implies

$$\varrho(\varphi(t, x, \omega), \varphi(t, y, \omega)) \leq c e^{at} \varrho(x, y) < c e^{a(t-\tau)} \varepsilon \leq c\varepsilon.$$

For fixed $\omega \in \pi_{\mathcal{U}} \mathcal{K}_Q$, $\tau > 0$, and $\varepsilon > 0$ consider the minimal number $n(K(\omega, \tau), e^{-a\tau}\varepsilon)$ of balls of radius $e^{-a\tau}\varepsilon$ necessary to cover the compact set $K(\omega, \tau)$. We have shown that every such ball $B(x, e^{-a\tau}\varepsilon)$ is contained in the Bowen-ball $B_\omega^\tau(x, c\varepsilon)$ which implies

$$r_{\text{span}}(\omega, \tau, c\varepsilon, K, Q) \leq n(e^{-a\tau}\varepsilon, K(\omega, \tau)).$$

Hence, we obtain

$$\begin{aligned}
\overline{r}_{\text{span}}(\tau, K, Q) &\leq \limsup_{\varepsilon \searrow 0} \sup_\omega (c\varepsilon)^d n(e^{-a\tau}\varepsilon, K(\omega, \tau)) \\
&= c^d \limsup_{\varepsilon \searrow 0} \sup_\omega e^{ad\tau} \mu_C\left(K(\omega, \tau), d, e^{-a\tau}\varepsilon\right) \\
&\leq c^d e^{ad\tau} \limsup_{\varepsilon \searrow 0} \mu_C\left(K, d, e^{-a\tau}\varepsilon\right) = c^d e^{ad\tau} \mu_C(K, d).
\end{aligned}$$

This implies $h_{\text{esc}}(K, Q) \leq ad < \infty$. To show that $h_{\text{esc}}(K, Q) > -\infty$, note that $\varepsilon^d r_{\text{span}}(\omega, \tau, \varepsilon, K, Q) = \mu_C\left(K(\omega, \tau), d, \varepsilon; \varrho_{\omega,\tau}\right)$ and hence

$$\overline{r}_{\text{span}}(\tau, K, Q) = \limsup_{\varepsilon \searrow 0} \sup_{\omega \in \pi_{\mathcal{U}} \mathcal{K}_Q} \mu_C\left(K(\omega, \tau), d, \varepsilon; \varrho_{\omega,\tau}\right).$$

Since $B_\omega^\tau(x, \varepsilon) \subset B(x, \varepsilon)$ for all $\varepsilon, \tau > 0$ and $(\omega, x) \in \mathcal{U} \times M$, we have

$$\mu_C\left(K(\omega, \tau), d, \varepsilon; \varrho_{\omega,\tau}\right) \geq \mu_C\left(K(\omega, \tau), d, \varepsilon; \varrho\right).$$

By assumption, minimal (τ, K, Q)-spanning sets are finite. If $\mathscr{S}$ is such a set, then $K = \bigcup_{\omega \in \mathscr{S}} K(\omega, \tau)$ and hence

$$\mu_C\left(K, d, \varepsilon; \varrho\right) \leq \sum_{\omega \in \mathscr{S}} \mu_C\left(K(\omega, \tau), d, \varepsilon; \varrho\right)$$

$$\leq r_{\mathrm{inv}}(\tau, K, Q) \sup_{\omega \in \pi_{\mathscr{U}} \mathscr{K}_Q} \mu_C\left(K(\omega, \tau), d, \varepsilon; \varrho\right).$$

Altogether, we obtain

$$\overline{r}_{\mathrm{span}}(\tau, K, Q) \geq \limsup_{\varepsilon \searrow 0} \frac{\mu_C(K, d, \varepsilon; \varrho)}{r_{\mathrm{inv}}(\tau, K, Q)} = \frac{\mu_C(K, d; \varrho)}{r_{\mathrm{inv}}(\tau, K, Q)}.$$

Using (ii) and $\mu_C(K, d; \varrho) \geq \mu_H(K, d; \varrho) > 0$, we can conclude that

$$h_{\mathrm{esc}}(K, Q) \geq -h_{\mathrm{inv}}(K, Q) \geq -h_{\mathrm{inv}}(Q) > -\infty,$$

which finishes the proof. $\qquad\qquad\qquad\qquad\qquad\qquad\qquad\qquad\qquad\qquad\qquad\square$

Remark 6.1. Another property of $h_{\mathrm{esc}}(K, Q)$ that can easily be seen is its invariance with respect to bi-Lipschitz state transformations. (We leave the proof of this fact to the reader.)

An Estimate for the Hausdorff Measure of a Bowen-Ball

In this subsection, we present a result of Gelfert [49, 50], which gives an estimate for the outer Hausdorff measure of a Bowen-ball. Gelfert proved the result for classical dynamical systems, but the generalization to control systems is straightforward. This result is based on techniques that are also used in Franz [44] and Gu [54], and an essential argument in the proof is that Bowen-balls can be approximated by ellipsoids in corresponding tangent spaces. Therefore, we first need to introduce some concepts related to ellipsoids in Euclidean spaces.

Let $\mathscr{E}$ be an ellipsoid in a d-dimensional Euclidean space X. Then the lengths of the half-axes of $\mathscr{E}$ are denoted by

$$\sigma_1(\mathscr{E}) \geq \cdots \geq \sigma_d(\mathscr{E}) \geq 0.$$

Analogously to the definition of the singular value function (cf. Sect. A.1), we put

$$\alpha_r(\mathscr{E}) := \begin{cases} \sigma_1(\mathscr{E})\sigma_2(\mathscr{E})\cdots\sigma_r(\mathscr{E}) & \text{for } r > 0 \\ 1 & \text{for } r = 0 \end{cases}, \quad r = 0, 1, \ldots, d.$$

The following lemma gives an estimate of the number of metric balls necessary to cover an ellipsoid.

Lemma 6.1. *Let $\mathcal{E}$ be an ellipsoid in a d-dimensional Euclidean space X and $\zeta > 0$. Then $\mathcal{E}$ can be covered by $\lfloor 2^r (\alpha_r(\mathcal{E})/\zeta^r) \rfloor$ balls of radii $\zeta \sqrt{r+1}$, where*

$$
r = \begin{cases} 0 & \text{for } \zeta > \sigma_1(\mathcal{E}), \\ l & \text{for } \sigma_{l+1}(\mathcal{E}) \leq \zeta \leq \sigma_l(\mathcal{E}), \ l \in \{1, \dots, d-1\}, \\ d & \text{for } \zeta \leq \sigma_d(\mathcal{E}). \end{cases}
$$

Proof. If $\zeta > \sigma_1(\mathcal{E})$, the lemma claims that $\mathcal{E}$ can be covered by one ball of radius ζ which is obviously true. In the other cases, let us assume without loss of generality that $X = \mathbb{R}^d$ and that $\mathcal{E}$ is given by

$$
\mathcal{E} = \left\{ x \in \mathbb{R}^d : \sum_{i=1}^d \left(\frac{x_i}{\sigma_i(\mathcal{E})} \right)^2 \leq 1 \right\}.
$$

Assume that $\sigma_{l+1}(\mathcal{E}) \leq \zeta \leq \sigma_l(\mathcal{E})$ for some $l \in \{1, \dots, d-1\}$, define r as in the assertion, and set $\mathcal{E}_0 := \mathcal{E} \cap \mathbb{R}^r$, $\rho := \sigma_{r+1}(\mathcal{E})$. The ellipsoid $\mathcal{E}_0$ is contained in the parallelepiped $P := \prod_{i=1}^r [-\sigma_i(\mathcal{E}), \sigma_i(\mathcal{E})]$. Cover P by N cubes of side length 2ζ, where $N := \prod_{i=1}^r (\lfloor \sigma_i(\mathcal{E})/\zeta \rfloor + 1)$. Since $\sigma_i(\mathcal{E})/\zeta \geq 1$ for $i = 1, \dots, r$, it follows that $N \leq 2^r \prod_{i=1}^r (\sigma_i(\mathcal{E})/\zeta)$. Let $B'(\rho)$ be the metric ball of radius ρ centered at the origin in $\mathbb{R}^{d-r}$. Then $\mathcal{E}$ is contained in $\mathcal{E}_0 \times B'(\rho)$, and hence we can cover $\mathcal{E}$ by N sets of the form $K \times B'(\rho)$, where K is a cube of side length 2ζ. Each of these Cartesian products is contained in a ball of radius $\zeta \sqrt{r+1}$ which gives

$$
n\left(\zeta \sqrt{r+1}, \mathcal{E}\right) \leq 2^r \prod_{i=1}^r \frac{\sigma_i(\mathcal{E})}{\zeta} = 2^r \frac{\alpha_r(\mathcal{E})}{\zeta^r}.
$$

This implies the desired estimate. Finally, if $\zeta \leq \sigma_d(\mathcal{E})$, the ellipsoid $\mathcal{E}$ is contained in $\prod_{i=1}^d [-\sigma_i(\mathcal{E}), \sigma_i(\mathcal{E})]$ and hence, $\prod_{i=1}^d (\lfloor \sigma_i(\mathcal{E})/\zeta \rfloor + 1) \leq 2^d \prod_{i=1}^d (\sigma_i(\mathcal{E})/\zeta)$ cubes of side length 2ζ or the same number of balls of radius $\sqrt{d+1}\zeta$ are sufficient to cover $\mathcal{E}$. We conclude that

$$
n\left(\zeta \sqrt{r+1}, \mathcal{E}\right) \leq 2^d \prod_{i=1}^d \frac{\sigma_i(\mathcal{E})}{\zeta} = 2^d \frac{\alpha_d(\mathcal{E})}{\zeta^r},
$$

which finishes the proof. $\square$

Now consider system Σ and let $Q \subset M$ be a compact controlled invariant set. Let $E \to \mathcal{Q}$, $E = \bigcup_{(\omega, x) \in \mathcal{Q}} \{\omega\} \times E_{\omega, x}$, be a subbundle of rank n, $0 < n \leq d$, of the extended tangent bundle (6.2). Then for $(\omega, x) \in \mathcal{Q}$, $\tau \in \mathbb{R}$, and $i = 1, \dots, n$ we define

$$
\sigma_i^E(\omega, x, \tau) := \sigma_i \left(d_x \varphi_{\tau, \omega} |_{E_{\omega, x}} : E_{\omega, x} \to d_x \varphi_{\tau, \omega} E_{\omega, x} \right),
$$

where $\sigma_i(\cdot)$ denotes the i-th singular value, and for $r \in \{0, 1, \ldots, n\}$ we also consider the singular value function of order r:

$$\alpha_r^E(\omega, x, \tau) := \alpha_r \left(d_x \varphi_{\tau,\omega}|_{E_{\omega,x}} : E_{\omega,x} \to d_x \varphi_{\tau,\omega} E_{\omega,x} \right).$$

For the proof of the estimate for the Hausdorff measures of Bowen-balls we use a lemma which can be found in Gelfert [49, Lemma 5.2.1] and can be regarded as a generalization or analog of the Taylor formula in $\mathbb{R}^d$.

Lemma 6.2. *Let (M, g) be a Riemannian $\mathscr{C}^3$-manifold, $U \subset M$ an open set, $\phi :$ $U \to M$ a $\mathscr{C}^1$-mapping and $Q \subset U$ a compact set with $\inf_{x \in Q} |\det d_x \phi| > 0$. Then there is $\varepsilon_0 > 0$ such that for all $x \in Q$ and $y \in B(x, \varepsilon_0)$ with $\phi(y) \in B(\phi(x), \varepsilon_0)$ the mapping $d_y \phi : T_y M \to T_{\phi(y)} M$ is invertible and the inequality*

$$\left| \exp_x^{-1}(y) - (d_x \phi)^{-1} \exp_{\phi(x)}^{-1}(\phi(y)) \right| \le \eta(x) \cdot \varepsilon_0$$

holds with

$$\eta(x) := \sup_{y: \phi(y) \in B(\phi(x), \varepsilon_0)} \left\| \tau_y^x \circ (d_y \phi)^{-1} \circ \tau_{\phi(x)}^{\phi(y)} - (d_x \phi)^{-1} \right\|,$$

where τ_y^x is the isometric operator defined by parallel transport along the shortest geodesic from y to x.[3] Hence, for every $y \in B(x, \varepsilon_0)$ it holds that

$$y = \exp_x \left((d_x \phi)^{-1} \exp_{\phi(x)}^{-1}(\phi(y)) + w(y) \right)$$

for a tangent vector $w(y)$ with $|w(y)| \le \eta(x) \varepsilon_0$.

Proof. Choose $x \in Q$ arbitrarily. The map $d_x \phi$ is by assumption an isomorphism. From the inverse function theorem the existence of numbers $\varepsilon_x, \delta_x > 0$ follows such that ϕ maps $B(x, \varepsilon_x)$ diffeomorphically into $B(\phi(x), \delta_x)$. Moreover, these constants can be chosen such that the relation $d_z \phi^{-1} = (d_{\phi^{-1}(z)} \phi)^{-1}$ holds for all $z \in B(\phi(x), \delta_x)$. By compactness of Q we find $\varepsilon_0, \delta_0 > 0$ such that this statement holds for all $x \in Q$ with $\varepsilon_x \le \varepsilon_0$ and $\delta_x \le \delta_0$. The Taylor formula (as can be found in Noack [87, Proposition 2.1]) applied to ϕ^{-1} yields for $y = \phi(x)$ and all $w \in B(y, \delta_x)$ the inequality

$$\left| \exp_{\phi^{-1}(y)}^{-1}(\phi^{-1}(w)) - d_y \phi^{-1} \exp_y^{-1}(w) \right|$$

$$\le \sup_{\widetilde{w} \in B(y, \delta_x)} \left\| \tau_{\phi^{-1}(\widetilde{w})}^{\phi^{-1}(y)} \circ d_{\widetilde{w}} \phi^{-1} \circ \tau_y^{\widetilde{w}} - d_y \phi^{-1} \right\| \cdot \left| \exp_y^{-1}(w) \right|.$$

---

[3]The operator $\tau_y^x : T_y M \to T_x M$ is defined via the solutions of the differential equation $(DX/dt)(t) \equiv 0$, where D/dt denotes the covariant derivative along the shortest geodesic $\gamma : [0, 1] \to M$ from y to x, and $\tau_y^x(v) := X(1)$ if $X(0) = v$.

With $x = \phi^{-1}(y)$ and $z := \phi^{-1}(w)$ we obtain

$$\left| \exp_x^{-1}(z) - (d_x \phi)^{-1} \exp_{\phi(x)}^{-1}(\phi(z)) \right|$$

$$\leq \sup_{\widetilde{w} \in B(y, \delta_x)} \left\| \tau_{\phi^{-1}(\widetilde{w})}^x \circ (d_{\phi^{-1}(\widetilde{w})} \phi)^{-1} \circ \tau_{\phi(x)}^{\widetilde{w}} - (d_x \phi)^{-1} \right\| \cdot \left| \exp_{\phi(x)}^{-1}(\phi(z)) \right|.$$

With $\eta(x) := \sup_{w : \phi(w) \in B(\phi(x), \varepsilon_0)} \| \tau_w^x \circ (d_w \phi)^{-1} \circ \tau_{\phi(x)}^{\phi(w)} - (d_x \phi)^{-1} \|$ the assertion follows. $\square$

Now we can give the proof of the volume lemma.

Lemma 6.3 (Gelfert-Franz Volume Lemma). *Let $Q \subset M$ be a compact controlled invariant set and $E \to \mathcal{Q}$ a subbundle of (6.2) of rank n, $0 < n \leq d$. Furthermore, let $\omega \in \pi_{\mathcal{U}} \mathcal{Q}$ and $\tau > 0$ be such that*

$$\inf_{x \in Q : (\omega, x) \in \mathcal{Q}} \alpha_n^E(\omega, x, \tau) > 1.$$

Then there is $\tilde{\varepsilon} = \tilde{\varepsilon}(\omega, \tau) > 0$ such that for all $x \in Q$ with $(\omega, x) \in \mathcal{Q}$ and all $\varepsilon \in (0, \tilde{\varepsilon})$ it holds that

$$\mu_H \left(B_\omega^\tau(x, \varepsilon), d, 8\sqrt{d}\varepsilon \right) \leq \left(2^4 \sqrt{d}\varepsilon \right)^d \alpha_n^E(\omega, x, \tau)^{-1}.$$

Proof. Let $E^1 := E$. If $n < d$, we choose another subbundle E^0 of rank $d - n$, which is complementary and orthogonal to E^1, that is, $E^0 \oplus E^1 = \bigcup_{(\omega, x) \in \mathcal{Q}} \{\omega\} \times T_x M$ and $E_{\omega, x}^0 \perp E_{\omega, x}^1$. If $n = d$, then $E^1 = \bigcup_{(\omega, x) \in \mathcal{Q}} \{\omega\} \times T_x M$ and it is not necessary to consider another subbundle. The projections onto E^0 and E^1 are denoted by

$$\pi_i : \bigcup_{(\omega, x) \in \mathcal{Q}} \{\omega\} \times T_x M \to E^i, \quad (\omega, v) \mapsto \pi_i(\omega)v, \quad i = 0, 1.$$

We consider the map $\phi := \varphi_{\tau, \omega}$. Since ϕ is of class $\mathcal{C}^1$ and $\inf_{x \in Q} |\det d_x \phi| > 0$ (ϕ is a $\mathcal{C}^1$-diffeomorphism), there are $\theta, \eta > 0$ such that the inequalities

$$\theta \geq \left(\inf_{x \in Q} \sigma_d(d_x \phi) \right)^{-1}, \quad \eta \leq \left(\sup_{x \in Q} \sigma_1(d_x \phi) \right)^{-1}$$

hold. (Note that $|\det d_x \phi| = \prod_{i=1}^d \sigma_i(d_x \phi)$.) Since

$$\sup_{x \in Q} \sigma_1 \left((d_x \phi)^{-1} \right) = \sup_{x \in Q} \frac{1}{\sigma_d(d_x \phi)},$$

$$\inf_{x \in Q} \sigma_d \left((d_x \phi)^{-1} \right) = \inf_{x \in Q} \frac{1}{\sigma_1(d_x \phi)},$$

we have

$$\sup_{x \in Q} \sigma_1((\mathrm{d}_x \phi)^{-1}) \leq \theta \quad \text{and} \quad \eta \leq \inf_{x \in Q} \sigma_d((\mathrm{d}_x \phi)^{-1}). \tag{6.3}$$

From the assumptions of the lemma and elementary transformations we obtain

$$\inf_{x \in Q:\, (\omega,x) \in \mathscr{Q}} \alpha_n^{E^1}(\omega, x, \tau) > 1$$

$$\Leftrightarrow \quad \sup_{x \in Q:\, (\omega,x) \in \mathscr{Q}} \alpha_n \left((\mathrm{d}_x \phi)^{-1} |_{\mathrm{d}_x \phi E^1_{\omega,x}} \right) < 1$$

and therefore

$$\inf_{x \in Q:\, (\omega,x) \in \mathscr{Q}} \sigma_1^{E^1}(\omega, x, \tau) > 1$$

$$\Leftrightarrow \quad \sup_{x \in Q:\, (\omega,x) \in \mathscr{Q}} \sigma_n \left((\mathrm{d}_x \phi)^{-1} |_{\mathrm{d}_x \phi E^1_{\omega,x}} \right) < 1. \tag{6.4}$$

(Note that $\alpha_n^{E^1}(\omega, x, \tau) = \sigma_1^{E^1}(\omega, x, \tau) \cdots \sigma_n^{E^1}(\omega, x, \tau)$ and $\sigma_1(\cdot)$ is the greatest singular value.) We choose $\tilde{\varepsilon} > 0$ small enough such that the following conditions are satisfied:

(1) $\tilde{\varepsilon}$ is smaller than ε_0 from Lemma 6.2 applied to ϕ and the compact set Q;
(2) For every $x \in Q$ we have

$$\left\| \tau_y^x \circ (\mathrm{d}_y \phi)^{-1} \circ \tau_{\phi(x)}^{\phi(y)} - (\mathrm{d}_x \phi)^{-1} \right\| \leq \eta \tag{6.5}$$

for all $y \in B(x, \tilde{\varepsilon})$ with $\phi(y) \in B(\phi(x), \tilde{\varepsilon})$;
(3) The inequality

$$\varrho \left(\exp_y(v_1), \exp_y(v_2) \right) \leq 2|v_1 - v_2| \tag{6.6}$$

holds for all $y \in Q$ and $v_1, v_2 \in B(0_y, \tilde{\varepsilon})$ with $|v_1 - v_2| \leq (\theta + \eta)(2\sqrt{d} + 1)\tilde{\varepsilon}$.

Now we fix a point $x \in Q$ with $(\omega, x) \in \mathscr{Q}$ and a number $\varepsilon \in (0, \tilde{\varepsilon})$. Obviously, we have $B_\omega^\tau(x, \varepsilon) \subset B(x, \varepsilon)$. By applying Lemma 6.2 to $\phi = \varphi_{\tau,\omega}$, with (6.5) we get

$$\exp_x^{-1} \left(B_\omega^\tau(x, \varepsilon) \right) \subset (\mathrm{d}_x \phi)^{-1} B(0_{\phi(x)}, \varepsilon) + B(0_x, \eta\varepsilon). \tag{6.7}$$

With the chosen splitting of $\bigcup_{(\omega,x) \in \mathscr{Q}} \{\omega\} \times T_x M$ we obtain

$$\exp_x^{-1} \left(B_\omega^\tau(x, \varepsilon) \right) \subset \pi_1(\omega)(\mathrm{d}_x \phi)^{-1} B(0_{\phi(x)}, \varepsilon) \oplus \pi_0(\omega) B(0_x, \varepsilon) + B(0_x, \eta\varepsilon). \tag{6.8}$$

By Proposition A.1 the set $\mathscr{E}_1 := \pi_1(\omega)(d_x\phi)^{-1}B(0_{\phi(x)}, \varepsilon)$ is an ellipsoid in $E^1_{\omega,x}$ with half-axes of lengths

$$\sigma_i(\mathscr{E}_1) = \sigma_i\left((d_x\phi)^{-1}|_{d_x\phi E^1_{\omega,x}}\right)\varepsilon, \quad i = 1,\dots,n. \tag{6.9}$$

The set $\mathscr{E}_0 := \pi_0(\omega)(\exp_x^{-1}(B(x,\varepsilon)))$ is an ε-ball in $E^0_{\omega,x}$ and hence an ellipsoid with half-axes of lengths $\sigma_1(\mathscr{E}_0) = \cdots = \sigma_{d-n}(\mathscr{E}_0) = \varepsilon$. The set $\mathscr{E}_1 + (B(0_x, \eta\varepsilon) \cap E^1_{\omega,x})$ is easily seen to be contained in an ellipsoid $\mathscr{E}_1'$ with half-axes of lengths

$$\sigma_i(\mathscr{E}_1') = \left(1 + \frac{\eta\varepsilon}{\sigma_n(\mathscr{E}_1)}\right)\sigma_i(\mathscr{E}_1), \quad i = 1,\dots,n.$$

From (6.4) and (6.9) we conclude that

$$\sigma_n(\mathscr{E}_1) < \varepsilon.$$

With $\varepsilon + \eta\varepsilon = (1 + (\eta\varepsilon)/\varepsilon)\varepsilon$ it follows that $\mathscr{E}_0 + (B(0_x, \eta\varepsilon) \cap E^0_{\omega,x})$ is contained in an ellipsoid $\mathscr{E}_0' \subset E^0_{\omega,x}$ with half-axes of lengths

$$\sigma_i(\mathscr{E}_0') = \left(1 + \frac{\eta\varepsilon}{\sigma_n(\mathscr{E}_1)}\right)\varepsilon, \quad i = 1,\dots,d-n.$$

We set

$$\zeta := \left(1 + \frac{\eta\varepsilon}{\sigma_n(\mathscr{E}_1)}\right)\sigma_n(\mathscr{E}_1).$$

By Lemma 6.1 we can cover $\mathscr{E}_1'$ with $N_1 := \lfloor 2^n\alpha_n(\mathscr{E}_1')/\zeta^n \rfloor$ balls of radii $\zeta\sqrt{n+1}$ and $\mathscr{E}_0'$ with $N_0 := \lfloor 2^{d-n}\alpha_{d-n}(\mathscr{E}_0')/\zeta^{d-n} \rfloor$ balls of radii $\zeta\sqrt{d-n+1}$. From (6.8) it follows that

$$\exp_x^{-1}\left(B^\tau_\omega(x,\varepsilon)\right) \subset \mathscr{E}_1 \oplus \mathscr{E}_0 + B(0_x, \eta\varepsilon) \subset \mathscr{E}_1' \oplus \mathscr{E}_0',$$

and hence the set $\exp_x^{-1}(B^\tau_\omega(x,\varepsilon))$ can be covered with $N_1 N_0$ balls of radii $2\zeta\sqrt{d}$. (The product of a Euclidean ball of radius $\zeta\sqrt{n+1}$ and one of radius $\zeta\sqrt{d-n+1}$ is contained in a ball of radius $((\zeta\sqrt{n+1})^2 + (\zeta\sqrt{d-n+1})^2)^{1/2} = \zeta\sqrt{d+2} \le 2\zeta\sqrt{d}$.) From (6.3) and (6.7) it follows that each of these balls lies in the ball of radius $(\theta + \eta)(2\sqrt{d} + 1)\varepsilon$ and center 0_x. This is shown as follows: By (6.7) each $v \in \exp_x^{-1}(B^\tau_\omega(x,\varepsilon))$ can be written as $v = w_1 + w_2$ with $|w_1| \le \sigma_1((d_x\phi)^{-1})\varepsilon$, $|w_2| \le \eta\varepsilon$. Hence, $|v| \le \varepsilon(\theta + \eta)$. Now consider a ball with radius $2\zeta\sqrt{d}$ around v. Since

$$2\zeta\sqrt{d} = 2\sqrt{d}(\sigma_n(\mathscr{E}_1) + \eta\varepsilon) \le 2\sqrt{d}\varepsilon(\sigma_1((d_x\phi)^{-1}) + \eta) \le 2\sqrt{d}\varepsilon(\theta + \eta),$$

we find that the norm of such a vector is bounded by $\varepsilon(\theta + \eta) + 2\sqrt{d}\varepsilon(\theta + \eta) = (\theta + \eta)(2\sqrt{d} + 1)\varepsilon$. If one maps this cover with $\exp_x$ down to the manifold, then, by (6.6), $B_\omega^\tau(x, \varepsilon)$ is covered by $N_1 N_0$ balls of radii $4\zeta\sqrt{d}$. Hence, we obtain

$$\mu_H\left(B_\omega^\tau(x, \varepsilon), d, 4\zeta\sqrt{d}\right) \leq N_1 N_0 \left(4\zeta\sqrt{d}\right)^d.$$

This implies

$$\mu_H\left(B_\omega^\tau(x, \varepsilon), d, 4\zeta\sqrt{d}\right) \leq \frac{2^n \alpha_n(\mathscr{E}_1') }{\zeta^n} \frac{2^{d-n}\alpha_{d-n}(\mathscr{E}_0')}{\zeta^{d-n}}\left(4\zeta\sqrt{d}\right)^d$$

$$\leq 2^d \frac{\left(1 + \frac{\eta\varepsilon}{\sigma_n(\mathscr{E}_1)}\right)^d \alpha_n(\mathscr{E}_1)\alpha_{d-n}(\mathscr{E}_0)}{\left(1 + \frac{\eta\varepsilon}{\sigma_n(\mathscr{E}_1)}\right)^d \sigma_n(\mathscr{E}_1)^d}\left(4\zeta\sqrt{d}\right)^d$$

$$= \left(2^3\sqrt{d}\right)^d \alpha_n\left((d_x\phi)^{-1}|_{d_x\phi E_{\omega,x}^1}\right)$$

$$\left(\frac{\zeta}{\sigma_n\left((d_x\phi)^{-1}|_{d_x\phi E_{\omega,x}^1}\right)}\right)^d$$

$$= \left(2^3\sqrt{d}\varepsilon\right)^d \alpha_n^{E^1}(\omega, x, \tau)^{-1}$$

$$\left(\frac{\sigma_n\left((d_x\phi)^{-1}|_{d_x\phi E_{\omega,x}^1}\right) + \eta}{\sigma_n\left((d_x\phi)^{-1}|_{d_x\phi E_{\omega,x}^1}\right)}\right)^d.$$

By the choice of η we have

$$\eta \leq \inf_{y \in Q} \sigma_d\left((d_y\phi)^{-1}\right) \leq \sigma_d\left((d_x\phi)^{-1}\right) \leq \sigma_n\left((d_x\phi)^{-1}|_{d_x\phi E_{\omega,x}^1}\right).$$

Hence,

$$\mu_H\left(B_\omega^\tau(x, \varepsilon), d, 4\zeta\sqrt{d}\right) \leq \left(2^4\sqrt{d}\varepsilon\right)^d \alpha_n^{E^1}(\omega, x, \tau)^{-1}.$$

We can assume that $\eta \leq 1$ and hence, with (6.4) we get

$$\zeta = \varepsilon\left(\sigma_n\left((d_x\phi)^{-1}|_{d_x\phi E_{\omega,x}^1}\right) + \eta\right) < 2\varepsilon.$$

This gives

$$\mu_H\left(B_\omega^\tau(x,\varepsilon),d,8\sqrt{d}\varepsilon\right) \le \mu_H\left(B_\omega^\tau(x,\varepsilon),d,4\zeta\sqrt{d}\right),$$

which concludes the proof.   □

The Lower Bound Theorem

Now we are in position to formulate our main result which gives a lower estimate of $h_{\mathrm{inv}}(K,Q)$ in terms of a volume growth rate and the escape entropy.

Theorem 6.1. *Consider system Σ and let (K,Q) be an admissible pair such that $\mu_H(K,d) > 0$ and Q is compact and controlled invariant with $h_{\mathrm{inv}}(Q) < \infty$. Let $E \to \mathcal{Q}$ be a subbundle of (6.2) of rank n and assume that there is $\tau_0 > 0$ such that for all $\omega \in \pi_\mathcal{U} \mathcal{Q}$ and $\tau \ge \tau_0$*

$$\inf_{x\in Q:\ (\omega,x)\in\mathcal{Q}} \alpha_n^E(\omega,x,\tau) > 1.$$

Then it holds that

$$h_{\mathrm{inv}}(K,Q) \ge \limsup_{\tau\to\infty} \frac{1}{\tau} \inf_{(\omega,x)\in\mathcal{K}_Q} \log \alpha_n^E(\omega,x,\tau) - h_{\mathrm{esc}}(K,Q).$$

Proof. Fix an arbitrary $\tau \ge \tau_0$ and a minimal (τ,K,Q)-spanning set $\mathcal{S}$. Since $h_{\mathrm{inv}}(Q) < \infty$ by assumption, $\mathcal{S}$ is finite. Moreover, we can assume that $\mathcal{S} \subset \pi_\mathcal{U}\mathcal{K}_Q \subset \pi_\mathcal{U}\mathcal{Q}$. For each $\omega \in \mathcal{S}$ Lemma 6.3 yields an $\tilde{\varepsilon}(\omega) > 0$ such that for all $x \in Q$ with $(\omega,x) \in \mathcal{Q}$ and $\varepsilon \in (0,\tilde{\varepsilon}(\omega))$ the estimate

$$\mu_H\left(B_\omega^\tau(x,\varepsilon),d,8\sqrt{d}\varepsilon\right) \le \left(2^4\sqrt{d}\varepsilon\right)^d \alpha_n^E(\omega,x,\tau)^{-1}$$

holds. Let $\tilde{\varepsilon} := \min_{\omega\in\mathcal{S}}\tilde{\varepsilon}(\omega)$. For each $\varepsilon \in (0,\tilde{\varepsilon})$ and $\omega \in \mathcal{S}$ let $S_{\omega,\tau,\varepsilon}$ be a maximal $(\omega,\tau,\varepsilon)$-separated subset of $K(\omega,\tau)$. Then, since $K = \bigcup_{\omega\in\mathcal{S}} K(\omega,\tau)$ and $K(\omega,\tau) \subset \bigcup_{x\in S_{\omega,\tau,\varepsilon}} B_\omega^\tau(x,\varepsilon)$, for each $\varepsilon \in (0,\tilde{\varepsilon})$ we obtain

$$\mu_H\left(K,d,8\sqrt{d}\varepsilon\right) \le \sum_{\omega\in\mathcal{S}} \mu_H\left(K(\omega,\tau),d,8\sqrt{d}\varepsilon\right)$$

$$\le r_{\mathrm{inv}}(\tau,K,Q) \max_{\omega\in\mathcal{S}} \mu_H\left(K(\omega,\tau),d,8\sqrt{d}\varepsilon\right)$$

$$\le r_{\mathrm{inv}}(\tau,K,Q) \max_{\omega\in\mathcal{S}} \sum_{x\in S_{\omega,\tau,\varepsilon}} \mu_H\left(B_\omega^\tau(x,\varepsilon),d,8\sqrt{d}\varepsilon\right)$$

$$\le r_{\mathrm{inv}}(\tau,K,Q) \max_{\omega\in\mathcal{S}} \sum_{x\in S_{\omega,\tau,\varepsilon}} \left(2^4\sqrt{d}\varepsilon\right)^d \alpha_n^E(\omega,x,\tau)^{-1}$$

and

$$\max_{\omega \in \mathscr{S}} \sum_{x \in S_{\omega,\tau,\varepsilon}} \varepsilon^d \alpha_n^E(\omega, x, \tau)^{-1}$$

$$\leq \max_{\omega \in \mathscr{S}} \left(\varepsilon^d r_{\mathrm{sep}}(\omega, \tau, \varepsilon, K, Q) \sup_{x \in K(\omega,\tau)} \alpha_n^E(\omega, x, \tau)^{-1} \right)$$

$$\leq \left(\sup_{\omega \in \pi_{\mathscr{U}} \mathscr{K}_Q} \varepsilon^d r_{\mathrm{sep}}(\omega, \tau, \varepsilon, K, Q) \right) \left(\sup_{(\omega,x) \in \mathscr{K}_Q} \alpha_n^E(\omega, x, \tau)^{-1} \right).$$

With $\gamma := (2^4 \sqrt{d})^{-d}$ this implies the estimate

$$r_{\mathrm{inv}}(\tau, K, Q) \geq \gamma \mu_H \left(K, d, 8\sqrt{d}\,\varepsilon \right) \overline{r}_{\mathrm{sep}}(\tau, \varepsilon, K, Q)^{-1} \inf_{(\omega,x) \in \mathscr{K}_Q} \alpha_n^E(\omega, x, \tau).$$

Applying the logarithm to this inequality yields

$$\log r_{\mathrm{inv}}(\tau, K, Q) \geq \log \left(\gamma \mu_H \left(K, d, 8\sqrt{d}\,\varepsilon \right) \right)$$

$$- \log \overline{r}_{\mathrm{sep}}(\tau, \varepsilon, K, Q) + \inf_{(\omega,x) \in \mathscr{K}_Q} \log \alpha_n^E(\omega, x, \tau).$$

Since this holds for all $\varepsilon \in (0, \tilde{\varepsilon})$, we also get

$$\log r_{\mathrm{inv}}(\tau, K, Q) \geq \lim_{\varepsilon \searrow 0} \log \left(\gamma \mu_H \left(K, d, 8\sqrt{d}\,\varepsilon \right) \right)$$

$$- \limsup_{\varepsilon \searrow 0} \log \overline{r}_{\mathrm{sep}}(\tau, \varepsilon, K, Q) + \inf_{(\omega,x) \in \mathscr{K}_Q} \log \alpha_n^E(\omega, x, \tau)$$

$$= \log \left(\gamma \mu_H \left(K, d \right) \right) - \log \overline{r}_{\mathrm{sep}}(\tau, K, Q)$$

$$+ \inf_{(\omega,x) \in \mathscr{K}_Q} \log \alpha_n^E(\omega, x, \tau).$$

Dividing by τ and sending τ to infinity yields

$$h_{\mathrm{inv}}(K, Q) \geq \limsup_{\tau \to \infty} \left[-\frac{1}{\tau} \log \overline{r}_{\mathrm{sep}}(\tau, K, Q) + \frac{1}{\tau} \inf_{(\omega,x) \in \mathscr{K}_Q} \log \alpha_n^E(\omega, x, \tau) \right]$$

$$\geq \limsup_{\tau \to \infty} \frac{1}{\tau} \inf_{(\omega,x) \in \mathscr{K}_Q} \log \alpha_n^E(\omega, x, \tau) - \limsup_{\tau \to \infty} \frac{1}{\tau} \log \overline{r}_{\mathrm{sep}}(\tau, K, Q).$$

This finishes the proof. $\square$

The generalized Liouville formula (Proposition A.5) together with the fact that

$$\alpha_n^E(\omega, x, \tau) = \left| \det d_x \varphi_{\tau,\omega} \right|_{E_{\omega,x}} : E_{\omega,x} \to d_x \varphi_{\tau,\omega} E_{\omega,x} \right|$$

immediately gives the following corollary.

Corollary 6.1. *Assume that the subbundle $E \to \mathcal{Q}$ in Theorem 6.1 is invariant under the differential* $d\varphi_{(\cdot,\cdot)}$, *that is,*

$$d_x \varphi_{t,\omega} E_{\omega,x} = E_{\Theta_t \omega, \varphi_{t,\omega}(x)} \quad \text{for all } t \geq 0, \ (\omega, x) \in \mathcal{Q}.$$

Then

$$h_{\text{inv}}(K, Q) \geq \limsup_{\tau \to \infty} \frac{1}{\tau} \inf_{(\omega,x) \in \mathcal{K}_Q} \int_0^\tau \text{tr} \left[\nabla F_{\omega(s)}(\varphi_{s,\omega}(x)) \circ Q(\Theta_s \omega, \varphi_{s,\omega}(x)) \right] ds$$
$$- h_{\text{esc}}(K, Q),$$

where $Q(\omega, x) : T_x M \to E_{\omega,x}$ is the orthogonal projection.

6.3   The Second Lower Bound Theorem

In this section, we assume that $\Sigma = (\mathbb{R}, M, \mathbb{R}^m, \mathcal{U}, \varphi)$ is a control-affine system such that (M, g) is a Riemannian $\mathcal{C}^3$-manifold. The Riemannian distance on M is denoted by ϱ and the Riemannian volume by vol. Recall from Sect. 1.3 that $\mathcal{U}$ becomes a compact metrizable space with the weak*-topology and the associated control flow is a continuous dynamical system.

Uniformly Hyperbolic Sets

The following definition introduces hyperbolic sets for control-affine systems.

Definition 6.2. Assume that $Q \subset M$ is a compact set which is controlled invariant in forward and in backward time for Σ, that is, for every $x \in Q$ there exists $\omega \in \mathcal{U}$ with $\varphi(\mathbb{R}, x, \omega) \subset Q$. Define the *full time lift of Q* by

$$\mathcal{Q} := \{(\omega, x) \in \mathcal{U} \times M : \varphi(\mathbb{R}, x, \omega) \subset Q\}.$$

Let $\Phi : \mathbb{R} \times (\mathcal{U} \times M) \to \mathcal{U} \times M$ denote the control flow of Σ. Further assume that for each $(\omega, x) \in \mathcal{Q}$ the tangent space $T_x M$ can be written as a direct sum

$$T_x M = E_{\omega,x}^- \oplus E_{\omega,x}^+$$

of subspaces such that the following statements hold:

(1) For all $t \in \mathbb{R}$ and $(\omega, x) \in \mathscr{Q}$ we have

$$d_x \varphi_{t,\omega} E^-_{\omega,x} = E^-_{\Phi_t(\omega,x)} \quad \text{and} \quad d_x \varphi_{t,\omega} E^+_{\omega,x} = E^+_{\Phi_t(\omega,x)};$$

(2) There are constants $c, \lambda > 0$ such that

$$|d_x \varphi_{t,\omega}(v)| \leq c^{-1} e^{-\lambda t} |v| \quad \text{for all } t \geq 0, \ (\omega, x) \in \mathscr{Q}, \ v \in E^-_{\omega,x},$$

and

$$|d_x \varphi_{t,\omega}(v)| \geq c e^{\lambda t} |v| \quad \text{for all } t \geq 0, \ (\omega, x) \in \mathscr{Q}, \ v \in E^+_{\omega,x};$$

(3) $E^-_{\omega,x}$ and $E^+_{\omega,x}$ vary continuously with (ω, x) (where on $\mathscr{U}$ we consider the weak*-topology). That is, the projections $\pi^{\pm}_{\omega,x} : T_x M \to E^{\pm}_{\omega,x}$ with respect to the decomposition $E^-_{\omega,x} \oplus E^+_{\omega,x}$ depend continuously on (ω, x).

Then Q is called *uniformly hyperbolic*.

Remark 6.2. With regard to the classical hyperbolic theory for continuous-time systems (flows), one might wonder if the existence of a nontrivial uniformly hyperbolic set according to the above definition is possible, since a one-dimensional center subbundle is missing. However, note that for a nonautonomous differential equation a central direction, in which the Lyapunov exponent vanishes, not necessarily exists. One can think of a uniformly hyperbolic set Q in this context as a set which arises by a time-dependent perturbation of a flow around a hyperbolic equilibrium point. In Sects. 7.3 and 7.4 examples for such sets are given.

Lemma 6.4. *For a uniformly hyperbolic set Q the following statements hold:*

(i) *The full time lift $\mathscr{Q}$ of Q is compact and invariant under the control flow Φ, that is, $\Phi_t(\mathscr{Q}) = \mathscr{Q}$ for all $t \in \mathbb{R}$.*

(ii) *Property (3) in Definition 6.2 follows from the first two properties. Moreover, the dimensions of $E^-_{\omega,x}$ and $E^+_{\omega,x}$, respectively, are locally constant on $\mathscr{Q}$.*

(iii) *If M is compact and Q is a chain control set, the dimensions of $E^{\pm}_{\omega,x}$ are constant on $\mathscr{Q}$.*

(iv) *There exists $\alpha > 0$ such that for every $(\omega, x) \in \mathscr{Q}$ the angle between any $v \in E^-_{\omega,x}$ and $w \in E^+_{\omega,x}$ is at least α. More precisely,*

$$\inf_{(\omega,x) \in \mathscr{Q}} \inf_{(v,w) \in E^-_{\omega,x} \times E^+_{\omega,x}} \arccos \frac{g_x(v,w)}{|v||w|} =: \alpha > 0.$$

(v) *The subspaces $E^{\pm}_{\omega,x}$ (more precisely, $\{\omega\} \times E^{\pm}_{\omega,x}$) are the fibers of subbundles $E^{\pm} \to \mathscr{Q}$ of the vector bundle*

$$\bigcup_{(\omega,x)\in\mathscr{Q}} \{\omega\} \times T_x M \to \mathscr{Q}, \quad (\omega,v) \mapsto (\omega, \pi_{TM}(v)),$$

and this bundle decomposes into the Whitney sum of E^- and E^+.

Proof.   (i) This is proved in the same manner as Proposition 1.10.

(ii) For all $(\omega, x) \in \mathscr{Q}$ and $v \in E_{\omega,x}^-$ it holds that

$$|\mathrm{d}_x\varphi_{t,\omega}(v)| \le c^{-1}\mathrm{e}^{-\lambda t}|v| \quad \text{for all } t \ge 0, \tag{6.10}$$

and these inequalities characterize the subspace $E_{\omega,x}^-$. Indeed, for any $v \in T_x M$, $v = v^- \oplus v^+$, $v^\pm \in E_{\omega,x}^\pm$, we have

$$\mathrm{d}_x\varphi_{t,\omega}(v) = \mathrm{d}_x\varphi_{t,\omega}(v^-) + \mathrm{d}_x\varphi_{t,\omega}(v^+),$$

and if $v^+ \ne 0$ we see that $|\mathrm{d}_x\varphi_{t,\omega}(v)|$ converges to ∞ for $t \to \infty$, which implies that the inequalities (6.10) cannot be satisfied. Now let (ω_m, x_m) be a sequence in $\mathscr{Q}$ with $(\omega_m, x_m) \to (\omega, x)$. By taking a subsequence if necessary[4] we can assume that $\dim E_{\omega_m,x_m}^- = \text{const.} =: d^-$ and that we can choose an orthonormal basis $(\xi_m^{(1)}, \ldots, \xi_m^{(d^-)})$ in each E_{ω_m,x_m}^- such that $\xi_m^{(i)} \to \xi^{(i)} \in T_x M$, $i = 1, \ldots, d^-$.

By continuity of the map $(x, \omega) \mapsto \mathrm{d}_x\varphi_{t,\omega}$ (see Theorem 1.1) the inequalities (6.10) for $\xi = \xi_m^{(i)}$ imply that

$$|\mathrm{d}_x\varphi_{t,\omega}(\xi^{(i)})| \le c^{-1}\mathrm{e}^{-\lambda t}|\xi^{(i)}| = c^{-1}\mathrm{e}^{-\lambda t}.$$

It follows that $(\xi^{(1)}, \ldots, \xi^{(d^-)})$ is an orthonormal system in $E_{\omega,x}^-$ and hence $\dim E_{\omega,x}^- \ge d^-$. Similarly, vectors $v \in E_{\omega,x}^+$ are characterized by the inequalities

$$|\mathrm{d}_x\varphi_{t,\omega}(v)| \ge c\mathrm{e}^{\lambda t}|v| \quad \text{for all } t \ge 0. \tag{6.11}$$

Indeed, substituting $\tilde{v} = \mathrm{d}_x\varphi_{t,\omega}(v)$ gives $v = \mathrm{d}_{\varphi(t,x,\omega)}\varphi_{-t,\Theta_t\omega}(\tilde{v})$ and hence we find that the inequalities

$$|\mathrm{d}_{\varphi(t,x,\omega)}\varphi_{-t,\Theta_t\omega}(\tilde{v})| \le c^{-1}\mathrm{e}^{-\lambda t}|\tilde{v}|$$

---

[4]It is sufficient to consider subsequences, because if the continuity statement would not hold, there would be a sequence $(\omega_m, x_m) \to (\omega, x)$ such that for any subsequence (ω_{n_m}, x_{n_m}) the convergence statement $E_{\omega_{n_m},x_{n_m}}^\pm \to E_{\omega,x}^\pm$ fails.

hold for all $(\omega, x) \in \mathcal{Q}$, $t \geq 0$, and $\tilde{v} \in E^+_{\Phi_t(\omega, x)}$. Another substitution shows that in fact for all $(\omega, x) \in \mathcal{Q}$ and $v \in E^+_{\omega,x}$ we have

$$|\mathrm{d}_x \varphi_{-t,\omega}(v)| \leq c^{-1} \mathrm{e}^{-\lambda t} |v| \quad \text{for all } t \geq 0.$$

Here we used that $\Phi_t(\mathcal{Q}) = \mathcal{Q}$ for all $t \in \mathbb{R}$. Similarly, it is shown that on the subspaces $E^-_{\omega,x}$ we have an expansion in backward time. With the same arguments as above this implies that the subspace $E^+_{\omega,x}$ is characterized by (6.11), and it follows that $\dim E^+_{\omega,x} \geq d^+ := \dim M - d^-$. Since $E^-_{\omega,x} \oplus E^+_{\omega,x} = T_x M$, this implies $\dim E^\pm_{\omega,x} = d^\pm$ which shows that the dimensions are locally constant. Since we can find orthonormal bases of the spaces $E^\pm_{\omega,x}$, depending continuously on (ω, x), the continuity statement follows.

(iii) First notice that any chain control set is controlled invariant in forward and backward time by Definition 1.15. Proposition 1.24 (iv) shows that the full time lift $\mathcal{Q}$ is a maximal invariant chain transitive set for the control flow. Then Proposition B.1 guarantees that $\mathcal{Q}$ is connected and hence the assertion follows from statement (ii).

(iv) This is an immediate consequence of item (i) and continuity of $(\omega, x) \mapsto E^\pm_{\omega,x}$.

 (v) For the proof that $E^- \to \mathcal{Q}$ is a subbundle we only have to show that the set $\bigcup_{(\omega,x) \in \mathcal{Q}} \{\omega\} \times E^-_{\omega,x}$ is closed in $E := \bigcup_{(\omega,x) \in \mathcal{Q}} \{\omega\} \times T_x M$. Therefore, consider a sequence (ω_n, v_n) with $(\omega_n, \pi_{TM}(v_n)) \in \mathcal{Q}$ and $v_n \in E^-_{\omega_n, \pi_{TM}(v_n)}$, which converges to some (ω, v) with $(\omega, \pi_{TM}(v)) \in \mathcal{Q}$. Then we have to show that $v \in E^-_{\omega,x}$, $x := \pi_{TM}(v)$. But this immediately follows from continuity of $E^-_{(\cdot,\cdot)}$. It is obvious that the same reasoning works for E^+ and that $E^- \oplus E^+ = E$. $\square$

The Bowen–Ruelle–Liu Volume Lemma

We further need one major result about uniformly hyperbolic sets of control-affine systems, namely the *Bowen–Ruelle volume lemma*. This result, which gives an estimate of the volumes of Bowen-balls, is well-known in the context of classical hyperbolic systems (diffeomorphisms or flows). A nonautonomous version of the lemma was given by Liu [78, Lemma 3.3] for random dynamical systems which arise by small random perturbations of Axiom A diffeomorphisms.[5] However, by inspecting Liu's proof, one sees that his arguments can also be applied in a deterministic context, for a hyperbolic set of a discrete-time skew-product

---

[5]This proof is mainly based on arguments applied in Qian and Zhang [92] to prove a Bowen–Ruelle volume lemma for hyperbolic endomorphisms.

system with compact base space.[6] Given a control-affine system with a uniformly hyperbolic set, one obtains such a skew product system by time-discretization of the control flow, and one finds that this system satisfies all assumptions necessary for the application of the volume lemma.

Lemma 6.5 (Bowen–Ruelle–Liu Volume Lemma). *Assume that the vector fields* $f_0, f_1, \ldots, f_m$ *of the control-affine system are of class* $\mathscr{C}^2$ *and let* $Q \subset M$ *be a compact set, controlled invariant in forward and backward time, which is uniformly hyperbolic. Then for every sufficiently small* $\varepsilon > 0$ *there is a constant* $C_\varepsilon \geq 1$ *such that*

$$\mathrm{vol}\left(B_\omega^\tau(x, \varepsilon)\right) \leq C_\varepsilon \left|\det \mathrm{d}_x \varphi_{\tau,\omega} : E_{\omega,x}^+ \to E_{\Phi_\tau(\omega,x)}^+\right|^{-1}$$

holds for all $\tau \geq 0$ *and* $(\omega, x) \in \mathcal{Q}$.

Proof. First of all, note that by Theorem 1.1 the assumption that the vector fields $f_0, f_1, \ldots, f_m$ are of class $\mathscr{C}^2$ guarantees that the first and second derivatives of φ with respect to the state variable exist and are continuous as functions of (t, x, ω). This implies that for any time-discretization of the control flow the global bounds and Lipschitz constants as defined in (A.1)–(A.5) of the proof of [78, Lemma 3.3] are finite and the hence rest of the proof works. To be more precise, let us discretize the control flow by setting

$$\Phi_k^D(\omega, x) := (\Theta_{\tau_0 k}\omega, \varphi(k\tau_0, x, \omega)) =: (\Theta_k^D \omega, \varphi^D(k, x, \omega))$$

to obtain a discrete-time skew product system $\Phi^D : \mathbb{Z} \times (\mathscr{U} \times M) \to \mathscr{U} \times M$. Here $\tau_0 > 0$ is a time step which has to be chosen sufficiently large so that we find a number $\mu \in (e^{-\lambda\tau_0}, 1)$ with

$$c^{-1}e^{-\lambda(k\tau_0)} \leq \mu^k \quad \text{for all } k \geq 1.$$

Then it is clear that $\mathcal{Q}$ is a hyperbolic set for Φ^D. In particular, it holds that

$$\left|\mathrm{d}_x \varphi_{k,\omega}^D(v)\right| \leq \mu^k |v| \quad \text{for all } k \geq 0, \ (\omega, x) \in \mathcal{Q}, \ v \in E_{\omega,x}^-,$$

and

$$\left|\mathrm{d}_x \varphi_{k,\omega}^D(v)\right| \geq \mu^{-k} |v| \quad \text{for all } k \geq 0, \ (\omega, x) \in \mathcal{Q}, \ v \in E_{\omega,x}^+.$$

---

[6]A *skew product system* is a dynamical system of the form $\Phi : \mathbb{T} \times (B \times X) \to B \times X$, $\Phi(t, (b, x)) = (\theta_t(b), \varphi(t, b, x))$, where θ is a dynamical system on the base space B and φ a cocycle over θ, that is, $\varphi(t + s, b, x) \equiv \varphi(t, \theta_s(b), \varphi(s, b, x))$.

This allows to apply the arguments of Liu's proof. Precisely, for the volumes of the Bowen-balls

$$B_\omega^k(x, \varepsilon; \varphi^D) := \left\{ y \in M \ : \ \max_{0 \le i \le k-1} \varrho\left(\varphi^D(i, x, \omega), \varphi^D(i, y, \omega)\right) < \varepsilon \right\}$$

we find that for $\varepsilon > 0$ sufficiently small there is a constant $\tilde{C}_\varepsilon \ge 1$ with

$$\tilde{C}_\varepsilon^{-1} \le \mathrm{vol}(B_\omega^k(x, \varepsilon; \varphi^D))a_{k\tau_0}(\omega, x) \le \tilde{C}_\varepsilon$$

for all $k \ge 0$ and $(\omega, x) \in \mathcal{Q}$, where $a_t(\omega, x) := |\det \mathrm{d}_x \varphi_{t,\omega} : E_{\omega,x}^+ \to E_{\Phi_t(\omega,x)}^+|$. (Note that a is a multiplicative cocycle over the control flow.) Writing each $t \ge 0$ as $t = k_t \tau_0 + r_t$ with $k_t \in \mathbb{Z}_+$ and $r_t \in [0, \tau_0)$, we find that

$$a_t(\omega, x) = a_{r_t}(\Phi_{k_t \tau_0}(\omega, x))a_{k_t \tau_0}(\omega, x) \le \underbrace{\left[\max_{(r,(\omega,x)) \in [0,\tau_0] \times \mathcal{Q}} a_r(\omega, x) \right]}_{=:\alpha} \cdot a_{k_t \tau_0}(\omega, x).$$

Since $B_\tau^\omega(x, \varepsilon) \subset B_{k_\tau}^\omega(x, \varepsilon; \varphi^D)$, we have

$$\mathrm{vol}(B_\tau^\omega(x, \varepsilon))a_\tau(\omega, x) \le \mathrm{vol}(B_{k_\tau}^\omega(x, \varepsilon; \varphi^D))\alpha a_{k_\tau \tau_0}(\omega, x) \le (\alpha\tilde{C}_\varepsilon) =: C_\varepsilon$$

for all $\tau \ge 0$ and $(\omega, x) \in \mathcal{Q}$, which concludes the proof. $\qquad\qquad\square$

The Lower Bound Theorem

In order to formulate the lower bound theorem for the invariance entropy of a uniformly hyperbolic set, we need to introduce another version of escape entropy. We abstain from giving this quantity a name.

Definition 6.3. Let $Q \subset M$ be a compact and controlled invariant set and $K \subset Q$ compact. Then (K, Q) is an admissible pair and we define

$$\hat{h}_{\mathrm{esc}}(K, Q) := \lim_{\varepsilon \searrow 0} \limsup_{\tau \to \infty} \frac{1}{\tau} \log \left(\sup_{\omega \in \pi_{\mathcal{U}} \mathcal{K}_Q} r_{\mathrm{sep}}(\omega, \tau, \varepsilon, K, Q) \right).$$

The quantity $\hat{h}_{\mathrm{esc}}(K, Q)$ is better behaved than $h_{\mathrm{esc}}(K, Q)$, as the following proposition shows.

Proposition 6.2. *The following assertions hold:*

(i) $\hat{h}_{\mathrm{esc}}(K, Q) \in [0, \infty)$.
(ii) It holds that

$$\hat{h}_{\mathrm{esc}}(K, Q) = \lim_{\varepsilon \searrow 0} \limsup_{\tau \to \infty} \frac{1}{\tau} \log \left(\sup_{\omega \in \pi_{\mathcal{U}} \mathcal{K}_Q} r_{\mathrm{span}}(\omega, \tau, \varepsilon, K, Q) \right).$$

(iii) $\hat{h}_{\mathrm{esc}}(K, Q)$ is invariant with respect to $\mathscr{C}^0$-state equivalence and hence metric-independent.

Proof. (i) Since $r_{\mathrm{sep}}(\omega, \tau, \varepsilon, K, Q) \geq 1$ if $K(\omega, \tau) \neq \emptyset$, and for each $x \in K$ there is $\omega \in \pi_{\mathcal{U}} \mathcal{Q}$ with $(\omega, x) \in \mathcal{K}_Q$, and hence $K(\omega, \tau) \neq \emptyset$ for all $\tau > 0$, we find

$$\frac{1}{\tau} \log \left(\sup_{\omega \in \pi_{\mathcal{U}} \mathcal{K}_Q} r_{\mathrm{sep}}(\omega, \tau, \varepsilon, K, Q) \right) \geq 0,$$

implying $\hat{h}_{\mathrm{esc}}(K, Q) \geq 0$. Finiteness of $\hat{h}_{\mathrm{esc}}(K, Q)$ is proved with the same arguments that are used to show finiteness of $h_{\mathrm{esc}}(K, Q)$.

(ii) From Proposition 6.1 we know that

$$r_{\mathrm{span}}(\omega, \tau, \varepsilon, K, Q) \leq r_{\mathrm{sep}}(\omega, \tau, \varepsilon, K, Q) \leq r_{\mathrm{span}}\left(\omega, \tau, \frac{\varepsilon}{2}, K, Q\right).$$

This immediately implies the assertion.

(iii) Consider another control-affine system of the form $\Sigma' = (\mathbb{R}, N, \mathbb{R}^m, \mathcal{U}, \psi)$. Further suppose that $h : M \to N$ is a homeomorphism which relates the transition maps φ and ψ, that is, $h(\varphi(t, x, \omega)) \equiv \psi(t, h(x), \omega)$. Then (K', Q') with $K' := h(K)$ and $Q' := h(Q)$ is an admissible pair for Σ' such that Q' is compact and controlled invariant. If ϱ' is a metric on N, then the restriction of h to a δ_0-neighborhood $N_{\delta_0}(Q)$ of Q is uniformly continuous for sufficiently small $\delta_0 > 0$. Hence, for given $\varepsilon > 0$ we find $\delta \in (0, \delta_0)$ such that for every minimal (ω, τ, δ)-spanning set E of $K(\omega, \tau)$, we have that $E' := h(E)$ is a $(\omega, \tau, \varepsilon)$-spanning set for $K'(\omega, \tau) = h(K(\omega, \tau))$. This implies the assertion. $\square$

Using Lemma 6.5, we are able to prove the following result.

Theorem 6.2. *Assume that the vector fields $f_0, f_1, \ldots, f_m$ are of class $\mathscr{C}^2$ and let Q be a compact set, controlled invariant in forward and backward time, which is uniformly hyperbolic and satisfies $h_{\mathrm{inv}}(Q) < \infty$. Then for each compact set $K \subset Q$ of positive volume we have*

$$h_{\mathrm{inv}}(K, Q) \geq \limsup_{\tau \to \infty} \inf_{(\omega, x) \in \mathcal{K}_Q} \frac{1}{\tau} \int_0^\tau \mathrm{tr}\left[\nabla F_{\omega(s)}(\varphi_{s,\omega}(x)) \circ Q(\Phi_s(\omega, x)) \right] ds$$

$$- \hat{h}_{\mathrm{esc}}(K, Q), \tag{6.12}$$

where $Q(\omega, x) : T_x M \to E^+_{\omega,x}$ denotes the orthogonal projection. In the case $K = Q$, the limit superior and the infimum can be interchanged.

Proof. Choose ε small enough according to Lemma 6.5. For fixed $\tau > 0$ let $\mathscr{S} \subset \mathscr{U}$ be a minimal (τ, K, Q)-spanning set. For each $\omega \in \pi_{\mathscr{U}} \mathscr{K}_Q$ choose a maximal $(\omega, \tau, \varepsilon)$-separated subset $E_{\omega, \tau, \varepsilon}$ of $K(\omega, \tau)$. Then $E_{\omega, \tau, \varepsilon}$ is also a $(\omega, \tau, \varepsilon)$-spanning set, and hence $K(\omega, \tau) \subset \bigcup_{x \in E_{\omega, \tau, \varepsilon}} B_{\omega}^{\tau}(x, \varepsilon)$, which implies

$$\mathrm{vol}(K(\omega, \tau)) \leq \sum_{x \in E_{\omega, \tau, \varepsilon}} \mathrm{vol}(B_{\omega}^{\tau}(x, \varepsilon))$$

$$\leq \#E_{\omega, \tau, \varepsilon} \cdot C_{\varepsilon} \cdot \sup_{x \in K(\omega, \tau)} \left| \det d_x \varphi_{\tau, \omega} : E_{\omega, x}^+ \to E_{\Phi_\tau(\omega, x)}^+ \right|^{-1}.$$

Since $\mathrm{vol}(K) \leq r_{\mathrm{inv}}(\tau, K, Q) \max_{\omega \in \mathscr{S}} \mathrm{vol}(K(\omega, \tau))$, we obtain

$$r_{\mathrm{inv}}(\tau, K, Q)$$

$$\geq \mathrm{vol}(K) \cdot \min_{\omega \in \mathscr{S}} \left((\#E_{\omega, \tau, \varepsilon})^{-1} \cdot C_{\varepsilon}^{-1} \cdot \inf_{x \in K(\omega, \tau)} \left| \det d_x \varphi_{\tau, \omega} \big|_{E_{\omega, x}^+} \right| \right)$$

$$\geq \mathrm{const} \cdot \left(\inf_{\omega \in \pi_{\mathscr{U}} \mathscr{K}_Q} (\#E_{\omega, \tau, \varepsilon})^{-1} \right) \cdot \left(\inf_{(\omega, x) \in \mathscr{K}_Q} \left| \det d_x \varphi_{\tau, \omega} \big|_{E_{\omega, x}^+} \right| \right).$$

Since ε does not depend on τ and $\#E_{\omega, \tau, \varepsilon} = r_{\mathrm{sep}}(\omega, \tau, \varepsilon, K, Q)$, we can conclude that $h_{\mathrm{inv}}(K, Q)$ is bounded from below by

$$\limsup_{\tau \to \infty} \frac{1}{\tau} \log \left[\left(\inf_{\omega \in \pi_{\mathscr{U}} \mathscr{K}_Q} (\#E_{\omega, \tau, \varepsilon})^{-1} \right) \cdot \left(\inf_{(\omega, x) \in \mathscr{K}_Q} \left| \det d_x \varphi_{\tau, \omega} \big|_{E_{\omega, x}^+} \right| \right) \right]$$

$$\geq \limsup_{\tau \to \infty} \inf_{(\omega, x) \in \mathscr{K}_Q} \frac{1}{\tau} \log \left| \det d_x \varphi_{\tau, \omega} \big|_{E_{\omega, x}^+} \right|$$

$$- \limsup_{\tau \to \infty} \frac{1}{\tau} \log \sup_{\omega \in \pi_{\mathscr{U}} \mathscr{K}_Q} r_{\mathrm{sep}}(\omega, \tau, \varepsilon, K, Q).$$

Now, taking ε arbitrarily small and applying the generalized Liouville formula (Proposition A.5), inequality (6.12) follows. The function $(\tau, (\omega, x)) \mapsto \log | \det d_x \varphi_{\tau, \omega} |_{E_{\omega, x}^+} |$ is a continuous additive cocycle over the control flow. Indeed, continuity follows from continuity of $d_x \varphi_{\tau, \omega}$ (see Theorem 1.1) and continuity of $(\omega, x) \mapsto E_{\omega, x}^+$. The validity of the cocycle equation follows from the cocycle property of φ, the chain rule, invariance of $E \to \mathscr{Q}$, and the product rule for determinants. Moreover, $\mathscr{Q}$ is a compact invariant set for the control flow (see Lemma 6.4 (i)). Hence, by Theorem B.2, we can interchange the limit superior with the infimum in the case that $K = Q$. $\qquad\square$

Remark 6.3. There should be no difficulty in transferring the proofs of this chapter to the discrete-time setting. However, in view of Lemma 6.2, it is clear that one has to assume that the time-t-maps of the system are locally invertible in order to obtain a discrete-time version of Theorem 6.1. Also note that the volume lemma of Gelfert and Franz (Lemma 6.3) has been proved by Gelfert for discrete- and continuous-time classical dynamical systems simultaneously, and the Bowen–Ruelle volume lemma has been proved by Liu for discrete-time systems anyway.

6.4 Comments and Bibliographical Notes

The Lemma 6.1 on coverings of ellipsoids by metric balls can be found in Douady and Oesterlé [38, Lemme 1] or Temam [108, Chap. V, Lemma 3.1]. The results of Sect. 6.2 have appeared before in Kawan [65]. Escape rates for classical dynamical systems have been studied, for instance, in [35, 46, 115]. There are many open questions about the escape entropies $h_{\mathrm{esc}}(K, Q)$ and $\hat{h}_{\mathrm{esc}}(K, Q)$. In particular, one is interested in criteria which guarantee that these numbers are less than or equal to zero, since in this case the invariance entropy is bounded below by the volume growth rate on the unstable bundle. In [65] it is shown that $h_{\mathrm{esc}}(K, Q) \leq 0$ for uniformly expanding systems and for inhomogeneous bilinear systems under a weak hyperbolicity condition (see also Propositions 7.4 and 7.5). In other cases, no results in this direction have been obtained so far. Hyperbolic sets for skew-product systems are also studied in the context of random dynamical systems, where they are called *random hyperbolic sets*. We refer to Gundlach and Kifer [55] for an excellent survey paper on this topic. In a deterministic context, hyperbolic sets of the above type have been considered in Meyer and Sell [81] (for discrete-time systems), where they are called *skew hyperbolic sets*. Similar ideas can be found in Stoffer [104]. Furthermore, hyperbolic control and chain control sets for control-affine systems have been considered in Colonius and Du [21]. The original version of the Bowen–Ruelle volume lemma, which plays an important role in the ergodic theory of Axiom A systems, first appeared in Bowen and Ruelle [14, Lemma 4.2] for systems of class $\mathscr{C}^2$. A slightly different version with weaker differentiability assumptions can be found in Fried and Shub [45].

Chapter 7
Examples

In this chapter, the theory developed so far is applied to specific classes of systems. In particular, we obtain formulas and/or estimates for the invariance entropy of control sets of scalar control-affine systems, uniformly expanding systems, inhomogeneous bilinear systems given by differential equations, and projective systems (which are control-affine systems on real projective space $\mathbb{P}^d$ induced by bilinear systems on $\mathbb{R}^{d+1}$).

7.1 One-Dimensional Control-Affine Systems

Consider a scalar control-affine system $\Sigma = (\mathbb{R}, \mathbb{R}, \mathbb{R}^m, \mathcal{U}, \varphi)$ given by differential equations with right-hand side $F(x, u) = f_0(x) + \sum_{i=1}^m u_i f_i(x)$, whose compact and convex control range $\Omega \subset \mathbb{R}^m$ satisfies $0 \in \mathrm{int}\,\Omega$.

By Sect. 1.4 these assumptions guarantee that $\mathcal{U}$ becomes a compact metrizable space with the weak*-topology of $L^\infty(\mathbb{R}, \mathbb{R}^m) = L^1(\mathbb{R}, \mathbb{R}^m)^*$ and that the associated control flow

$$\Phi(t, (\omega, x)) = (\Theta_t \omega, \varphi(t, x, \omega)), \quad \Phi : \mathbb{R} \times (\mathcal{U} \times \mathbb{R}) \to \mathcal{U} \times \mathbb{R},$$

is continuous. If D is a control set of Σ, then, by Proposition 1.11, $Q := \mathrm{cl}\,D$ is controlled invariant. If such Q is compact, then, Proposition 1.10 guarantees that $\mathcal{Q} = \{(\omega, x) \in \mathcal{U} \times \mathbb{R} : \varphi(\mathbb{R}_+, x, \omega) \subset Q\}$ is a compact forward-invariant set for Φ.

Before we state the main results of this section, we collect some facts about one-dimensional continuous-time systems which can be found in Colonius and Kliemann [25, Chap. 8]. First, we define the Lyapunov spectrum of Σ over a set $Q \subset \mathbb{R}$.

C. Kawan, *Invariance Entropy for Deterministic Control Systems*, Lecture Notes in Mathematics 2089, DOI 10.1007/978-3-319-01288-9_7,
© Springer International Publishing Switzerland 2013

Definition 7.1. For $(\omega, x) \in \mathcal{U} \times \mathbb{R}$, the *Lyapunov exponent* $\lambda(\omega, x)$ is defined by

$$\lambda(\omega, x) := \limsup_{t \to \infty} \frac{1}{t} \log |D\varphi_{t,\omega}(x)|.$$

For a compact controlled invariant subset $Q \subset \mathbb{R}$, the *Lyapunov spectrum* over Q is given by

$$\mathrm{Sp}_{\mathrm{Ly}}(Q) = \{\lambda(\omega, x) \, : \, (\omega, x) \in \mathcal{Q}\}.$$

The statement of the next proposition is part of Theorem 8.1.1 in Colonius and Kliemann [25].

Proposition 7.1. *A point $x \in \mathbb{R}$ is in the interior of some control set if and only if there are $u_\pm \in \Omega$ with $F(x, u_+) > 0$ and $F(x, u_-) < 0$.*

Proof. If x is an element of the interior of a control set, then approximate controllability implies that one can reach points which are on the left and points which are on the right of x. Hence, there must be $u_\pm$ as in the assertion. On the other hand, if there are such $u_\pm$, we can consider the maximal connected set A with $x \in A$ such that for every $y \in A$ there are $u_\pm(y)$ with $F(y, u_+(y)) > 0$ and $F(y, u_-(y)) < 0$. By continuity, A is open. Since Ω is connected, the intermediate value theorem implies the existence of $u_0(y) \in \Omega$ with $F(y, u_0(y)) = 0$ for every $y \in A$. Hence, A is controlled invariant. Taking $y_1, y_2 \in A$ and assuming without loss of generality that $y_1 < y_2$, there are $u_+(y) \in \Omega$ such that $F(y, u_+(y)) \geq \alpha$ for all $y \in [y_1, y_2]$ and some $\alpha > 0$. This easily implies that one can get from y_1 to y_2 using a piecewise constant control function. Hence, the system is controllable on A which implies the existence of a control set $D \supset A$. $\qquad\square$

The following proposition is part of Theorem 8.1.2 in Colonius and Kliemann [25]. We omit its proof.

Proposition 7.2. *Let $D \subset \mathbb{R}$ be a control set with nonempty interior such that* cl D *is a compact chain control set. Assume that for every $y \in$ cl D with $F(y, u) = 0$ for some $u \in \Omega$ there is a sequence $(y_n, u_n) \in$ int $D \times \Omega$ with $F(y_n, u_n) = 0$ for all $n \geq 1$ and $(y_n, u_n) \to (y, u)$ for $n \to \infty$. Then the Lyapunov spectrum over* cl D *is given by*

$$\mathrm{Sp}_{\mathrm{Ly}}(\mathrm{cl}\, D) = \left[\inf_{y \in \mathrm{cl}\, D} \min_{u:\, F(y,u)=0} \frac{\partial F}{\partial x}(y, u), \; \sup_{y \in \mathrm{cl}\, D} \max_{u:\, F(y,u)=0} \frac{\partial F}{\partial x} F(y, u) \right].$$

Theorem 7.1. *Let $D \subset \mathbb{R}$ be a bounded control set of Σ with nonempty interior such that $Q := \mathrm{cl}\, D$ is a chain control set. Further assume that for each $(y, u) \in Q \times \Omega$ with $F(y, u) = 0$ there exists a sequence $(y_n, u_n) \in$ int $D \times$ int Ω with $F(y_n, u_n) = 0$ for all $n \geq 1$ and $(y_n, u_n) \to (y, u)$. Then for every compact set $K \subset D$ with nonempty interior it holds that*

$$h_{\mathrm{inv}}(K, Q) = \max\left\{0, \inf \mathrm{Sp}_{\mathrm{Ly}}(Q)\right\}.$$

Proof. From Corollary 4.4 (iii) and the Liouville formula it follows that

$$h_{\mathrm{inv}}(K, Q) \geq \inf_{(\omega, x) \in \mathscr{Q}} \limsup_{\tau \to \infty} \frac{1}{\tau} \log |D\varphi_{\tau, \omega}(x)|$$

which implies

$$h_{\mathrm{inv}}(K, Q) \geq \max\left\{0, \inf_{(\omega, x) \in \mathscr{Q}} \lambda(\omega, x)\right\} = \max\left\{0, \inf \mathrm{Sp}_{\mathrm{Ly}}(Q)\right\}.$$

To prove the upper bound, let $(u^*, x^*) \in \mathrm{int}\,\Omega \times \mathrm{int}\,D$ be an arbitrary equilibrium pair. We show that (u^*, x^*) is regular and hence Corollary 5.2 can be applied. Indeed, consider the derivatives

$$A := \frac{\partial F}{\partial x}(x^*, u^*) \in \mathbb{R}^{1 \times 1}, \quad B := \frac{\partial F}{\partial u}(x^*, u^*) \in \mathbb{R}^{1 \times m}.$$

Regularity of (u^*, x^*) is (by the Kalman rank condition) equivalent to controllability of the pair (A, B) which is equivalent to B having full rank, or equivalently, to $B \neq 0$, which follows by definition. If we assume that $B = (f_1(x^*), \ldots, f_m(x^*)) = 0$, then $F(x^*, u^*) = f_0(x^*) = 0$ follows. But this implies $\mathcal{O}^+(x^*) = \{x^*\}$ and hence contradicts approximate controllability on D. Since (u^*, x^*) was chosen arbitrarily, we obtain

$$\ ^{.}h_{\mathrm{inv}}(K, Q) \leq \max\left\{0, \inf \lambda(u, x)\right\}, \tag{7.1}$$

where the infimum runs over all $(u, x) \in \mathrm{int}\,\Omega \times \mathrm{int}\,D$ with $F(x, u) = 0$. By Proposition 7.2 the minimum of the Lyapunov spectrum $\mathrm{Sp}_{\mathrm{Ly}}(Q)$ is given by

$$\inf_{y \in Q} \min_{u \in \Omega:\ F(y, u) = 0} \frac{\partial F}{\partial x}(y, u) = \inf_{(x, u) \in (Q \times \Omega) \cap F^{-1}(0)} \lambda(u, x).$$

By continuous differentiability of F, the assumption about the equilibria pairs, and (7.1), the inequality $h_{\mathrm{inv}}(K, Q) \leq \max\{0, \inf \mathrm{Sp}_{\mathrm{Ly}}(Q)\}$ follows. $\qquad\square$

If the system Σ has only one control vector field, we can give a formula in terms of the right-hand side vector fields.

Theorem 7.2. *Assume that the right-hand side of system Σ has the form $F(x, u) = f_0(x) + u f_1(x)$. Let $D \subset \mathbb{R}$ be a bounded control set of Σ with nonempty interior. Moreover, assume that Σ is locally accessible on $Q := \mathrm{cl}\,D$. Then for every compact set $K \subset D$ with nonempty interior it holds that*

$$h_{\mathrm{inv}}(K, Q) = h_{\mathrm{inv,out}}(K, Q) = \max\left\{0, \min_{x \in Q}\left[f_0'(x) - \frac{f_1'(x)}{f_1(x)} f_0(x)\right]\right\}. \tag{7.2}$$

Proof. The proof proceeds in three steps.

Step 1. From approximate controllability it follows that D is connected. Thus, Q is a compact interval. In order to show that formula (7.2) makes sense, we have to prove that $f_1(x) \neq 0$ for all $x \in Q$: Assume to the contrary that $f_1(x^*) = 0$ for some $x^* \in Q$. From Proposition 7.1 and the intermediate value theorem it follows that for every $x \in Q$ there exists $u_x \in \Omega$ with $f_0(x) + u_x f_1(x) = 0$. Hence, $f_0(x^*) = 0$, which implies $\varphi(t, x^*, \omega) = x^*$ for all $t \in \mathbb{R}$ and $\omega \in \mathcal{U}$ and therefore contradicts local accessibility on Q.

Step 2. Now we prove the lower bound for $h_{\mathrm{inv,out}}(K, Q)$, using the estimate (4.12). To this end, define a $\mathscr{C}^1$-function on a neighborhood of Q by $\alpha(x) := -\ln |f_1(x)|$. Since $f_1(x) \neq 0$ on Q implies $f_1(x) \neq 0$ on a neighborhood of Q, the definition of α is correct. Because

$$(f_0 + u f_1)\alpha(x) = -\frac{f_1'(x)}{f_1(x)} f_0(x) - u f_1'(x),$$

we obtain

$$h_{\mathrm{inv}}(K, N_\varepsilon(Q)) \geq \max \left\{ 0, \inf_{x \in N_\varepsilon(Q)} \left[f_0'(x) - \frac{f_1'(x)}{f_1(x)} f_0(x) \right] \right\}.$$

Letting ε go to zero, the desired estimate follows.

Step 3. We prove the upper bound for $h_{\mathrm{inv}}(K, Q)$, using Corollary 5.2. To this end, let $x \in \mathrm{int}\, D$. Then, by Proposition 7.1, there exist $u_\pm \in \Omega$ such that $f_0(x) + u_- f_1(x) < 0$ and $f_0(x) + u_+ f_1(x) > 0$. Since Ω is connected, the intermediate value theorem gives $f_0(x) + u_x f_1(x) = 0$ for some u_x which lies in the interior of the interval with endpoints u_- and u_+, and hence $u_x \in \mathrm{int}\, \Omega$. Since $f_1(x) \neq 0$, u_x is unique, namely $u_x = -f_0(x)/f_1(x)$. The linearization of Σ at the equilibrium pair (u_x, x) is controllable which in this case is equivalent to $f_1(x) \neq 0$ by the Kalman rank condition. Corollary 5.2 yields

$$h_{\mathrm{inv}}(K, Q) \leq \max \left\{ 0, f_0'(x) + u_x f_1'(x) \right\} = \max \left\{ 0, f_0'(x) - \frac{f_0(x)}{f_1(x)} f_1'(x) \right\}.$$

The point x was chosen arbitrarily in $\mathrm{int}\, D$ and thus we get

$$
\begin{aligned}
h_{\mathrm{inv}}(K, Q) &\leq \inf_{x \in \mathrm{int}\, D} \max \left\{ 0, f_0'(x) - \frac{f_0(x)}{f_1(x)} f_1'(x) \right\} \\
&= \min_{x \in Q} \max \left\{ 0, f_0'(x) - \frac{f_0(x)}{f_1(x)} f_1'(x) \right\} \\
&= \max \left\{ 0, \min_{x \in Q} \left[f_0'(x) - \frac{f_1'(x)}{f_1(x)} f_0(x) \right] \right\}.
\end{aligned}
$$

Interchanging the minimum and the maximum here is possible, since for every continuous function $g : \mathbb{R} \to \mathbb{R}$ the inequality $\min_{x \in Q} \max\{0, g(x)\} \geq \max\{0, \min_{x \in Q} g(x)\}$ trivially holds, and on the other hand compactness of Q gives $x^* \in Q$ with

$$\max\{0, \min_{x \in Q} g(x)\} = \max\{0, g(x^*)\} \geq \min_{x \in Q} \max\{0, g(x)\}.$$

Since $h_{\text{int,out}}(K, Q) \leq h_{\text{inv}}(K, Q)$, the assertion of the theorem follows. $\square$

Remark 7.1. If Q is a chain control set in the above theorem, the expression

$$\min_{x \in Q} \left[f_0'(x) - \frac{f_1'(x)}{f_1(x)} f_0(x) \right]$$

coincides with $\inf \text{Sp}_{\text{Ly}}(Q)$. This follows from Proposition 7.2 and the fact that for an equilibrium pair (u_x, x), $u_x = -f_0(x)/f_1(x)$, one has

$$\lambda(u_x, x) = \frac{\partial F}{\partial x}(x, u_x) = f_0'(x) - \frac{f_1'(x)}{f_1(x)} f_0(x).$$

Example 7.1. Consider a planar bilinear system $\Sigma = (\mathbb{R}, \mathbb{R}^2, \mathbb{R}, \mathcal{U}, \varphi)$ given by differential equations

$$\dot{x}(t) = (A_0 + \omega(t) A_1) x(t), \quad \omega \in \mathcal{U}. \tag{7.3}$$

Let $A_0 = (a_{ij}^0)$, $A_1 = (a_{ij}^1)$, and $A(u) = A_0 + u A_1$. Consider the projection Σ' of Σ to the unit circle $S^1 \subset \mathbb{R}^2$, given by

$$\dot{s}(t) = (A(\omega(t)) - s(t)^T A(\omega(t)) s(t) I) s(t), \quad \omega \in \mathcal{U}. \tag{7.4}$$

Let $D \subset S^1$ be a control set of system Σ' with nonempty interior which is not the whole circle and assume that Σ' is locally accessible. We want to compute $h_{\text{inv}}(K, Q)$ for every compact set $K \subset D$ with nonempty interior. To this end, we describe system Σ' in polar coordinates. By writing $s(t) = (\cos \varphi(t), \sin \varphi(t))$, a simple calculation leads to the equations

$$\dot{\varphi}(t) = f_0(\varphi(t)) + \omega(t) f_1(\varphi(t)), \quad \omega \in \mathcal{U},$$

where $f_0, f_1 : [0, 2\pi) \to \mathbb{R}$ are given by

$$f_k(\varphi) = (a_{22}^k - a_{11}^k) \sin \varphi \cos \varphi - a_{12}^k \sin^2 \varphi + a_{21}^k \cos^2 \varphi, \quad k = 0, 1.$$

For the derivatives f_k' we get

$$f_k'(\varphi) = (a_{22}^k - a_{11}^k) \cos(2\varphi) - (a_{12}^k + a_{21}^k) \sin(2\varphi).$$

By Theorem 7.2 we obtain

$$h_{\mathrm{inv}}(K, Q) = \max\left\{0, \min_{\varphi \in Q}\left[f_0'(\varphi) - \frac{f_1'(\varphi)}{f_1(\varphi)}f_0(\varphi)\right]\right\}.$$

The next example provides an application of this formula.

Example 7.2. We consider the scalar second-order equations

$$\ddot{y}(t) + 2b\dot{y}(t) - (1 + \omega(t))y(t) = 0, \quad \omega \in \mathscr{U},$$

with $b > 0$ and

$$\mathscr{U} = \{\omega \in L^{\infty}(\mathbb{R}, \mathbb{R}) \; : \; \omega(t) \in [-\rho, \rho] \text{ a.e.}\},$$

where $0 < \rho < b^2 + 1$. These equations describe the linearization of a controlled damped mathematical pendulum at the unstable position, where the control acts as a reset force. The corresponding first-order system is the following bilinear system:

$$\dot{x}(t) = \underbrace{\begin{pmatrix} 0 & 1 \\ 1 & -2b \end{pmatrix}}_{=:A_0} x(t) + \omega(t)\underbrace{\begin{pmatrix} 0 & 0 \\ 1 & 0 \end{pmatrix}}_{=:A_1} x(t), \quad \omega \in \mathscr{U}.$$

The eigenvalues of the matrix A_0 are given by

$$\lambda_{\pm} = -b \pm \sqrt{b^2 + 1}.$$

Since $b > 0$, λ_- is negative and λ_+ is positive. Hence, the uncontrolled system $\dot{x} = A_0 x$ has one stable and one unstable direction. From the preceding example it follows that the projected system on S^1 is given by

$$\dot{\varphi} = (-2b\sin\varphi\cos\varphi - \sin^2\varphi + \cos^2\varphi) + \omega(t)\cos^2\varphi, \quad \omega \in \mathscr{U}.$$

From Proposition 7.1 it follows that the control sets on S^1 consist of equilibria. Hence, in order to determine these sets, we have to find the zeros of the right-hand side. To this end, we divide by $\cos^2\varphi$ (which is possible for $2\varphi \notin \{\pi, 3\pi\}$). This yields

$$\tan^2\varphi + 2b\tan\varphi - (1 + u) = 0 \quad \Leftrightarrow \quad \tan\varphi = -b \pm \sqrt{b^2 + 1 + u}.$$

Hence, we obtain the solutions

$$\varphi_{1,\pm} = \arctan\left(-b \pm \sqrt{b^2 + 1 + u}\right) \in \left(-\frac{\pi}{2}, \frac{\pi}{2}\right)$$

and, by π-periodicity of the tangent function,

$$\varphi_{2,\pm} = \pi + \arctan\left(-b \pm \sqrt{b^2 + 1 + u}\right) \in \left(\frac{\pi}{2}, \frac{3\pi}{2}\right).$$

The solutions are real numbers, since

$$b^2 + 1 + u \in \left[b^2 + 1 - \rho, b^2 + 1 + \rho\right] \subset (0, 2(b^2 + 1)).$$

Hence, in $(-\pi/2, \pi/2)$ we obtain the following two intervals of equilibria which are the closures of control sets:

$$Q_{1,-} = \left[\arctan\left(-b - \sqrt{b^2 + 1 + \rho}\right), \arctan\left(-b - \sqrt{b^2 + 1 - \rho}\right)\right],$$

$$Q_{1,+} = \left[\arctan\left(-b + \sqrt{b^2 + 1 - \rho}\right), \arctan\left(-b + \sqrt{b^2 + 1 + \rho}\right)\right],$$

and in $(\pi/2, (3\pi)/2)$ we obtain the sets $Q_{2,\pm} = \pi + Q_{1,\pm}$. Applying the result from the preceding example we can compute the invariance entropy of these control sets. An elementary computation gives

$$h_{\mathrm{inv}}(K, Q_{i,\pm}) = \max\left\{0, \min_{\varphi \in Q_{i,\pm}} (-2b - 2\tan\varphi)\right\}, \quad i = 1, 2.$$

Hence, we obtain

$$h_{\mathrm{inv}}(K, Q_{i,-}) = \max\left\{0, \min_{u \in [-\rho, \rho]} \left(2\sqrt{b^2 + 1 - u}\right)\right\} = 2\sqrt{b^2 + 1 - \rho},$$

$$h_{\mathrm{inv}}(K, Q_{i,+}) = 0.$$

The sets $Q_{i,+}$ are easily seen to be invariant control sets, while the sets $Q_{i,-}$ are the closures of open, variant control sets. Figure 7.1 illustrates the situation. The set $Q_{1,-}$ contains the point $\varphi_0 = \arctan(-b - \sqrt{b^2 + 1})$ which is an equilibrium for the constant zero control function, that is, the vectors $\pm(\cos(\varphi_0), \sin(\varphi_0))$ are eigenvectors of A_0 corresponding to the stable eigenvalue $\lambda_- = -b - \sqrt{b^2 + 1}$. On int $Q_{1,-}$ and int $Q_{2,-}$ the projected system is controllable. This implies that from any point in the interior of the cone over $Q_{1,-}$ and $Q_{2,-}$, $C := \pi^{-1}(Q_{1,-} \cup Q_{2,-})$, where

$$\pi(x) = \frac{x}{|x|}, \quad \pi : \mathbb{R}^2 \setminus \{0\} \to S^1,$$

it is possible to steer to the stable axis, that is, the eigenspace of A_0 corresponding to λ_-. Hence, here it is possible to stabilize the system. It is easily seen that outside of C stabilization is not possible.

Fig. 7.1 The control sets on S^1

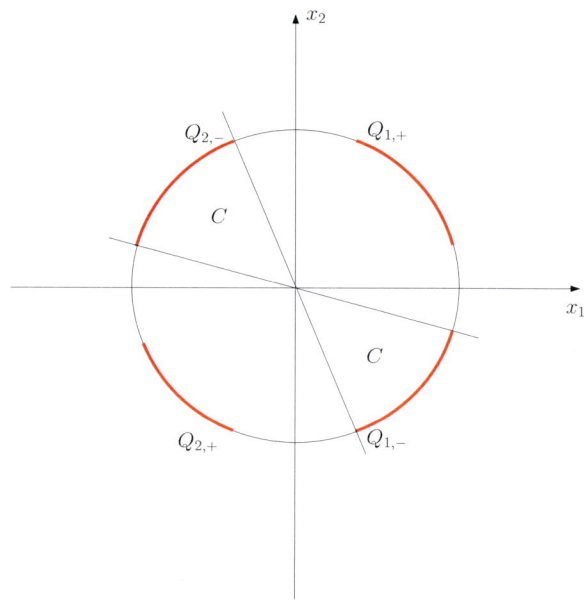

7.2 Uniformly Expanding Systems

In this section, we consider continuous-time topological systems $\Sigma = (\mathbb{R}, X, U, \mathscr{U}, \varphi)$ which have the property that the maps $\varphi_{t,\omega} : X \to X$ increase distances on a given subset Q of the state space as described precisely in the following definition.

Definition 7.2. Fix a metric ϱ on the state space X of a continuous-time topological time-invariant system $\Sigma = (\mathbb{R}, X, U, \mathscr{U}, \varphi)$, and let $Q \subset X$ be a nonempty set. We say that Σ is *uniformly expanding on Q (with respect to ϱ)* if there are constants $C, \lambda > 0$ such that

$$\varrho(\varphi(\tau, x_1, \omega), \varphi(\tau, x_2, \omega)) \geq C e^{\lambda \tau} \varrho(x_1, x_2)$$

whenever $x_1, x_2 \in Q$, $\omega \in \mathscr{U}$, and $\tau \geq 0$ with $\varphi(t, x_i, \omega) \in Q$ for all $t \in [0, \tau]$ and $i = 1, 2$. The constant λ is called an *expansion factor*.

Our first main result gives a lower bound for the invariance entropy of an admissible pair (K, Q), when Σ is uniformly expanding on Q.

Theorem 7.3. *Let (K, Q) be an admissible pair for Σ such that Q is bounded and Σ is uniformly expanding on Q with expansion factor λ. Then*

$$h_{\mathrm{inv}}(K, Q) \geq \lambda \cdot \overline{\dim}_C(K).$$

Proof. As in the proof of Theorem 3.3, we may assume that $r_{\mathrm{inv}}(\tau, K, Q) < \infty$ for all $\tau > 0$. Under this assumption, let $\mathscr{S} = \{\omega_1, \ldots, \omega_n\}$ be a minimal (τ, K, Q)-spanning set for some $\tau > 0$. Define

$$K_j := \{x \in K \mid \varphi([0, \tau], x, \omega_j) \subset Q\}, \quad j = 1, \ldots, n.$$

Then $\{K_1, \ldots, K_n\}$ is a cover of K and by minimality $K_j \neq \emptyset$ for $j = 1, \ldots, n$. Let $x, y \in K_j$ for some $j \in \{1, \ldots, n\}$. Then, since Σ is uniformly expanding on Q and Q is bounded, it follows that

$$C e^{\lambda \tau} \varrho(x, y) \leq \varrho(\varphi(\tau, x, \omega_j), \varphi(\tau, y, \omega_j)) \leq \operatorname{diam} Q < \infty,$$

which implies

$$\varrho(x, y) \leq \frac{\operatorname{diam} Q}{C} e^{-\lambda \tau}.$$

Let $\alpha := \operatorname{diam}(Q)/C$. Then K_j is contained in the ball with radius $\alpha e^{-\lambda \tau}$ centered at any point in K_j, and hence

$$r_{\mathrm{inv}}(\tau, K, Q) = n \geq n(\alpha e^{-\lambda \tau}, K), \tag{7.5}$$

where $n(\alpha e^{-\lambda \tau}, K)$ denotes the minimal number of $\alpha e^{-\lambda \tau}$-balls necessary to cover K (cf. Sect. B.3). It holds that $\log([\alpha e^{-\lambda \tau}]^{-1}) = \lambda \tau - \log \alpha$, and thus

$$\tau = \frac{\log(\alpha^{-1} e^{\lambda \tau}) + \log \alpha}{\lambda} = \frac{\log(\alpha^{-1} e^{\lambda \tau})}{\lambda} \left(1 + \frac{\log \alpha}{\log(\alpha^{-1} e^{\lambda \tau})} \right). \tag{7.6}$$

Note that

$$\lim_{\tau \to \infty} \left(1 + \frac{\log \alpha}{\log(\alpha^{-1} e^{\lambda \tau})} \right) = 1. \tag{7.7}$$

Hence, we obtain

$$\begin{aligned} h_{\mathrm{inv}}(K, Q) &= \limsup_{\tau \to \infty} \frac{1}{\tau} \log r_{\mathrm{inv}}(\tau, K, Q) \\ &\overset{(7.5)}{\geq} \limsup_{\tau \to \infty} \frac{1}{\tau} \log n(\alpha e^{-\lambda \tau}, K) \\ &= \lambda \limsup_{\tau \to \infty} \frac{\log n(\alpha e^{-\lambda \tau}, K)}{\lambda \tau} \end{aligned}$$

$$\stackrel{(7.6)}{=} \lambda \limsup_{\tau \to \infty} \frac{\log n(\alpha e^{-\lambda \tau}, K)}{\log(\alpha^{-1} e^{\lambda \tau}) \left(1 + \frac{\log \alpha}{\log(\alpha^{-1} e^{\lambda \tau})}\right)}$$

$$\stackrel{(7.7)}{=} \lambda \limsup_{\tau \to \infty} \frac{\log n(\alpha e^{-\lambda \tau}, K)}{\log(\alpha^{-1} e^{\lambda \tau})} = \lambda \cdot \overline{\dim}_C(K),$$

which concludes the proof.   □

Remark 7.2. There is no problem to define also uniformly expanding discrete-time systems and to prove an analogous theorem as the preceding one. However, in the discrete-time case, the upper capacitive dimension has to be replaced by the lower capacitive dimension $\underline{\dim}_C(K)$, since the expression

$$\limsup_{\tau \to \infty} \frac{\log n(\alpha e^{-\lambda \tau}, K)}{\log(\alpha^{-1} e^{\lambda \tau})}$$

in general does not give the upper capacitive dimension if τ is an integer variable.

Example 7.3. Consider a one-dimensional continuous-time linear system $\Sigma = (\mathbb{R}, \mathbb{R}, U, \mathscr{U}, \varphi)$. By Proposition 1.3 we can write $\varphi_{t,0}(x) = e^{at} x$ for some $a \in \mathbb{R}$. If $|\varphi_{1,0}| = |e^a| > 1$, it easily follows that the system is uniformly expanding on every set $Q \subset \mathbb{R}$ with expansion factor a. Together with the upper estimate (4.3), we obtain the following result (without the assumption that $|\varphi_{1,0}| > 1$): If (K, Q) is an admissible pair for Σ with Q compact, then

$$h_{\mathrm{inv,out}}(K, Q) = \max\{0, a\} \cdot \overline{\dim}_C(K). \tag{7.8}$$

Remark 7.3. From formula (7.8) it follows that in general $h_{\mathrm{inv,out}}(K, Q) \neq \sup_{j \in \mathbb{Z}_+} h_{\mathrm{inv,out}}(K_j, Q)$ if $K = \bigcup_{j \in \mathbb{Z}_+} K_j$. As a counter-example, consider the linear system Σ given by the differential equations

$$\dot{x}(t) = x(t) + \omega(t), \quad \omega \in \mathscr{U},$$

with control range $\Omega = [-1, 1]$. Then $Q := [-1, 1]$ is controlled invariant, since every point $x \in Q$ becomes an equilibrium for the constant control function $\omega_x(t) :\equiv -x$. The set $K := \{0\} \cup \{1, 1/2, 1/3, \ldots\}$ is compact, and by Boichenko et al. [9, Chap. III, Example 2.2.2] its upper capacitive dimension is $1/2$. Now let $K_0 := \{0\}$ and $K_j := \{1/j\}$ for every $j \in \mathbb{N}$. Then $K = \bigcup_{j \in \mathbb{Z}_+} K_j$, but

$$h_{\mathrm{inv,out}}(K, Q) \stackrel{(7.8)}{=} \frac{1}{2} \neq 0 = \underbrace{\sup_{j \in \mathbb{Z}_+} h_{\mathrm{inv,out}}(K_j, Q)}_{=0},$$

which proves the claim.

The following proposition provides a simple condition for the right-hand side of a smooth system given by differential equations, which guarantees that the system is uniformly expanding on a set Q.

Proposition 7.3. *Let* $\Sigma = (\mathbb{R}, M, \mathbb{R}^m, \mathscr{U}, \varphi)$ *be a smooth system given by differential equations with right-hand side* F *and a compact control range* $\Omega \subset \mathbb{R}^m$. *Moreover, assume that* (M, g) *is a complete Riemannian* $\mathscr{C}^3$-*manifold. Let* $Q \subset M$ *be a bounded set and* $\rho > 0$ *a real number such that*

$$\lambda_{\min}(S\nabla F_u(x)) \geq \rho \quad \text{for all } (x, u) \in \mathrm{gh}(Q) \times \Omega,$$

where $\lambda_{\min}(\cdot)$ *denotes the minimal eigenvalue and* $\mathrm{gh}(Q)$ *the union of the images of all shortest geodesics joining two points in* Q.[1] *Then* Σ *is uniformly expanding on* Q *with expansion factor* ρ.

Proof. We subdivide the proof into three steps. First we prove expansiveness for constant control functions, then for piecewise constant ones and finally, for arbitrary admissible control functions.

Step 1. Let $x_1, x_2 \in Q$, $\tau > 0$, and $\omega \in \mathscr{U}$ a constant control function, say $\omega(t) \equiv u_0 \in \Omega$. Assume further that $\varphi([0, \tau], x_j, \omega) \subset Q$ for $j = 1, 2$. In order to prove expansiveness with expansion factor ρ, we show the following:

$$\forall \delta \in (0, \rho) : \; \varrho(\varphi(\tau, x_1, \omega), \varphi(\tau, x_2, \omega)) \geq e^{\delta\tau}\varrho(x_1, x_2). \tag{7.9}$$

To this end, we consider the time-reversed system

$$\dot{x}(t) = f(x(t)), \quad f(x) :\equiv -F(x, u_0).$$

The time-t-map of the corresponding flow is denoted by $\phi^t : M \to M$. It obviously holds that $\phi^t \equiv \varphi_{-t,\omega}$. Hence, (7.9) is equivalent to

$$\forall \delta \in (0, \rho) : \; \varrho(x_1, x_2) \leq e^{-\delta\tau}\varrho(\phi^{-\tau}(x_1), \phi^{-\tau}(x_2)).$$

With the substitution $\tilde{x}_j := \phi^{-\tau}(x_j) = \varphi(\tau, x_j, \omega)$, $j = 1, 2$, we obtain the equivalent statement

$$\forall \delta \in (0, \rho) : \; \varrho(\phi^\tau(\tilde{x}_1), \phi^\tau(\tilde{x}_2)) \leq e^{-\delta\tau}\varrho(\tilde{x}_1, \tilde{x}_2). \tag{7.10}$$

In order to prove (7.10), we introduce for every $\sigma > 0$ the set

$$A(\sigma) := \bigcup_{t \in [0,\sigma]} \phi^t(\mathrm{gh}(Q))$$

---

[1] The letters "gh" are supposed to stand for "geodesic hull", though we mean something slightly different here.

and show that the following statement holds:

$$\forall \delta \in (0, \rho) \; \exists \sigma > 0 \; \forall (x, u) \in A(\sigma) \times \Omega : \; \lambda_{\min}(S \nabla F_u(x)) \geq \delta. \qquad (7.11)$$

To this end, first note that $\mathrm{gh}(Q)$ is relatively compact which follows from the assumption that (M, g) is complete and $\mathrm{gh}(Q)$ is bounded, since obviously $\operatorname{diam} \mathrm{gh}(Q) = \operatorname{diam} Q$. For every $\delta \in (0, \rho)$ we find a neighborhood W of $\operatorname{cl} \mathrm{gh}(Q)$ such that $\lambda_{\min}(S \nabla F_u(x)) \geq \delta$ holds for all $(x, u) \in W \times \Omega$ which follows from the fact that the map $(x, u) \mapsto \lambda_{\min}(S \nabla F_u(x))$ is uniformly continuous on the compact set $\operatorname{cl} \mathrm{gh}(Q) \times \Omega$. Hence, it suffices to show that σ can be chosen small enough such that $A(\sigma) \subset W$. Assume to the contrary that there is no such σ. Then we find sequences $(t_n)_{n \in \mathbb{N}}$ in $(0, \infty)$ with $t_n \to 0$ and $(x_n)_{n \in \mathbb{N}}$ in $\mathrm{gh}(Q)$ converging to some point $x \in \operatorname{cl} \mathrm{gh}(Q)$ such that $\phi^{t_n}(x_n) \in M \setminus W$ for all $n \in \mathbb{N}$. By continuity, the contradiction $x = \phi^0(x) \in (M \setminus W) \cap \operatorname{cl} \mathrm{gh}(Q) = \emptyset$ follows. Hence, (7.11) is proved. Now let $t \in [0, \sigma]$ for some $\sigma = \sigma(\delta)$. Let $c : [0, 1] \to M$ be a shortest geodesic from $\tilde{x}_1$ to $\tilde{x}_2$, which exists by completeness of (M, g). Then we have

$$\varrho(\phi^t(\tilde{x}_1), \phi^t(\tilde{x}_2)) \leq \mathscr{L}(\phi^t \circ c) = \int_0^1 \left| \frac{\mathrm{d}}{\mathrm{d}s}(\phi^t \circ c)(s) \right| \mathrm{d}s$$

$$= \int_0^1 |\mathrm{d}_{c(s)}\phi^t(\dot{c}(s))| \mathrm{d}s \leq \int_0^1 \|\mathrm{d}_{c(s)}\phi^t\| \cdot |\dot{c}(s)| \mathrm{d}s$$

$$\leq \left(\max_{\xi \in c([0,1])} \|\mathrm{d}_\xi \phi^t\| \right) \int_0^1 |\dot{c}(s)| \mathrm{d}s$$

$$\leq \left(\max_{\xi \in \operatorname{cl} \mathrm{gh}(Q)} \|\mathrm{d}_\xi \phi^t\| \right) \varrho(\tilde{x}_1, \tilde{x}_2).$$

By the Wazewski inequality (Proposition A.4), we obtain the estimate

$$\|\mathrm{d}_\xi \phi^t\| \leq \exp\left(t \sup_{s \in [0,t]} \lambda(s) \right),$$

where $\lambda(s)$ denotes the maximal eigenvalue of $S \nabla f(\phi^s(\xi)) = -S \nabla F_{u_0}(\phi^s(\xi))$. Since $\phi^s(\xi) \in A(\sigma)$ for all $s \in [0, \sigma]$, (7.11) implies

$$\|\mathrm{d}_\xi \phi^t\| \leq \exp\left(t \sup_{s \in [0,t]} [-\lambda_{\min}(S \nabla F_{u_0}(\phi^s(\xi)))] \right)$$

$$\leq \exp\left(t \sup_{s \in [0,t]} (-\delta) \right) = \exp(-\delta t).$$

Hence, for all $t \in [0, \sigma]$ we have

$$\varrho(\phi^t(\tilde{x}_1), \phi^t(\tilde{x}_2)) \leq \mathrm{e}^{-\delta t} \varrho(\tilde{x}_1, \tilde{x}_2).$$

An inductive argument (using the flow property) shows that the same estimate holds for arbitrary $\sigma > 0$. Hence, we have proved (7.10).

Step 2. Let $x_1, x_2 \in Q$, $\omega \in \mathcal{U}$ a control function, and $\tau > 0$ a time such that $\varphi([0, \tau], x_j, \omega) \subset Q$, $j = 1, 2$, and such that there exists a partition $0 = t_0 < t_1 < \cdots < t_n = \tau$ with $\omega(t)$ constant on each of the intervals $[t_{j-1}, t_j)$, $j = 1, \ldots, n$. Then, by Step 1, we have

$$\varrho(\varphi_{t,\omega}(x_1), \varphi_{t,\omega}(x_2)) \geq e^{t\rho} \varrho(x_1, x_2) \quad \text{for all } t \in [t_0, t_1].$$

Now let $t \in [t_1, t_2]$. Then the cocycle property implies

$$\varrho(\varphi_{t,\omega}(x_1), \varphi_{t,\omega}(x_2))$$
$$= \varrho(\varphi(t - t_1, \varphi(t_1, x_1, \omega), \Theta_{t_1}\omega), \varphi(t - t_1, \varphi(t_1, x_2, \omega), \Theta_{t_1}\omega)).$$

Since $\Theta_{t_1}\omega$ is constant on $[0, t_2 - t_1]$, again Step 1 implies

$$\varrho(\varphi_{t,\omega}(x_1), \varphi_{t,\omega}(x_2)) \geq e^{(t-t_1)\rho} \varrho(\varphi_{t_1,\omega}(x_1), \varphi_{t_1,\omega}(x_2))$$
$$\geq e^{(t-t_1)\rho} e^{t_1\rho} \varrho(x_1, x_2) = e^{t\rho} \varrho(x_1, x_2).$$

Inductively we obtain the assertion for all $t \in [0, \tau]$.

Step 3. Expansiveness for arbitrary admissible control functions now can easily be concluded using the approximation result of Proposition 1.6 (approximation of measurable control functions by piecewise constant ones). $\quad\square$

If Σ is expanding on a compact controlled invariant set Q, then it is possible to show that the escape entropy, introduced in Sect. 6.2, of any admissible pair (K, Q) is nonpositive.

Proposition 7.4. *Let* $\Sigma = (\mathbb{R}, M, \mathbb{R}^m, \mathcal{U}, \varphi)$ *be a smooth system given by differential equations with right-hand side* F *and compact control range* $\Omega \subset \mathbb{R}^m$. *Moreover, assume that* (M, g) *is a Riemannian* $\mathscr{C}^3$-*manifold. If* (K, Q) *is an admissible pair for* Σ, *with* Q *being compact and controlled invariant, such that* Σ *is uniformly expanding on* Q, *then* $h_{\mathrm{esc}}(K, Q) \leq 0$.

Proof. For arbitrary $\omega \in \pi_{\mathcal{U}} \mathcal{K}_Q$, $\tau > 0$, and $\varepsilon > 0$ let S be an $(\omega, \tau, \varepsilon)$-separated subset of $K(\omega, \tau)$. Choose distinct points $x_1, x_2 \in S$ and let $s = s(x_1, x_2) \in [0, \tau]$ such that $\varrho(\varphi(s, x_1, \omega), \varphi(s, x_2, \omega)) = \varrho_{\omega,\tau}(x_1, x_2)$. Then, using the cocycle property of φ, we find

$$\varrho\left(\varphi(\tau, x_1, \omega), \varphi(\tau, x_2, \omega)\right)$$
$$= \varrho\left(\varphi(\tau - s, \varphi(s, x_1, \omega), \Theta_s\omega), \varphi(\tau - s, \varphi(s, x_2, \omega), \Theta_s\omega)\right)$$
$$\geq C e^{\lambda(\tau-s)} \varrho\left(\varphi(s, x_1, \omega), \varphi(s, x_2, \omega)\right)$$
$$\geq C e^{\lambda(\tau-s)} \varepsilon \geq C\varepsilon.$$

Hence, $\varphi_{\tau,\omega}(S)$ is a $C\varepsilon$-separated subset of Q, which has the same cardinality as S (since $\varphi_{\tau,\omega}$ is injective). By compactness, we can cover Q with finitely many balls $B(x_1, \delta), \ldots, B(x_n, \delta)$ of a fixed radius $\delta > 0$ such that $\exp_{x_i}^{-1}(B(x_i, \delta)) = B(0_{x_i}, \delta)$ and

$$\varrho\left(\exp_{x_i}(v), \exp_{x_i}(w)\right) \leq 2|v - w| \quad \text{for all } v, w \in B(0_{x_i}, \delta) \tag{7.12}$$

for $i = 1, \ldots, n$. The latter is possible because $d_{0_x} \exp_x = \mathrm{id}_{T_x M}$ for all $x \in M$. Then

$$\#\varphi_{\tau,\omega}(S) \leq \sum_{i=1}^{n} \#\left(\varphi_{\tau,\omega}(S) \cap B(x_i, \delta)\right)$$

$$\leq n \max_{1 \leq i \leq n} \underbrace{\#\exp_{x_i}^{-1}\left(\varphi_{\tau,\omega}(S) \cap B(x_i, \delta)\right)}_{=:N_i}.$$

By (7.12) the set $\exp_{x_i}^{-1}(\varphi_{\tau,\omega}(S) \cap B(x_i, \delta))$ is a $(C\varepsilon)/2$-separated subset of $B(0_{x_i}, \delta)$. Hence, $B(0_{x_i}, \delta + (C\varepsilon)/4)$ contains N_i disjoint balls of radii $(C\varepsilon)/4$. Letting $d = \dim M$, this implies

$$\left(\delta + \frac{C\varepsilon}{4}\right)^d \geq N_i \left(\frac{C\varepsilon}{4}\right)^d \quad \Rightarrow \quad N_i \leq \left(\frac{4\delta + C\varepsilon}{C\varepsilon}\right)^d.$$

Hence, we get

$$\#S = \#\varphi_{\tau,\omega}(S) \leq n \left(\frac{4\delta + C\varepsilon}{C\varepsilon}\right)^d,$$

and therefore we obtain

$$\overline{r}_{\mathrm{sep}}(\tau, K, Q) = \limsup_{\varepsilon \searrow 0} \overline{r}_{\mathrm{sep}}(\tau, \varepsilon, K, Q)$$

$$\leq \limsup_{\varepsilon \searrow 0} n \left(\frac{4\delta + C\varepsilon}{C}\right)^d = n \left(\frac{4\delta}{C}\right)^d,$$

which implies the assertion. $\square$

The following theorem provides the best estimates for the invariance entropy of a uniformly expanding control set that we can give using the methods developed in Chaps. 4–6.

Theorem 7.4. *Let $\Sigma = (\mathbb{R}, M, \mathbb{R}^m, \mathcal{U}, \varphi)$ be a smooth control-affine system with right-hand side F and a compact and convex control range Ω with nonempty interior, satisfying the Lie algebra rank condition on a control set $D \subset M$ with nonempty interior and compact closure $Q := \mathrm{cl}\, D$. Furthermore, assume that Σ is* *uniformly expanding on Q with respect to some Riemannian metric. Then for every compact set $K \subset D$ it holds that*

$$h_{\mathrm{inv}}(K, Q) \leq \inf_{(\omega, x)} \limsup_{\tau \to \infty} \frac{1}{\tau} \int_0^\tau \mathrm{div}\, F_{\omega(s)}(\varphi(s, x, \omega))\mathrm{d}s,$$

where the infimum is taken over all $(\omega, x) \in \mathrm{int}\, \mathscr{U} \times \mathrm{int}\, D$ such that $\varphi(\mathbb{R}_+, x, \omega)$ is contained in a compact subset of $\mathrm{int}\, D$. Moreover, if K has positive volume, then

$$h_{\mathrm{inv}}(K, Q) \geq \inf_{(\omega, x) \in \mathscr{Q}} \limsup_{\tau \to \infty} \frac{1}{\tau} \int_0^\tau \mathrm{div}\, F_{\omega(s)}(\varphi(s, x, \omega))\mathrm{d}s.$$

Proof. First note that the assumptions on Σ guarantee that we can use the approximation results of Sect. 5.2 (using Proposition 5.12). In particular, we can use Proposition 5.11 with $k = d := \dim M$. To this end, we have to check that the assumptions (i) and (ii) in Proposition 5.11 are satisfied. So let $(\varphi(\cdot, x, \omega), \omega(\cdot))$ be a periodic controlled trajectory with $(x, \omega) \in \mathrm{int}\, D \times \mathrm{int}\, \mathscr{U}$ and let $\tau \geq 0$ and $v \in T_x M \setminus \{0_x\}$. Then there exists a geodesic γ, defined on a small interval $[0, \alpha]$ such that $\gamma(0) = x$ and $\dot{\gamma}(0) = v$. By continuous dependence on initial conditions, there is $\delta \in (0, \alpha]$ such that $\varphi([0, \tau], \gamma(\delta), \omega)$ is still contained in D. This implies

$$\varrho(\varphi(\tau, x, \omega), \varphi(\tau, \gamma(\delta), \omega)) \geq C\mathrm{e}^{\lambda \tau} \varrho(x, \gamma(\delta)) = C\mathrm{e}^{\lambda \tau} \delta |v|. \tag{7.13}$$

Moreover, we have

$$\varrho(\varphi(\tau, x, \omega), \varphi(\tau, \gamma(\delta), \omega)) \leq \mathscr{L}\left(\varphi_{\tau, \omega} \circ \gamma|_{[0, \delta]}\right)$$

$$= \int_0^\delta \left|\mathrm{d}_{\gamma(s)}\varphi_{\tau, \omega}(\dot{\gamma}(s))\right| \mathrm{d}s$$

$$\leq \delta \max_{s \in [0, \delta]} \left|\mathrm{d}_{\gamma(s)}\varphi_{\tau, \omega}(\dot{\gamma}(s))\right|.$$

Combining this estimate with (7.13), we find

$$\max_{s \in [0, \delta]} \left|\mathrm{d}_{\gamma(s)}\varphi_{\tau, \omega}(\dot{\gamma}(s))\right| \geq C\mathrm{e}^{\lambda \tau} |v|.$$

We may choose δ arbitrarily small, which gives $|\mathrm{d}_x\varphi_{\tau, \omega}(v)| \geq C\mathrm{e}^{\lambda \tau} |v|$. This implies

$$\limsup_{\tau \to \infty} \frac{1}{\tau} \log |\mathrm{d}_x\varphi_{\tau, \omega}(v)| \geq \lambda.$$

Hence, all Lyapunov exponents at (ω, x) are positive. It remains to show that for every controlled trajectory $(\varphi(\cdot, x, \omega), \omega(\cdot))$ such that $\omega \in \mathrm{int}\, \mathscr{U}$ and $\varphi(\mathbb{R}_+, x, \omega)$ is contained in a compact subset of $\mathrm{int}\, D$ there exists $t_0 \geq 0$ with $a_t^d(\omega, x) \geq 0$ for all $t \geq t_0$. First note that $a_t^d(\omega, x) = \log |\det \mathrm{d}_x\varphi_{t, \omega}|$. With the same arguments as above it follows that $|d_x\varphi_{t,\omega}(v)| \geq Ce^{\lambda t}|v|$ for all nonzero $v \in T_xM$. This implies $|\det d_x\varphi_{t,\omega}| \geq (Ce^{\lambda t})^d$ and hence

$$a_t^d(\omega, x) \geq \lambda dt + \log(C)d \rightarrow \infty \quad \text{for } t \rightarrow \infty.$$

Therefore, the assumptions of Proposition 5.11 are satisfied and the upper estimate follows (using the Liouville formula). The lower bound follows from Corollary 4.4 (iii) (using Proposition 1.11, which guarantees that Q is controlled invariant). $\square$

Remark 7.4. An alternative proof of the lower estimate can be given by combining Corollary 6.1 with Proposition 7.4.

Remark 7.5. The expression

$$C := \inf_{(\omega,x)\in\mathcal{Q}} \limsup_{\tau\rightarrow\infty} \frac{1}{\tau} \int_0^\tau \operatorname{div} F_{\omega(s)}(\varphi(s, x, \omega))ds$$

can be rewritten in terms of the Lyapunov exponents of the invariant probability measures of the control flow Φ restricted to the compact set $\mathcal{Q}$. Using methods from Gelfert [50] one can show that

$$C = \inf_{\mu\in\mathcal{E}(\mathcal{Q})} (\nu_1(\mu) + \cdots + \nu_d(\mu)),$$

where $\mathcal{E}(\mathcal{Q})$ denotes the set of all ergodic invariant measures of $\Phi|\mathcal{Q}$ and $\nu_1(\mu), \ldots, \nu_d(\mu)$ the associated (μ-almost everywhere on $\mathcal{Q}$ defined) Lyapunov exponents.

Remark 7.6. It is clear that Theorem 7.4 also holds under weaker assumptions. Indeed, notice that for the lower bound we do not need smoothness, the Lie algebra rank condition, or uniform expansiveness. For the upper bound, instead of uniform expansiveness, it is sufficient to assume that all Lyapunov exponents of trajectories in D are positive.

7.3 Inhomogeneous Bilinear Systems Revisited

Let $\Sigma = (\mathbb{R}, \mathbb{R}^d, \mathbb{R}^{m_1+m_2}, \mathcal{U} \times \mathcal{V}, \varphi)$ be an inhomogeneous bilinear system given by differential equations

$$\dot{x}(t) = \underbrace{\left[A_0 + \sum_{i=1}^{m_1} \omega_i(t)A_i\right]}_{=:A(\omega(t))} x(t) + B\mu(t), \quad (\omega, \mu) \in \mathcal{U} \times \mathcal{V}.$$

The control range has the form $\Omega = \Omega_1 \times \Omega_2$ with $\Omega_1 \subset \mathbb{R}^{m_1}$ and $\Omega_2 \subset \mathbb{R}^{m_2}$ and we assume that both Ω_1 and Ω_2 are compact and convex. (Recall that this implies weak*-compactness of $\mathscr{U}$ and $\mathscr{U} \times \mathscr{V}$.) The transition map φ has the form

$$\varphi(t, x, (\omega, \mu)) = \Lambda_\omega(t, 0)x + \int_0^t \Lambda_\omega(t, s) B\mu(s)\mathrm{d}s.$$

For the last term in this expression we also briefly write $\varphi^s(t, \omega, \mu)$.

The following theorem yields a slight improvement of the estimate of Sect. 3.2. However, the methods used in the proof are different; they are a variation of those applied in Chap. 6 for nonlinear systems. We recall the notations

$$\mathscr{K}_Q := \{((\omega, \mu), x) \, : \, \varphi(\mathbb{R}_+, x, (\omega, \mu)) \subset Q, \ x \in K\}$$

and

$$K((\omega, \mu), \tau) = \{x \in K \, : \, \varphi([0, \tau], x, (\omega, \mu)) \subset Q\},$$

where (K, Q) is an admissible pair. Furthermore, we recall the definition of the Bowen-balls:

$$B_{(\omega,\mu)}^\tau(x, \varepsilon) = \left\{ y \in \mathbb{R}^d \, : \, \max_{t \in [0,\tau]} |\varphi(t, x, (\omega, \mu)) - \varphi(t, y, (\omega, \mu))| < \varepsilon \right\}$$

$$= \left\{ y \in \mathbb{R}^d \, : \, \max_{t \in [0,\tau]} |\Lambda_\omega(\tau, 0)(x - y)| < \varepsilon \right\}.$$

Theorem 7.5. *Let (K, Q) be an admissible pair for Σ such that K has positive Lebesgue measure and Q is compact and controlled invariant. Then, for the homogeneous system, there exists a (maximal) invariant subbundle $\mathscr{W}^+ \subset \mathscr{U} \times \mathbb{R}^d$ on which the system is uniformly expanding such that the estimate*

$$h_{\mathrm{inv}}(K, Q) \geq \limsup_{\tau \to \infty} \inf_{\omega \in \pi_{\mathscr{U}} \mathscr{K}_Q} \frac{1}{\tau} \int_0^\tau \mathrm{tr}\,[A(\omega(s)) \circ Q(\Theta_s\omega)]\,\mathrm{d}s \qquad (7.14)$$

holds, where $Q(\omega) : \mathbb{R}^d \to \mathscr{W}_\omega^+$ is the orthogonal projection and $\pi_{\mathscr{U}} : \mathscr{U} \times \mathscr{V} \times \mathbb{R}^d \to \mathscr{U}, (\omega, \mu, x) \mapsto \omega$. Moreover, it holds that

$$h_{\mathrm{inv}}(K, Q) \geq \inf_{\omega \in \pi_{\mathscr{U}} \mathscr{Q}} \limsup_{\tau \to \infty} \frac{1}{\tau} \int_0^\tau \mathrm{tr}\,[A(\omega(s)) \circ Q(\Theta_s\omega)]\,\mathrm{d}s. \qquad (7.15)$$

Proof. As in the proof of Theorem 3.3, we may assume that $r_{\mathrm{inv}}(\tau, K, Q) < \infty$ for all $\tau > 0$. Then we prove the theorem in two steps.

Step 1. Selgrade's theorem B.1 yields a decomposition $\mathscr{U} \times \mathbb{R}^d = \mathscr{W}^1 \oplus \cdots \oplus \mathscr{W}^r$ into subbundles, which are invariant under the control flow of the homogeneous system associated with Σ. We can define $\mathscr{W}^+$ as the Whitney sum of all those subbundles on which the system is uniformly expanding. Then (since the angles between the fibers of the subbundles stay bounded) it is easy to see that there are $C, \lambda > 0$ such that

$$|\Lambda_\omega(t, 0)x| \geq C e^{\lambda t}|x| \quad \text{for all } t \geq 0, \quad (\omega, x) \in \mathscr{W}^+. \tag{7.16}$$

Let $\mathscr{W}^-$ be the sum of the remaining subbundles. For each $\omega \in \mathscr{U}$ let $\pi_\omega^+ : \mathbb{R}^d \to \mathscr{W}_\omega^+$ be the projection onto $\mathscr{W}_\omega^+$ along $\mathscr{W}_\omega^-$ and let d^+ be the rank of $\mathscr{W}^+$. By m^{d+} we denote the d^+-dimensional Lebesgue measure on any of the fibers $\mathscr{W}_\omega^+, \omega \in \mathscr{U}$. Let $\mathscr{S} \subset \pi_{\mathscr{U} \times \mathscr{V}} \mathscr{K}_Q$ (without loss of generality) be a minimal (τ, K, Q)-spanning set which is finite by assumption. Moreover,

$$K = \bigcup_{(\omega,\mu) \in \mathscr{S}} K((\omega, \mu), \tau).$$

Since K has positive Lebesgue measure, Lemma 3.3 yields a constant $\alpha > 0$ with

$$\alpha \leq \sum_{(\omega,\mu) \in \mathscr{S}} m^{d+} \left(\pi_\omega^+(K((\omega, \mu), \tau))\right)$$

$$\leq \#\mathscr{S} \cdot \sup_{(\omega,\mu) \in \pi_{\mathscr{U} \times \mathscr{V}} \mathscr{K}_Q} m^{d+} \left(\pi_\omega^+(K((\omega, \mu), \tau))\right).$$

Since $\#\mathscr{S} = r_{\mathrm{inv}}(\tau, K, Q)$, this implies

$$h_{\mathrm{inv}}(K, Q) \geq \limsup_{\tau \to \infty} \frac{1}{\tau} \log \frac{\alpha}{\sup_{(\omega,\mu)} m^{d+}(\pi_\omega^+(K((\omega, \mu), \tau)))}$$

$$= \limsup_{\tau \to \infty} \left[-\frac{1}{\tau} \log \sup_{(\omega,\mu) \in \pi_{\mathscr{U} \times \mathscr{V}} \mathscr{K}_Q} m^{d+}(\pi_\omega^+(K((\omega, \mu), \tau)))\right].$$

Step 2. Fix an arbitrary $\varepsilon > 0$. For each $\tau > 0$ and $(\omega, \mu) \in \pi_{\mathscr{U} \times \mathscr{V}} \mathscr{K}_Q$ select a maximal $((\omega, \mu), \tau, \varepsilon)$-separated set $S_{(\omega,\mu),\tau} \subset \pi_\omega^+(K((\omega, \mu), \tau)) \subset \mathscr{W}_\omega^+$. Then for all distinct $x_1, x_2 \in S_{(\omega,\mu),\tau}$ it holds that

$$\max_{t \in [0,\tau]} |\Lambda_\omega(t, 0)(x_1 - x_2)| = \max_{t \in [0,\tau]} |\varphi(t, x_1, (\omega, \mu)) - \varphi(t, x_2, (\omega, \mu))| \geq \varepsilon.$$

Since $x_1 - x_2 \in \mathscr{W}_\omega^+$, from (7.16) we obtain with the same arguments as in the proof of Proposition 7.4 that

$$|\Lambda_\omega(\tau, 0)(x_1 - x_2)| \geq C\varepsilon.$$

The set $\Lambda_\omega(\tau, 0)S_{(\omega,\mu),\tau}$ has the same cardinality as $S_{(\omega,\mu),\tau}$ and

$$S_{(\omega,\mu),\tau} \subset \left\{ \pi_\omega^+ x \ : \ \varphi(\tau, x, (\omega, \mu)) \in Q \right\}$$
$$= \left\{ \pi_\omega^+ x \ : \ \Lambda_\omega(\tau, 0)x \in Q - \varphi^s(\tau, \omega, \mu) \right\}.$$

Hence, we find

$$\Lambda_\omega(\tau, 0)S_{(\omega,\mu),\tau} \subset \left\{ \Lambda_\omega(\tau, 0)\pi_\omega^+ x \ : \ \Lambda_\omega(\tau, 0)x \in Q - \varphi^s(\tau, \omega, \mu) \right\}$$
$$= \left\{ \pi_{\Theta_\tau\omega}^+ \Lambda_\omega(\tau, 0)x \ : \ \Lambda_\omega(\tau, 0)x \in Q - \varphi^s(\tau, \omega, \mu) \right\}$$
$$= \left\{ \pi_{\Theta_\tau\omega}^+ y \ : \ y \in Q - \varphi^s(\tau, \omega, \mu) \right\}$$
$$= \pi_{\Theta_\tau\omega}^+ Q - \pi_{\Theta_\tau\omega}^+ \varphi^s(\tau, \omega, \mu).$$

By Lemma B.1 the projection π_ω^+ depends continuously on ω in the weak*-topology and hence, by compactness of $\mathscr{U}$, there is $R > 0$ such that each of the sets $\pi_\omega^+ Q$, $\omega \in \mathscr{U}$, is contained in a ball in $\mathscr{W}_\omega^+$ of radius R centered at the origin. Therefore,

$$\Lambda_\omega(\tau, 0)S_{(\omega,\mu),\tau} + \pi_{\Theta_\tau\omega}^+ \varphi^s(\tau, \omega, \mu) \subset B(0, R) \subset \mathscr{W}_{\Theta_\tau\omega}^+.$$

The set on the left-hand side has the same cardinality as $S_{(\omega,\mu),\tau}$ and each two distinct points in this set have distance $\geq C\varepsilon$. This implies that there is a global bound $N \in \mathbb{N}$ on $\#S_{(\omega,\mu),\tau}$, that is,

$$\#S_{(\omega,\mu),\tau} \leq N \quad \text{for all } (\omega, \mu) \in \pi_{\mathscr{U}\times\mathscr{V}}\mathscr{K}_{\mathcal{2}}, \ \tau > 0. \tag{7.17}$$

Since $S_{(\omega,\mu),\tau}$ is maximal $((\omega, \mu), \tau, \varepsilon)$-separated in $\pi_\omega^+(K((\omega, \mu), \tau))$, we have

$$\pi_\omega^+(K((\omega, \mu), \tau)) \subset \bigcup_{x \in S_{(\omega,\mu),\tau}} B_{(\omega,\mu)}^{\tau,+}(x, \varepsilon),$$

where

$$B_{(\omega,\mu)}^{\tau,+}(x, \varepsilon) := B_{(\omega,\mu)}^\tau(x, \varepsilon) \cap \mathscr{W}_\omega^+.$$

Hence, we find

$$m^{d+}(\pi_\omega^+(K((\omega, \mu), \tau))) \leq \sum_{x \in S_{(\omega,\mu),\tau}} m^{d+}\left(B_{(\omega,\mu)}^{\tau,+}(x, \varepsilon) \right)$$
$$\leq \#S_{(\omega,\mu),\tau} \cdot \max_x m^{d+}\left(B_{(\omega,\mu)}^{\tau,+}(x, \varepsilon) \right)$$
$$\overset{(7.17)}{\leq} N \cdot \max_x m^{d+}\left(B_{(\omega,\mu)}^{\tau,+}(x, \varepsilon) \right).$$

We have

$$B^{\tau,+}_{(\omega,\mu)}(x,\varepsilon) = \left\{ y \in \mathscr{W}^+_\omega \; : \; \max_{t\in[0,\tau]} |\Lambda_\omega(t,0)(x-y)| < \varepsilon \right\}$$

$$\subset \left\{ y \in \mathscr{W}^+_\omega \; : \; |\Lambda_\omega(\tau,0)(x-y)| < \varepsilon \right\}.$$

Since Lebesgue measure is invariant under translations, this implies

$$m^{d+}\left(B^{\tau,+}_{(\omega,\mu)}(x,\varepsilon) \right) \leq m^{d+}\left(\{ y \in \mathscr{W}^+_\omega \; : \; |\Lambda_\omega(\tau,0)y| < \varepsilon \} \right)$$

$$\leq m^{d+}\left(\Lambda_\omega(\tau,0)^{-1}\left(B(0,\varepsilon) \cap \mathscr{W}^+_{\Theta_\tau\omega} \right) \right)$$

$$= C(\varepsilon) \left| \det \Lambda_\omega(\tau,0)|_{\mathscr{W}^+_\omega} \right|^{-1},$$

where $C(\varepsilon)$ is the volume of the d^+-dimensional Euclidean ε-ball. Thus, with Step 1 we obtain

$$h_{\mathrm{inv}}(K,Q) \geq \limsup_{\tau\to\infty} \left[-\frac{1}{\tau}\log \sup_{(\omega,\mu)\in\pi_{\mathscr{U}\times\mathscr{V}}\mathscr{K}_Q} m^{d+}(\pi^+_\omega(K((\omega,\mu),\tau))) \right]$$

$$\geq \limsup_{\tau\to\infty} \left[-\frac{1}{\tau}\log \sup_{(\omega,\mu)\in\pi_{\mathscr{U}\times\mathscr{V}}\mathscr{K}_Q} N \cdot \max_x m^{d+}\left(B^{\tau,+}_{(\omega,\mu)}(x,\varepsilon) \right) \right]$$

$$\geq \limsup_{\tau\to\infty} \left[-\frac{1}{\tau}\log \sup_{(\omega,\mu)\in\pi_{\mathscr{U}\times\mathscr{V}}\mathscr{K}_Q} NC(\varepsilon) \left| \det \Lambda_\omega(\tau,0)|_{\mathscr{W}^+_\omega} \right|^{-1} \right]$$

$$= \limsup_{\tau\to\infty} \left[-\frac{1}{\tau}\log \sup_{\omega\in\pi_{\mathscr{U}}\mathscr{K}_Q} \left| \det \Lambda_\omega(\tau,0)|_{\mathscr{W}^+_\omega} \right|^{-1} \right]$$

$$= \limsup_{\tau\to\infty} \inf_{\omega\in\pi_{\mathscr{U}}\mathscr{K}_Q} \frac{1}{\tau}\log \left| \det \Lambda_\omega(\tau,0)|_{\mathscr{W}^+_\omega} \right|.$$

From invariance of $\mathscr{W}^+$ it follows with the generalized Liouville formula (Proposition A.5) that

$$\left| \det \Lambda_\omega(\tau,0)|_{\mathscr{W}^+_\omega} \right| = \exp\left(\int_0^\tau \mathrm{tr}\,[A(\omega(s)) \circ Q(\Theta_s\omega)]\,\mathrm{d}s \right).$$

This proves (7.14). To show (7.15), recall from the proof of Theorem 3.3 that $(\tau,\omega) \mapsto \log|\det \Lambda_\omega(\tau,0)|_{\mathscr{W}^+_\omega}|$ is an additive cocycle over the shift flow on $\mathscr{U}$. Using Theorem B.2 and the fact that $\pi_{\mathscr{U}}\mathscr{Q}$ is a compact shift-invariant set, we obtain

$$h_{\mathrm{inv}}(K, Q) \geq \limsup_{\tau\to\infty} \inf_{\omega\in\pi_{\mathcal{U}}\,\mathcal{K}_Q} \frac{1}{\tau} \int_0^\tau \mathrm{tr}\left[A(\omega(s)) \circ Q(\Theta_s\omega)\right] \mathrm{d}s$$

$$\geq \limsup_{\tau\to\infty} \inf_{\omega\in\pi_{\mathcal{U}}\,\mathcal{Q}} \frac{1}{\tau} \int_0^\tau \mathrm{tr}\left[A(\omega(s)) \circ Q(\Theta_s\omega)\right] \mathrm{d}s$$

$$= \inf_{\omega\in\pi_{\mathcal{U}}\,\mathcal{Q}} \limsup_{\tau\to\infty} \frac{1}{\tau} \int_0^\tau \mathrm{tr}\left[A(\omega(s)) \circ Q(\Theta_s\omega)\right] \mathrm{d}s.$$

This finishes the proof.   □

Corollary 7.1. *Under the assumptions of Theorem 7.5, assume that the homogeneous system is weakly contracting on $\mathscr{W}^-$, that is, there is $\tau_1 \geq 0$ with*

$$|\Lambda_\omega(\tau,0)x| \leq |x| \quad \text{for all } (\omega,x) \in \mathscr{W}^-,\ \tau \geq \tau_1.$$

Then

$$h_{\mathrm{inv}}(K, Q) \geq \inf_{\omega\in\pi_{\mathcal{U}}\,\mathcal{Q}} \limsup_{\tau\to\infty} \frac{1}{\tau} \log \left\| \Lambda_\omega(\tau,0)^\wedge \right\|,$$

where $\Lambda_\omega(\tau,0)^\wedge$ denotes the exterior power of $\Lambda_\omega(\tau,0)$ and $\|\cdot\|$ an arbitrary operator norm. If, additionally, Q is the closure of a control set D and Σ is locally accessible, then

$$h_{\mathrm{inv}}(K, Q) \leq \inf_\omega \limsup_{\tau\to\infty} \frac{1}{\tau} \log \left\| \Lambda_\omega(\tau,0)^\wedge \right\|,$$

where the infimum is taken over all $\omega \in \mathcal{U}$ such that there is $(x,\mu) \in Q \times \mathcal{V}$ with $\varphi(\mathbb{R}_+, x, (\omega,\mu))$ contained in a compact subset of $\mathrm{int}\,D$ and $(\omega,\mu) \in \mathrm{int}(\mathcal{U}\times\mathcal{V})$.

Proof. The proof is subdivided into two steps.

Step 1. For each $\omega \in \mathcal{U}$ let $\langle\cdot,\cdot\rangle_\omega$ be the inner product on $\mathbb{R}^d$ which coincides with the standard one on both $\mathscr{W}_\omega^+$ and $\mathscr{W}_\omega^-$, and satisfies $\mathscr{W}_\omega^+ \perp \mathscr{W}_\omega^-$. For all $\tau \geq 0$ and $\omega \in \mathcal{U}$ define the linear operator

$$L_{\tau,\omega} := \Lambda_\omega(\tau,0) : (\mathbb{R}^d, \langle\cdot,\cdot\rangle_\omega) \to (\mathbb{R}^d, \langle\cdot,\cdot\rangle_{\Theta_\tau\omega}).$$

We claim that $\mathscr{W}_\omega^+$ is an invariant subspace for the operator

$$L_{\tau,\omega}^* L_{\tau,\omega} : (\mathbb{R}^d, \langle\cdot,\cdot\rangle_\omega) \to (\mathbb{R}^d, \langle\cdot,\cdot\rangle_\omega),$$

where $L_{\tau,\omega}^*$ denotes the adjoint of $L_{\tau,\omega}$. Indeed, if $x \in \mathscr{W}_\omega^+$ and $y \in (\mathscr{W}_\omega^+)^\perp = \mathscr{W}_\omega^-$, then

$$\langle L_{\tau,\omega}^* L_{\tau,\omega} x, y \rangle_\omega = \langle L_{\tau,\omega} x, L_{\tau,\omega} y \rangle_{\Theta_\tau\omega} = 0$$

by choice of the inner products and invariance of the subbundles $\mathscr{W}^{\pm}$. We claim that for all sufficiently large $\tau > 0$ the eigenvalues of modulus greater than 1 of $L_{\tau,\omega}^{*}L_{\tau,\omega}$ are attained on the invariant subspace $\mathscr{W}_{\omega}^{+}$. From (7.16) we know that $|\Lambda_{\omega}(\tau,0)x| \geq C e^{\lambda\tau}|x|$ for all $(\omega, x) \in \mathscr{W}^{+}$ and $\tau \geq 0$. This implies for $x \in \mathscr{W}_{\omega}^{+}$ that

$$\langle L_{\tau,\omega}^{*}L_{\tau,\omega}x, x\rangle_{\omega} = |\Lambda_{\omega}(\tau,0)x|^{2} \geq C^{2}e^{2\lambda\tau}|x|^{2}.$$

Hence, every eigenvalue σ of $L_{\tau,\omega}^{*}L_{\tau,\omega}|_{\mathscr{W}_{\omega}^{+}}$ satisfies $\sigma^{1/2} \geq C e^{\lambda\tau}$. Consequently, for all $\tau \geq$ some τ_0 (which is independent of ω) these eigenvalues are greater than 1. The other eigenvalues of the self-adjoint operator $L_{\tau,\omega}^{*}L_{\tau,\omega}$ are attained on the orthogonal complement $\mathscr{W}_{\omega}^{-} = (\mathscr{W}_{\omega}^{+})^{\perp}$, where the weak contraction property holds. Hence, those eigenvalues are less than or equal to 1 for all $\tau \geq \tau_{*} :=$ $\max\{\tau_0, \tau_1\}$, which proves the claim. We thus obtain

$$\left|\det L_{\tau,\omega}|_{\mathscr{W}_{\omega}^{+}}\right| = \prod_{i=1}^{d}\max\{1, \sigma_i(L_{\tau,\omega})\} \quad \text{for all } \omega \in \mathscr{U}, \ \tau \geq \tau_{*}.$$

Since we chose the inner products $\langle\cdot,\cdot\rangle_{\omega}$ such that they all coincide with the standard one on the fibers $\mathscr{W}_{\omega}^{+}$, we have

$$\left|\det L_{\tau,\omega}|_{\mathscr{W}_{\omega}^{+}}\right| = \left|\det \Lambda_{\omega}(\tau,0)|_{\mathscr{W}_{\omega}^{+}}\right| \quad \text{for all } \omega \in \mathscr{U}.$$

Hence, we obtain from Theorem 7.5 that

$$h_{\text{inv}}(K, Q) \geq \inf_{\omega\in\pi\mathscr{U}\mathscr{Q}} \limsup_{\tau\to\infty} \frac{1}{\tau}\log \prod_{i=1}^{d}\max\{1, \sigma_i(L_{\tau,\omega})\}$$

$$= \inf_{\omega\in\pi\mathscr{U}\mathscr{Q}} \limsup_{\tau\to\infty} \frac{1}{\tau}\log \left\|L_{\tau,\omega}^{\wedge}\right\|_{\omega,\Theta_{\tau}\omega},$$

where for any $\omega_1, \omega_2 \in \mathscr{U}$, we denote by $\|\cdot\|_{\omega_1,\omega_2}$ the operator norm on

$$\mathscr{L}\left(\bigwedge(\mathbb{R}^{d}, \langle\cdot,\cdot\rangle_{\omega_1}), \bigwedge(\mathbb{R}^{d}, \langle\cdot,\cdot\rangle_{\omega_2})\right).$$

To complete the proof, it suffices to show that there are constants $c, C > 0$ with

$$c\|\cdot\|_{\omega_1,\omega_2} \leq \|\cdot\| \leq C\|\cdot\|_{\omega_1,\omega_2} \quad \text{for all } \omega_1, \omega_2 \in \mathscr{U}, \tag{7.18}$$

where the operator norm in the middle term is the one induced by the standard inner product on both the domain and the codomain. The inner product $\langle\cdot,\cdot\rangle_{\omega}$ is given by $\langle x, y\rangle_{\omega} = \langle\pi_{\omega}^{+}x, \pi_{\omega}^{+}y\rangle + \langle\pi_{\omega}^{-}x, \pi_{\omega}^{-}y\rangle$ and the projections $\pi_{\omega}^{\pm}$ depend continuously on ω by Lemma B.1. This implies that also the induced inner products on $\bigwedge(\mathbb{R}^d, \langle \cdot, \cdot \rangle_\omega)$ depend continuously on ω. Consequently, the operator norm $\| \cdot \|_{\omega_1, \omega_2}$ is continuous in (ω_1, ω_2). By compactness of $\mathscr{U}$, this implies (7.18).

Step 2. Now assume that Q is the closure of a control set and Σ is locally accessible. Since Σ is real-analytic, we can apply the results of Sect. 5.2 (using Proposition 5.12). In particular, we can apply Proposition 5.11, since the hyperbolicity assumption guarantees that every periodic controlled trajectory in $\mathrm{int}\, D \times \mathrm{int}(\mathscr{U} \times \mathscr{V})$ has exactly d^+ positive Lyapunov exponents, and if $(\varphi(\cdot, x, (\omega, \mu)), (\omega(\cdot), \mu(\cdot)))$ is a controlled trajectory in $\mathscr{Q}$, then

$$a_t^{d+}((\omega, \mu), x) = \log \prod_{i=1}^{d_+} \sigma_i(\Lambda_\omega(\tau, 0)) \to \infty \quad \text{for } t \to \infty.$$

Hence, using that

$$\|\Lambda_\omega(\tau, 0)^\wedge\| = \max_{0 \le k \le d} \sigma_1(\Lambda_\omega(\tau, 0)) \cdots \sigma_k(\Lambda_\omega(\tau, 0)) = \prod_{i=1}^{d_+} \sigma_i(\Lambda_\omega(\tau, 0))$$

for all sufficiently large τ, we obtain the upper bound.   □

Remark 7.7. The estimates of the preceding corollary are similar to the well-known integral formula for the topological entropy of a $\mathscr{C}^\infty$-map $f : M \to M$ on a compact Riemannian manifold (cf. Kozlovski [71]):

$$h_{\mathrm{top}}(f) = \lim_{n \to \infty} \frac{1}{n} \log \int \|(d_x f^n)^\wedge\| \, d\mathrm{vol}.$$

Remark 7.8. Existence of a unique control set with nonempty interior for the inhomogeneous bilinear system Σ is guaranteed under mild assumptions. Indeed, by Vera [109, Teorema 3.45] it is sufficient to assume that (A_0, B) is controllable and $0 \in \mathrm{int}\, \Omega_1$, $0 \in \mathrm{int}\, \Omega_2$, and Ω_2 is bounded and convex.

Finally, an alternative way how to obtain the lower bound in the preceding corollary is to use the following result in combination with Corollary 6.1.

Proposition 7.5. *Let (K, Q) be an admissible pair of Σ such that Q is compact and controlled invariant. Further assume that there exists a vector bundle decomposition*

$$\mathscr{U} \times \mathbb{R}^d = \mathscr{W}^+ \oplus \mathscr{W}^-, \quad d^\pm := \mathrm{rk}\, \mathscr{W}^\pm,$$

into subbundles $\mathscr{W}^+$ and $\mathscr{W}^-$, respectively, both invariant under the control flow of the associated homogeneous system, such that the following holds: There are constants $C, \lambda > 0$ with

$$|\Lambda_\omega(t, 0)x| \ge C e^{\lambda t}|x| \quad \text{for all } t \ge 0, \ (\omega, x) \in \mathscr{W}^+, \tag{7.19}$$

and for every $\delta > 0$ *there is* $D \geq 1$ *with*

$$|\Lambda_\omega(t,0)x| \leq De^{\delta t}|x| \quad \text{for all } t \geq 0, \ (\omega,x) \in \mathscr{W}^-. \tag{7.20}$$

Then $h_{\mathrm{esc}}(K,Q) \leq 0$.

Proof. Let $\pi^+(\omega)$ and $\pi^-(\omega)$ denote the corresponding projections onto the fibers $\mathscr{W}^+_\omega$ and $\mathscr{W}^-_\omega$ ($\omega \in \mathscr{U}$), respectively. Then

$$\pi^\pm(\Theta_t\omega)\Lambda_\omega(t,0) = \Lambda_\omega(t,0)\pi^\pm(\omega) \quad \text{for all } t \in \mathbb{R}, \ \omega \in \mathscr{U}. \tag{7.21}$$

For arbitrary $(\omega,\mu) \in \pi_{\mathscr{U}\times\mathscr{V}}\mathscr{K}_Q$, $\tau > 0$, and $\varepsilon > 0$ let S be a $((\omega,\mu),\tau,\varepsilon)$-separated subset of $K((\omega,\mu),\tau)$. Then it follows that

$$\varphi(t,x,(\omega,\mu)) - \varphi(t,y,(\omega,\mu)) \equiv \Lambda_\omega(t,0)(x-y),$$

and hence for each pair of distinct elements $x,y \in S$ we obtain

$$
\begin{aligned}
\varepsilon &\leq \max_{t\in[0,\tau]} |\Lambda_\omega(t,0)(x-y)| \\
&= \max_{t\in[0,\tau]} \left|\pi^+(\Theta_t\omega)\Lambda_\omega(t,0)(x-y) + \pi^-(\Theta_t\omega)\Lambda_\omega(t,0)(x-y)\right| \\
&\leq \max_{t\in[0,\tau]} \left(\left|\pi^+(\Theta_t\omega)\Lambda_\omega(t,0)(x-y)\right| + \left|\pi^-(\Theta_t\omega)\Lambda_\omega(t,0)(x-y)\right|\right) \\
&\leq \max_{t\in[0,\tau]} \left|\pi^+(\Theta_t\omega)\Lambda_\omega(t,0)(x-y)\right| + \max_{t\in[0,\tau]} |\pi^-(\Theta_t\omega)\Lambda_\omega(t,0)(x-y)| \\
&\overset{(7.21)}{=} \max_{t\in[0,\tau]} \left|\Lambda_\omega(t,0)\pi^+(\omega)(x-y)\right| + \max_{t\in[0,\tau]} |\Lambda_\omega(t,0)\pi^-(\omega)(x-y)|.
\end{aligned}
$$

Assume that the first maximum in the last term is attained at $s^+ \in [0,\tau]$ and the second one at $s^- \in [0,\tau]$. Then, using the cocycle property, we get

$$
\begin{aligned}
\left|\Lambda_\omega(\tau,0)\pi^+(\omega)(x-y)\right| &= \left|\Lambda_{\Theta_{s^+}\omega}(\tau-s^+,0)\Lambda_\omega(s^+,0)\pi^+(\omega)(x-y)\right| \\
&\overset{(7.19)}{\geq} Ce^{\lambda(\tau-s^+)}\left|\Lambda_\omega(s^+,0)\pi^+(\omega)(x-y)\right| \\
&\geq C\max_{t\in[0,\tau]}\left|\Lambda_\omega(t,0)\pi^+(\omega)(x-y)\right|
\end{aligned}
$$

and

$$
\begin{aligned}
\max_{t\in[0,\tau]} |\Lambda_\omega(t,0)\pi^-(\omega)(x-y)| &= |\Lambda_\omega(s^-,0)\pi^-(\omega)(x-y)| \\
&\overset{(7.20)}{\leq} De^{\delta s^-}|\pi^-(\omega)(x-y)| \\
&\leq De^{\delta\tau}|\pi^-(\omega)(x-y)|.
\end{aligned}
$$

Hence, altogether we obtain

$$\varepsilon \le C^{-1}\left|\Lambda_\omega(\tau,0)\pi^+(\omega)(x-y)\right| + De^{\delta\tau}\left|\pi^-(\omega)(x-y)\right|.$$

With $\gamma = \gamma(\tau) := \min\{C, D^{-1}e^{-\delta\tau}\}$ this gives

$$\left|\Lambda_\omega(\tau,0)\pi^+(\omega)(x-y)\right| + \left|\pi^-(\omega)(x-y)\right| \ge \gamma\varepsilon,$$

which implies

$$\left|\Lambda_\omega(\tau,0)\pi^+(\omega)(x-y)\right| \ge \frac{\gamma\varepsilon}{2} \quad \text{or} \quad \left|\pi^-(\omega)(x-y)\right| \ge \frac{\gamma\varepsilon}{2}.$$

Now we cover Q with sets $Q_1,\ldots,Q_n$ such that for each $i \in \{1,\ldots,n\}$ and $x, y \in S \cap Q_i$ it holds that $|\pi^-(\omega)(x-y)| < (\gamma\varepsilon)/2$. The sets $Q_1,\ldots,Q_n$ can be defined as follows: First cover $\pi^-(\omega)Q$ with a minimal collection of d^--dimensional balls of radii $(\gamma\varepsilon)/2$, say $B_1,\ldots,B_n$, $n = n(\omega,\varepsilon,\tau)$, and then define

$$Q_i := \pi^+(\omega)Q \oplus B_i, \quad i = 1,\ldots,n.$$

Then for each distinct $x, y \in S \cap Q_i$ it must hold that $\left|\Lambda_\omega(\tau,0)\pi^+(\omega)(x-y)\right| \ge (\gamma\varepsilon)/2$, so in particular $\pi^+(\omega)x \ne \pi^+(\omega)y$. Hence, the set $\Lambda_\omega(\tau,0)\pi^+(\omega)(S \cap Q_i)$ is $(\gamma\varepsilon)/2$-separated and has the same cardinality as $S \cap Q_i$. Using that $S \subset K((\omega,\mu),\tau)$, we obtain

$$\Lambda_\omega(\tau,0)\pi^+(\omega)(S \cap Q_i) = \pi^+(\Theta_\tau\omega)\Lambda_\omega(\tau,0)(S \cap Q_i)$$

$$\subset \pi^+(\Theta_\tau\omega)\left[\varphi(\tau, S \cap Q_i, (\omega,\mu)) - \int_0^\tau \Lambda_\omega(t,s)B\mu(s)ds\right]$$

$$\subset \underbrace{\pi^+(\Theta_\tau\omega)Q - \pi^+(\Theta_\tau\omega)\int_0^\tau \Lambda_\omega(t,s)B\mu(s)ds}_{=:b(\omega,\mu,\tau)}.$$

Since $\pi^+(\cdot)$ is continuous and $\mathscr{U}$ is compact in the weak*-topology, there is $R > 0$ such that each of the sets $\pi^+(\Theta_\tau\omega)Q - b(\omega,\mu,\tau)$ is contained in a d^+-dimensional ball with radius R contained in $\mathscr{W}_\omega^+$. Using a similar volume argument as in the proof of Proposition 7.4, one easily sees that the maximal cardinality of a $(\gamma\varepsilon)/2$-separated set contained in such a ball is approximately $((2R)/(\gamma\varepsilon))^{d^+}$. Also $\pi^-(\cdot)$ is continuous and hence there is $r > 0$ such that $\pi^-(\omega)Q$ is contained in a d^--dimensional ball with radius r centered at $0 \in \mathscr{W}_\omega^-$, which implies that n is approximately $((2r)/(\gamma\varepsilon))^{d^-}$. We thus obtain

$$\#S \le \sum_{i=1}^n \#(S \cap Q_i) = \sum_{i=1}^n \#\Lambda_\omega(\tau,0)\pi^+(\omega)(S \cap Q_i)$$

$$\le \text{const} \cdot \left(\frac{2r}{\gamma\varepsilon}\right)^{d^-}\left(\frac{2R}{\gamma\varepsilon}\right)^{d^+} = \frac{\text{const}}{\gamma(\tau)^d\varepsilon^d}.$$

This implies $\bar{r}_{\text{sep}}(\tau, K, Q) \le \text{const} \cdot \gamma(\tau)^{-d} = \text{const} \cdot \min\{C, D^{-1}e^{-\delta\tau}\}^{-d}$ and hence

$$h_{\text{esc}}(K, Q) \le \limsup_{\tau \to \infty} \frac{1}{\tau} \log\left(D^d e^{\delta d\tau}\right) = \delta d.$$

Since δ can be chosen arbitrarily small, we have $h_{\text{esc}}(K, Q) \le 0$. $\square$

7.4 Projective Systems

In this section, we apply the nonlinear theory developed in Chaps. 4–6 to continuous-time projective systems. These are the systems on d-dimensional real projective space $\mathbb{P}^d$ induced by bilinear systems on $\mathbb{R}^{d+1}$. The space $\mathbb{P}^d$ is defined as the quotient space of $\mathbb{R}^{d+1}\backslash\{0\}$ with respect to the equivalence relation

$$x \sim y \quad :\Leftrightarrow \quad \exists \alpha \in \mathbb{R}\backslash\{0\}: \ y = \alpha x.$$

That is, the equivalence classes

$$\mathbb{P}x := \{y \in \mathbb{R}^{d+1}\backslash\{0\} \ : \ x \sim y\}$$

are the lines through the origin in $\mathbb{R}^{d+1}$ minus the origin itself. With the quotient topology $\mathbb{P}^d$ becomes a compact topological space which can be endowed with the structure of a $\mathscr{C}^\omega$-manifold. The canonical projection $\pi : \mathbb{R}^{d+1}\backslash\{0\} \to \mathbb{P}^d$ is a submersion and its derivative satisfies

$$\mathrm{d}_x\pi = \alpha\mathrm{d}_{\alpha x}\pi \quad \text{for all } x \in \mathbb{R}^{d+1}\backslash\{0\}, \ \alpha \in \mathbb{R}\backslash\{0\}, \tag{7.22}$$

which follows by differentiation of the identity $\pi(x) \equiv \pi(\alpha x)$. For every subset $A \subset \mathbb{R}^{d+1}$ we denote by $\mathbb{P}A$ the set $\{\mathbb{P}x : x \in A\backslash\{0\}\}$. The following proposition summarizes some well-known properties of $\mathbb{P}^d$.

Proposition 7.6. *The following assertions hold:*

 (i) *The space $\mathbb{P}^d$ can be endowed with a $\mathscr{C}^\omega$-atlas such that it becomes a compact real-analytic manifold of dimension d.*
 (ii) *The space $\mathbb{P}^d$ is connected. For $d \ge 2$ its universal covering space is $S^d \subset \mathbb{R}^{d+1}\backslash\{0\}$ with the twofold covering projection $\pi : S^d \to \mathbb{P}^d$, $x \mapsto \mathbb{P}x$.*
 (iii) *The space $\mathbb{P}^1$ is homeomorphic to the circle S^1 and hence, its universal covering space is $\mathbb{R}$.*

(iv) *There exists a canonical Riemannian metric g on $\mathbb{P}^d$ which is defined by projection of the round metric on S^d.[2] More precisely, if g^r denotes the round metric on S^d, then g is given by*

$$g_{\mathbb{P}x} \left(d_x \pi(v), d_x \pi(w) \right) := g^r_x (v, w) \quad \text{for all } x \in S^d, \ v, w \in T_x S^d.$$

It follows from (7.22) that g is well-defined.

Every bilinear system $\Sigma = (\mathbb{R}, \mathbb{R}^{d+1}, \mathbb{R}^m, \mathcal{U}, \varphi)$ given by differential equations

$$\dot{x}(t) = \underbrace{\left[A_0 + \sum_{i=1}^{m} \omega_i(t) A_i \right]}_{=:A(\omega(t))} x(t), \quad \omega \in \mathcal{U},$$

induces a control-affine system on $\mathbb{P}^d$ with associated vector fields $f_0, f_1, \ldots, f_m$ given by

$$f_i(\mathbb{P}x) = d_x \pi(A_i x), \quad i = 0, 1, \ldots, m,$$

The vector fields f_i are real-analytic. The induced system on $\mathbb{P}^d$ is denoted by $\Sigma_{\mathbb{P}} = (\mathbb{R}, \mathbb{P}^d, \mathbb{R}^m, \mathcal{U}, \psi)$ and its right-hand side by $F = f_0 + \sum_i u_i f_i$. We further assume that the control range Ω is compact and convex with $0 \in \text{int } \Omega$. From the definition of the f_i it immediately follows that

$$\psi(t, \mathbb{P}x, \omega) \equiv \pi(\varphi(t, x, \omega)). \tag{7.23}$$

The system $\Sigma_{\mathbb{P}}$ has been studied in Colonius and Kliemann [25]. To formulate the main result about its controllability properties, we have to introduce the *semigroup* of the bilinear system Σ. To this end, denote by N the set $\{A_0 + \sum_{i=1}^{m} u_i A_i : u \in \Omega\}$ of possible constant right-hand sides of Σ. Then the semigroup $\mathscr{S}(\Sigma)$ is defined by

$$\mathscr{S}(\Sigma) := \left\{ \exp(t_n B_n) \cdots \exp(t_1 B_1) : t_j \geq 0, \ B_j \in N, \ j = 1, \ldots, n \in \mathbb{N} \right\}.$$

For each $t > 0$ also the subsets $\mathscr{S}_t(\Sigma)$ and $\mathscr{S}_{\leq t}(\Sigma)$ consisting of those elements of $\mathscr{S}(\Sigma)$ with $\sum t_j = t$ or $\sum t_j \leq t$, respectively, are of interest. A short version of the main result [25, Theorem 7.3.3] about the controllability properties of $\Sigma_{\mathbb{P}}$ reads as follows.

Theorem 7.6 (The Control Sets on Projective Space). *The following assertions are valid for the projective system $\Sigma_{\mathbb{P}}$ provided that local accessibility holds:*

---

[2]The *round metric* g^r on S^d is the Riemannian metric induced by the Euclidean metric of $\mathbb{R}^{d+1}$, that is, $g^r_x(v, w) = \langle v, w \rangle = \sum_i v_i w_i$ for all $x \in S^d$ and $v, w \in T_x S^d$, where $T_x S^d$ is identified with a linear subspace of $\mathbb{R}^{d+1}$.

(i) *There are k control sets $D_1, \ldots, D_k$ with nonempty interior and $1 \le k \le d + 1$. These control sets are called the main control sets of $\Sigma_{\mathbb{P}}$.*

(ii) *The main control sets can be ordered in such a way that D_k is closed and invariant, D_1 is open, and all other control sets are neither open nor closed.*

(iii) *For every $t > 0$, every $g \in \mathrm{int}\, \mathscr{S}_{\le t}(\Sigma)$, and every $\lambda \in \sigma(g)$, there is a main control set D_i such that the generalized eigenspace $E(\lambda)$ satisfies $\mathbb{P}E(\lambda) \subset \mathrm{int}\, D_i$. The interiors of the main control sets consist exactly of those elements $\mathbb{P}x \in \mathbb{P}^d$ such that x is an eigenvector for a real eigenvalue of some $g \in \mathscr{S}_t(\Sigma) \cap \mathrm{int}\, \mathscr{S}_{\le t+1}(\Sigma)$ for some $t > 0$.*

In order to find expressions for the lower and upper bounds of $h_{\mathrm{inv}}(K, Q)$ derived in Chaps. 4–6, for the projective system $\Sigma_{\mathbb{P}}$, there are several possibilities. If one wants to avoid computations in local coordinates on $\mathbb{P}^d$, one can use the map $\pi : S^d \to \mathbb{P}^d$ which relates the projective system and the system induced by Σ on S^d (to be introduced below). Another possibility is to describe $\mathbb{P}^d$ as a homogeneous space $\mathbb{P}^d = O(d + 1)/(O(d) \times O(1))$, using the fact that $O(d + 1)$ acts transitively on $\mathbb{P}^d$. In this case, we could use the elegant language of Lie groups. However, we do not want to introduce the necessary machinery and so we choose the first way.

Lemma 7.1. *Let f be a vector field on $\mathbb{P}^d$ and $\overline{f}$ a vector field on S^d such that f and $\overline{f}$ are related via the projection $\pi : S^d \to \mathbb{P}^d$, that is, $\mathrm{d}_x \pi\, \overline{f}(x) = f(\pi(x))$ for all $x \in S^d$. Then*

$$\nabla f(\mathbb{P}x)\mathrm{d}_x \pi(v) = \mathrm{d}_x \pi \left(\nabla \overline{f}(x)v \right) \quad \text{for all } x \in S^d, \ v \in T_x S^d.$$

Here, the covariant derivatives are the ones associated with the round metric on S^d and its projection to $\mathbb{P}^d$, respectively.

Proof. For every $x \in S^d$ we can find a chart (ϕ, U) of S^d around x such that $(\psi, V) := (\phi \circ (\pi|_U)^{-1}, \pi(U))$ is a chart of $\mathbb{P}^d$ around $\mathbb{P}x$, since π is a local diffeomorphism. Then it trivially follows that

$$\mathrm{d}\pi(\partial_i \phi) = \partial_i \psi \quad \text{for } i = 1, \ldots, d.$$

Therefore, using the local expressions of the objects involved, we find

$$\mathrm{d}_x \pi \left(\nabla \overline{f}(x)v \right) = v^i \left(\overline{f}^j(x)\Gamma_{ij}^k(x)\partial_k \psi(x) + \frac{\partial \overline{f}^j}{\partial \phi^i}(x)\partial_j \psi(x) \right),$$

where Γ_{ij}^k denote the Christoffel symbols on S^d with respect to (ϕ, U), and

$$\nabla f(\mathbb{P}x)\mathrm{d}_x \pi(v) = v^i \left(f^j(\mathbb{P}x)\tilde{\Gamma}_{ij}^k(\mathbb{P}x)\partial_k \psi(\mathbb{P}x) + \frac{\partial f^j}{\partial \psi^i}(\mathbb{P}x)\partial_j \psi(\mathbb{P}x) \right)$$

with $\tilde{\Gamma}_{ij}^k$ denoting the Christoffel symbols on $\mathbb{P}^d$ with respect to (ψ, V). From the assumption $d_x \pi \overline{f}(x) = f(\pi(x))$ it follows that $\overline{f}^j(x) = f^j(\mathbb{P}x)$ for $j = 1, \ldots, d$. Moreover,

$$\frac{\partial f^j}{\partial \psi^i}(\mathbb{P}x) = \frac{\partial (f^j \circ \psi^{-1})}{\partial x_i}(\psi(\pi(x)))$$

$$= \frac{\partial (f^j \circ \pi \circ \phi^{-1})}{\partial x_i}(\phi(x)) = \frac{\partial \overline{f}^j}{\partial \phi^i}(x).$$

From the definition of the Riemannian metric g on $\mathbb{P}^d$ it follows that $g_{ij}(\mathbb{P}x) = g_{ij}^r(x)$, where g_{ij} are the components of the metric g with respect to ψ and g_{ij}^r the components of the round metric g^r on S^d with respect to ϕ. This implies $\tilde{\Gamma}_{ij}^k(\mathbb{P}x) = \Gamma_{ij}^k(x)$, which concludes the proof. $\square$

To apply the above lemma, we introduce an intermediate system between the bilinear system Σ and the projective system $\Sigma_\mathbb{P}$, namely the system $\Sigma_S = (\mathbb{R}, S^d, \mathbb{R}^m, \mathcal{U}, \tilde{\psi})$ which is the control-affine system on S^d whose right-hand side vector fields g_i are given by

$$g_i(x) = d_x \tilde{\pi}(A_i x), \quad i = 0, 1, \ldots, m,$$

with $\tilde{\pi} : \mathbb{R}^{d+1} \setminus \{0\} \to S^d$, $\tilde{\pi}(x) = x/|x|$. More explicitly,

$$g_i(x) = \left(I - xx^T\right) A_i x = (A_i - (x^T A_i x)I)x.$$

We write $G = g_0 + \sum_i u_i g_i$ for the corresponding right-hand side. It is clear that the vector field g_i is related to f_i via $\pi : S^d \to \mathbb{P}^d$, $\pi(x) = \mathbb{P}x$. Hence, we can use Lemma 7.1 to compute the covariant derivatives of the vector fields f_i by computing the corresponding derivatives of the g_i's.

Lemma 7.2. *The covariant derivative of a vector field f on S^d of the form $f(x) = d_x \tilde{\pi}(Ax)$ with $A \in \mathbb{R}^{(d+1) \times (d+1)}$ is given by*

$$\nabla f(x)v = Q_x(A - x^T Ax I)v \quad \text{for all } x \in S^d, \ v \in T_x S^d,$$

where $Q_x := I - xx^T$ is the orthogonal projection onto $T_x S^d$. The symmetrized covariant derivative of f is given by

$$S\nabla f(x)v = (Q_x A^+ - x^T A^+ x I)v \quad \text{for all } x \in S^d, \ v \in T_x S^d,$$

where $A^+ := (1/2)(A + A^T)$.

Proof. By Gallot et al. [48, Proposition 2.56], $\nabla f(x)v$ is given by the orthogonal projection of $Df(x)v$ to $T_x S^d = x^\perp$, where $Df(x)$ denotes the Jacobi-matrix of f

at x, considered as a map on $\mathbb{R}^{d+1}$. An elementary computation gives

$$\mathrm{D}f(x)v = (A - x^T A x I - x x^T (A + A^T)) v \quad \text{for all } v \in T_x S^d.$$

With the orthogonal projection $Q_x = I - x x^T$ we obtain

$$Q_x \mathrm{D}f(x)v = (I - x x^T) (A - x^T A x I - x x^T (A + A^T)) v$$
$$= (I - x x^T) (A - x^T A x I) v.$$

Hence, $\nabla f(x)v = Q_x (A - x^T A x I)v$. The symmetrized covariant derivative of f is defined as

$$S\nabla f(x) = \frac{1}{2} \left(\nabla f(x) + \nabla f(x)^* \right).$$

The adjoint operator $\nabla f(x)^*$ is the unique endomorphism of $T_x S^d$ such that $\langle \nabla f(x)v, w \rangle = \langle v, \nabla f(x)^* w \rangle$ for all $v, w \in T_x S^d = x^\perp$. Since for $v, w \in x^\perp$ it holds that

$$\langle Q_x (A - x^T A x I)v, w \rangle = \langle \underbrace{v}_{=Q_x v}, (A^T - x^T A x I) \underbrace{Q_x w}_{=w} \rangle$$
$$= \langle v, Q_x (A^T - x^T A x I)w \rangle,$$

we have $\nabla f(x)^* v = Q_x (A^T - x^T A x I)v$ and thus

$$S\nabla f(x) = \frac{1}{2} \left[Q_x (A - x^T A x I) + Q_x (A^T - x^T A x I) \right]$$
$$= \frac{1}{2} Q_x \left[A + A^T - 2 x^T A x I \right].$$

Using that $x^T A x = x^T A^+ x$, we obtain

$$S\nabla f(x)v = Q_x \left[A^+ - x^T A^+ x I \right] = (Q_x A^+ - x^T A^+ x I)v,$$

which concludes the proof. $\square$

From the main results of Chap. 4 we now obtain the following estimates.

Theorem 7.7. *Let (K, Q) be an admissible pair for $\Sigma_\mathbb{P}$ such that Q is compact. Then*

$$h_{\text{inv,out}}(K, Q) \leq \max_{(\mathbb{P}x, u) \in Q \times \Omega} \lambda_{\max} \left(Q_{x/|x|} A(u)^+ - \frac{x^T A(u)^+ x}{|x|^2} I \right) \cdot \overline{\dim}_C(K).$$

$$(7.24)$$

If, in addition, K has positive volume, then

$$h_{\mathrm{inv,out}}(K, Q) \geq \min_{(\mathbb{P}x,u)\in Q\times\Omega} \left(\mathrm{tr}\, A(u) - (d+1) \cdot \frac{x^T A(u)x}{|x|^2} \right). \qquad (7.25)$$

Proof. From Lemma 7.1 it easily follows that

$$S\nabla f(\mathbb{P}x)\mathrm{d}_x \pi(v) \equiv \mathrm{d}_x \pi \left(S\nabla \overline{f}(x)v \right)$$

for related vector fields f on $\mathbb{P}^d$ and $\overline{f}$ on S^d. (Indeed, it holds that $\nabla f(\mathbb{P}x)^* \circ \mathrm{d}_x\pi \equiv \mathrm{d}_x\pi \circ \nabla\overline{f}(x)^*$, which follows from the fact that $\mathrm{d}_x\pi$ is an isometry.) This implies

$$\lambda_{\max}(S\nabla F_u(\mathbb{P}x)) = \lambda_{\max}\left(S\nabla G_u\left(\frac{x}{|x|} \right) \right).$$

Then Lemma 7.2 together with Corollary 4.1 gives the upper bound. For the lower bound, we have to compute the trace of $\nabla F_u(\mathbb{P}x)$ which coincides with the trace of $\nabla G_u(x/|x|)$. Choosing an orthonormal basis $v_1, \ldots, v_d$ of $T_x S^d$ for some $x \in S^d$, we find

$$\mathrm{tr}\, \nabla G_u(x) = \sum_{i=1}^{d} \langle Q_x(A(u) - x^T A(u)xI)v_i, v_i \rangle$$

$$= \sum_{i=1}^{d} \langle (A(u) - x^T A(u)xI)v_i, Q_x v_i \rangle$$

$$= \sum_{i=1}^{d} \langle (A(u) - x^T A(u)xI)v_i, v_i \rangle$$

$$= \mathrm{tr}(A(u) - x^T A(u)xI) - \underbrace{\langle (A(u) - x^T A(u)xI)x, x \rangle}_{=0}$$

$$= \mathrm{tr}\, A(u) - (d+1) \cdot x^T A(u)x,$$

which, together with Corollary 4.4 (ii) gives the desired result.   □

Our next aim is to apply the upper estimates of Chap. 5 to the main control sets of $\Sigma_{\mathbb{P}}$. The one-dimensional case can be treated with the results of Sect. 7.1. Hence, we may assume that $d \geq 2$. Notice that the basic assumptions of Sect. 5.2 are satisfied if we assume local accessibility for $\Sigma_{\mathbb{P}}$. Indeed, the system $\Sigma_{\mathbb{P}}$ is real-analytic and, by Proposition 5.6 (v) and Proposition 7.6 (ii), local accessibility implies strong accessibility. Alternatively, we can use Proposition 5.12.

Recall that each main control set is contained in a chain control set (see Proposition 1.24). The chain control sets of $\Sigma_{\mathbb{P}}$ can be described using Selgrade's theorem and the fact that their full time lifts to $\mathscr{U} \times \mathbb{P}^d$ are the maximal invariant chain transitive sets of the control flow. Indeed, by Selgrade's theorem B.1 there exists a unique vector bundle decomposition

$$\mathscr{U} \times \mathbb{R}^{d+1} = \mathscr{W}^1 \oplus \cdots \oplus \mathscr{W}^r$$

with subbundles $\mathscr{W}^i$ that are invariant under the control flow of Σ. Each $\mathscr{W}^i$ corresponds to a chain recurrent component $\mathscr{Q}_i$ of the control flow of $\Sigma_{\mathbb{P}}$ in the sense that

$$\mathscr{W}^i = \left\{ (\omega, x) \in \mathscr{U} \times \mathbb{R}^{d+1} \ : \ x \neq 0 \Rightarrow (\omega, \mathbb{P}x) \in \mathscr{Q}_i \right\}.$$

Each of the sets $\mathscr{Q}_i \subset \mathscr{U} \times \mathbb{P}^d$ projects onto a chain control set $Q_i \subset \mathbb{P}^d$ by Proposition 1.24 (iv), and we have the relations

$$\mathscr{Q}_i = \left\{ (\omega, \mathbb{P}x) \in \mathscr{U} \times \mathbb{P}^d \ : \ \psi(\mathbb{R}, \mathbb{P}x, \omega) \subset Q_i \right\}, \quad i = 1, \ldots, r. \qquad (7.26)$$

The *multiplicity* of the chain control set Q_i is defined as the rank of the corresponding subbundle $\mathscr{W}^i$. Before we formulate our main theorem, we prove a simple fact.

Proposition 7.7. *Assume that $\Sigma_{\mathbb{P}}$ is locally accessible and let $Q \subset \mathbb{P}^d$ be a chain control set of multiplicity one. Then Q contains only one main control set D and $\mathrm{cl}\, D = Q$.*

Proof. Assume to the contrary that Q contains two main control sets D_i and D_j, $i \neq j$. Take $\mathbb{P}x \in \mathrm{int}\, D_i$. Then one finds $t > 0$ and $g \in \mathscr{S}_t(\Sigma) \cap \mathrm{int}\, \mathscr{S}_{\leq t+1}(\Sigma)$ such that x is an eigenvector of g for a real eigenvalue λ. If $E(\lambda)$ denotes the generalized eigenspace of λ, then $\mathbb{P}E(\lambda) \subset \mathrm{int}\, D_i$ by Theorem 7.6 (iii). Since Q has multiplicity one, it follows that $\dim E(\lambda) = 1$. Hence, there exists another eigenvalue μ of g with $\mathbb{P}E(\mu) \subset \mathrm{int}\, D_j$. Again, $\dim E(\mu) = 1$ and μ is real. The piecewise constant periodic control function corresponding to g is denoted by ω_g. Then $\psi(\mathbb{R}, \mathbb{P}x, \omega_g) \subset Q$ and $\psi(\mathbb{R}, \mathbb{P}y, \omega_g) \subset Q$. This implies $(\omega_g, x) \in \mathscr{W}$ and $(\omega_g, y) \in \mathscr{W}$ or $x \in \mathscr{W}_{\omega_g}$, $y \in \mathscr{W}_{\omega_g}$, which is a contradiction, since $\mathscr{W}_{\omega_g}$ is one-dimensional and $y \notin \mathbb{R}x$. Hence, Q contains exactly one main control set D. It remains to show that $\mathrm{cl}\, D = Q$. By [25, Theorem 7.3.16]

$$Q = \mathrm{cl}\left\{ \mathbb{P}x \in \mathbb{P}^d \ : \ \exists t > 0, \ g \in \mathrm{int}\, \mathscr{S}_{\leq t} : \ \psi(\mathbb{R}, \mathbb{P}x, \omega_g) \subset \mathrm{cl}\, D \right\}.$$

This implies $Q = \mathrm{cl}\, \mathrm{int}\, D = \mathrm{cl}\, D$. $\square$

Remark 7.9. As we will see below, the chain control sets of multiplicity one are exactly the hyperbolic chain control sets. We remark that a more general result, proved in Colonius and Du [21, Theorem 3], states that every hyperbolic chain control set of a control-affine system is the closure of a control set, provided that local accessibility holds.

We also need the following lemma from [25, Lemma 7.3.2].

Lemma 7.3. *For some* $t > 0$ *let* $g \in \mathscr{S}_t(\Sigma)$. *Then there exist a decreasing sequence* $t_n \searrow t$ *and* $g_n \in \mathscr{S}_t(\Sigma) \cap \operatorname{int} \mathscr{S}_{\leq t_n}(\Sigma)$ *with* $\lim_{n \to \infty} g_n = g$.

Theorem 7.8. *Assume that* $\Sigma_{\mathbb{P}}$ *is locally accessible, let* Q *be a chain control set of multiplicity one, and* D *the corresponding main control set with* $\operatorname{cl} D = Q$. *Moreover, assume that every equilibrium in* $\operatorname{int} D$ *has positive Lyapunov exponents only. Then for every compact set* $K \subset D$ *with nonempty interior it holds that*

$$h_{\mathrm{inv}}(K, Q) = \inf_{(\omega, \mathbb{P}x) \in \mathscr{Q}} \limsup_{\tau \to \infty} \frac{1}{\tau} \int_0^\tau \operatorname{div} F_{\omega(s)}(\psi(s, \mathbb{P}x, \omega)) ds$$

$$= \inf_{(u, \mathbb{P}x) \in \Omega \times Q: \, F(\mathbb{P}x, u) = 0} \left[\operatorname{tr} A(u) - (d+1) \frac{x^T A(u)x}{|x|^2} \right]. \quad (7.27)$$

Proof. The theorem is proved in five steps.

Step 1. Note that the map $\alpha : \mathbb{R} \times (\mathscr{U} \times \mathbb{P}^d) \to \mathbb{R}$, defined by

$$\alpha_t(\omega, \mathbb{P}x) := \int_0^t \operatorname{div} F_{\omega(s)}(\psi(s, \mathbb{P}x, \omega)) ds = \log |\det d_{\mathbb{P}x} \psi_{t,\omega}|,$$

is an additive cocycle over the control flow Φ of $\Sigma_{\mathbb{P}}$. Writing $\varphi(t, x, \omega) = \varphi(t, \omega)x$ and using the expression for the divergence of $F_{\omega(s)}$ determined in the proof of Theorem 7.7 together with the Liouville formula, we obtain

$$\alpha_t(\omega, \mathbb{P}x) = \int_0^t \operatorname{div} F_{\omega(s)}(\psi(s, \mathbb{P}x, \omega)) ds$$

$$= \int_0^t \left[\operatorname{tr} A(\omega(s)) - (d+1) \frac{\langle A(\omega(s))\varphi(s, \omega)x, \varphi(s, \omega)x \rangle}{|\varphi(s, \omega)x|^2} \right] ds$$

$$= \log |\det \varphi(t, \omega)| - (d+1) \int_0^t \frac{\langle A(\omega(s))\varphi(s, \omega)x, \varphi(s, \omega)x \rangle}{|\varphi(s, \omega)x|^2} ds$$

$$= \log |\det \varphi(t, \omega)| - (d+1) \int_0^t \frac{d}{ds} \log |\varphi(s, \omega)x| ds$$

$$= \log |\det \varphi(t, \omega)| - (d+1)(\log |\varphi(t, \omega)x| - \log |x|).$$

Now let $\mathscr{W} \subset \mathscr{U} \times \mathbb{R}^{d+1}$ be the subbundle corresponding to the chain control set Q and let $(\omega, x) \in \mathscr{W}$ with $|x| = 1$. Writing

$$a(\omega, t) : \mathscr{W}_\omega \to \mathscr{W}_{\Theta_t \omega}, \quad y \mapsto \varphi(t, \omega)y,$$

for the maps between the one-dimensional fibers of $\mathcal{W}$, we find

$$\log|\varphi(t,\omega)x| = \log|a(\omega,t)x| = \log\|a(\omega,t)\|,$$

where $\|a(\omega,t)\|$ denotes the operator norm of $a(\omega,t)$. This gives

$$\alpha_t(\omega,\mathbb{P}x) = \log|\det\varphi(t,\omega)| - (d+1)\log\|a(\omega,t)\|.$$

So we see that $\alpha_t(\omega,\mathbb{P}x)$ is in fact independent of $\mathbb{P}x$ on $\mathcal{Q}$. Hence, in the following we sometimes only write $\alpha_t(\omega)$.

Step 2. By Lemma 6.4 (i), $\mathcal{Q}$ is a compact Φ-invariant set. Therefore, by Theorem B.2, the infimum

$$\kappa := \inf_{(\omega,\mathbb{P}x)\in\mathcal{Q}} \limsup_{t\to\infty} \frac{1}{t}\alpha_t(\omega,\mathbb{P}x)$$

is attained at some $(\omega,\mathbb{P}x) \in \mathcal{Q}$ as a limit. Our aim is to show that it is attained at an equilibrium pair. To this end, let $(t_n)_{n\geq 1}$ be any monotonically decreasing sequence of positive real numbers such that $T_n := \sum_{i=1}^{n} t_i \to \infty$ for $n \to \infty$ (for example, $t_n = 1/n$). Additionally, let $T_0 := 0$. Using additivity, for all $n \in \mathbb{N}$ we obtain

$$\kappa = \lim_{n\to\infty} \frac{1}{T_n}\alpha_{T_n}(\omega) = \lim_{n\to\infty} \frac{1}{T_n}\sum_{i=1}^{n}\alpha_{t_i}(\Theta_{T_{i-1}}\omega).$$

Moreover, we have

$$\frac{1}{T_n}\sum_{i=1}^{n}\alpha_{t_i}(\Theta_{T_{i-1}}\omega) = \sum_{i=0}^{n-1}\frac{t_{i+1}}{\sum_{j=1}^{n}t_j}\left[\frac{1}{t_{i+1}}\alpha_{t_{i+1}}(\Theta_{T_i}\omega)\right]$$

$$\geq \sum_{i=0}^{n-1}\frac{t_{i+1}}{\sum_{j=1}^{n}t_j}\left[\min_{0\leq i\leq n-1}\frac{1}{t_{i+1}}\alpha_{t_{i+1}}(\Theta_{T_i}\omega)\right]$$

$$= \min_{0\leq i\leq n-1}\frac{1}{t_{i+1}}\alpha_{t_{i+1}}(\Theta_{T_i}\omega).$$

For every $n \geq 1$ we extend $\Theta_{T_{n-1}}\omega|_{[0,t_n)}$ to a t_n-periodic function ω_n on $\mathbb{R}$. Let $x_n \in \mathcal{W}_{\omega_n}$ be an eigenvector with $|x_n| = 1$ of the corresponding matrix $g_n = \varphi(t_n,\omega_n)$. (Observe that this eigenvector for a real eigenvalue exists, since $\mathcal{W}$ is a subbundle of rank one.) Then

$$\lim_{t\to\infty}\frac{1}{t}\alpha_t(\omega_n) = \frac{1}{t_n}\alpha_{t_n}(\omega_n).$$

Since the cocycle α has the property that $\alpha_t(\mu_1) = \alpha_t(\mu_2)$ if μ_1 and μ_2 coincide on $[0, t]$, we have $\alpha_{t_n}(\omega_n) = \alpha_{t_n}(\Theta_{T_{n-1}}\omega)$ and we obtain

$$\kappa \geq \lim_{n\to\infty} \min_{1\leq i\leq n} \frac{1}{t_i}\alpha_{t_i}(\omega_i) = \inf_{n\in\mathbb{N}} \frac{1}{t_n}\alpha_{t_n}(\omega_n)$$

$$= \inf_{n\in\mathbb{N}} \lim_{t\to\infty} \frac{1}{t}\alpha_t(\omega_n, \mathbb{P}x_n) \geq \kappa.$$

Since the first number in the decreasing sequence t_n can be chosen arbitrarily small, this shows that κ can be approximated by a sequence $(1/\tau_n)\alpha_{\tau_n}(\mu_n, \mathbb{P}x_n)$ such that μ_n is τ_n-periodic and $\tau_n \leq 1/n$. In the following, we fix such a sequence and denote it by λ_n.

Step 3. Define

$$q(u, x) := \operatorname{tr} A(u) - (d+1)\frac{\langle A(u)x, x\rangle}{|x|^2}, \quad q : \mathbb{R}^m \times (\mathbb{R}^{d+1}\setminus\{0\}) \to \mathbb{R},$$

and note that q is continuous and by Step 1 satisfies

$$\lambda_n = \frac{1}{\tau_n}\int_0^{\tau_n} q_n(s)\mathrm{d}s \quad \text{with} \quad q_n(s) := q(\mu_n(s), \varphi(s, \mu_n)x_n).$$

We claim that for every $T > 0$ it holds that

$$\kappa = \lim_{n\to\infty} \frac{1}{T}\int_0^T q_n(s)\mathrm{d}s.$$

To prove this, fix $T > 0$ and $n \in \mathbb{N}$, and write $T = k_n\tau_n - r_n$ with $k_n \in \mathbb{N}$ and $r_n \in [0, \tau_n)$. Then

$$\lambda_n - \frac{1}{T}\int_0^T q_n(s)\mathrm{d}s$$

$$= \frac{1}{T+r_n}\int_0^{T+r_n} q_n(s)\mathrm{d}s - \frac{1}{T}\int_0^T q_n(s)\mathrm{d}s$$

$$= \left(\frac{1}{T+r_n} - \frac{1}{T}\right)\int_0^T q_n(s)\mathrm{d}s + \frac{1}{T+r_n}\int_T^{T+r_n} q_n(s)\mathrm{d}s,$$

where we used τ_n-periodicity of $q_n(\cdot)$. Since

$$q_n(s) \leq \max\left\{q(u, x) : (u, x) \in \Omega \times S^d\right\} < \infty$$

for almost all s, the difference $\lambda_n - (1/T)\int_0^T q_n(s)\mathrm{d}s$ tends to zero for $n \to \infty$. Since $\lambda_n \to \kappa$, this proves the claim.

Step 4. The sequence μ_n has a weak*-convergent subsequence with limit $\mu_* \in \mathcal{U}$. Hence, we may assume that $\mu_n \overset{\star}{\rightharpoonup} \mu_*$ and that also the corresponding eigenvectors x_n converge to some $x_* \in \mathcal{W}_{\mu_*}$. Fix $t \in \mathbb{R}$. Then for every $n \in \mathbb{N}$ we can write $t = \tau_n k(n) + r(n)$ with $k(n) \in \mathbb{Z}$ and $r(n) \in [0, \tau_n) \subset [0, 1/n)$ which gives

$$\psi(t, \mathbb{P}x_*, \mu_*) = \lim_{n \to \infty} \psi(t, \mathbb{P}x_n, \mu_n) = \lim_{n \to \infty} \psi(r(n), \mathbb{P}x_n, \mu_n)$$

$$= \psi(0, \mathbb{P}x_*, \mu_*) = \mathbb{P}x_*. \tag{7.28}$$

By continuity, we have $\varphi(s, \mu_n)x_n \to \varphi(s, \mu_*)x_*$ for $n \to \infty$ uniformly for $s \in [0, T]$. Moreover, q is uniformly continuous on the compact set $\Omega \times S^d$ which implies

$$\kappa = \lim_{n \to \infty} \frac{1}{T} \int_0^T q(\mu_n(s), \varphi(s, \mu_n)x_n)ds$$

$$= \lim_{n \to \infty} \frac{1}{T} \int_0^T q(\mu_n(s), \varphi(s, \mu_*)x_*)ds = \lim_{n \to \infty} \frac{1}{T} \int_0^T q(\mu_n(s), x_*)ds.$$

Note that

$$\int_0^T q(\mu_n(s), x_*)ds = \int_0^T \left[\operatorname{tr} A(\mu_n(s)) - (d+1)\langle A(\mu_n(s))x_*, x_* \rangle \right] ds$$

$$= \int_0^T \left[\operatorname{tr} A(\mu_n(s)) - (d+1)\langle A(\mu_n(s))x_*, x_* \rangle \right] ds$$

$$= T \left[\operatorname{tr} A_0 - (d+1)\langle A_0 x_*, x_* \rangle \right]$$

$$+ \sum_{i=1}^m \left[\operatorname{tr} A_i - (d+1)\langle A_i x_*, x_* \rangle \right] \int_0^T \mu_{n,i}(s)ds.$$

By weak*-convergence $\mu_n \overset{\star}{\rightharpoonup} \mu_*$ we obtain

$$\int_0^T \mu_{n,i}(s)ds = \int_0^T \langle \mu_n(s), e_i \rangle ds \to \int_0^T \langle \mu_*(s), e_i \rangle ds.$$

This gives

$$\kappa = \frac{1}{T} \int_0^T q(\mu_*(s), x_*)ds.$$

Now fix $T > 0$ and define

$$V := \{\mu_*(t) : t \in [0, T] \text{ and } x_* \text{ is an eigenvector of } A(\mu_*(t))\}.$$

By differentiating the identity (7.28) we find that the set $\{t \in [0,T] : \mu_*(t) \in V\}$ has Lebesgue measure T. Let

$$\gamma := \inf\{\operatorname{tr} A(v) - (d+1)\beta(v) : v \in V\},$$

where $\beta(v)$ denotes the eigenvalue of $A(v)$ such that for the corresponding eigenvector x_v (and for the constant control function $v \in \mathscr{U}$) one has $(v, x_v) \in \mathscr{W}$. One easily sees that necessarily $x_v = x_*$ for all $v \in V$. Existence of the infimum follows from compactness of Ω and continuity of q. There exists $v_* \in \Omega$ with

$$\gamma = \min_{v \in \operatorname{cl} V} q(v, x_*) = q(v_*, x_*).$$

This implies

$$\kappa = \frac{1}{T} \int_0^T q(\mu_*(s), x_*) \mathrm{d}s \geq q(v_*, x_*) = \lim_{t \to \infty} \frac{1}{t} \alpha_t(v_*, \mathbb{P}x_*) \geq \kappa.$$

This proves that κ is attained at the equilibrium pair (v_*, x_*) and hence gives the second equality in (7.27).

Step 5. We complete the proof by verifying the first equality in (7.27). The lower bound follows in the same way as in Theorem 7.4. For the upper bound, assume that $(u, \mathbb{P}x)$ is an equilibrium pair with $(u, x) \in \mathscr{W}$ and $|x| = 1$. Then x is an eigenvector of $A(u)$ and hence of $g_t := \exp(t A(u))$ for every $t \geq 0$. Fix $t_0 > 0$, let $g := g_{t_0}$, and denote by λ_* the corresponding eigenvalue of g. Then, by Lemma 7.3, there are sequences $g_n \to g$ and $t_n \searrow t_0$ such that $g_n \in \mathscr{S}_{t_0}(\Sigma) \cap \operatorname{int} \mathscr{S}_{\leq t_n}(\Sigma) \subset \mathscr{S}_{t_0}(\Sigma) \cap \operatorname{int} \mathscr{S}_{\leq t_0 + 1}(\Sigma)$ for all $n \in \mathbb{N}$. There also exists a corresponding sequence of eigenvalues λ_n such that $\lambda_n \to \lambda_*$ and a sequence of eigenvectors $x_n \to x$ with $|x_n| = 1$. Since $\mathscr{W}$ has rank one, these eigenvalues and eigenvectors are real. Hence, there is a periodic trajectory starting at $\mathbb{P}x_n$ of period t_0 corresponding to a piecewise constant control function ω_n. Furthermore, by Theorem 7.6 (iii), the points $\mathbb{P}x_n$ are contained in the interior of D. By slightly perturbing the control functions ω_n, we can achieve that $\omega_n \in \operatorname{int} \mathscr{U}$ without destroying the convergence statements. Then

$$\frac{1}{t_0} \alpha_{t_n}(\omega_n, \mathbb{P}x_n) = \frac{1}{t_0} (\log|\det g_n| - (d+1)\log|g_n x_n|)$$

$$\to \frac{1}{t_0}(\log|\det g| - (d+1)\log|gx|).$$

Now, with Proposition 5.9, we obtain

$$h_{\mathrm{inv}}(K, Q) \leq \frac{1}{t_0} \left(\log |\det g| - (d+1) \log |gx| \right)$$

$$= \frac{1}{t_0} \left(\log |\det \exp(t_0 A(u))| - (d+1) \log |\exp(t_0 A(u))x| \right)$$

$$= \mathrm{tr}\, A(u) - (d+1) x^T A(u) x.$$

Since the equilibrium pair $(u, \mathbb{P}x)$ was chosen arbitrarily, we can pass over to the infimum, which concludes the proof.   □

Let us finish this discussion with the questions whether the assumptions that all Lyapunov exponents be positive is realistic and if the above result can be generalized to other main control sets.

In the following, we fix a chain control set $Q = Q_i$ with its full time lift $\mathcal{Q}$. Then we can construct invariant subbundles of the vector bundle

$$\pi_{\mathcal{Q}}: \bigcup_{(\omega, \mathbb{P}x) \in \mathcal{Q}} \{\omega\} \times T_{\mathbb{P}x}\mathbb{P}^d \to \mathcal{Q}, \quad (\omega, v) \mapsto (\omega, \pi_{T\mathbb{P}^d}(v)), \qquad (7.29)$$

where $\pi_{T\mathbb{P}^d}: T\mathbb{P}^d \to \mathbb{P}^d$ is the base point projection. We define

$$\mathscr{V}^0 := \mathscr{W}^i, \quad \mathscr{V}^- := \mathscr{W}^1 \oplus \cdots \oplus \mathscr{W}^{i-1}, \quad \mathscr{V}^+ := \mathscr{W}^{i+1} \oplus \cdots \oplus \mathscr{W}^r.$$

Using the fact that the subbundles $\mathscr{W}^i$ are exponentially separated from each other (cf. Colonius and Kliemann [25, Theorem 5.1.4]), we may assume that the subbundles $\mathscr{W}^1, \ldots, \mathscr{W}^r$ are ordered by increasing growth rates. Then the growth rates in $\mathscr{V}^+$ dominate those in $\mathscr{V}^0$ and the growth rates in $\mathscr{V}^0$ dominate those in $\mathscr{V}^-$. More precisely: Let P^0, P^+ and P^- denote the projections from $\mathscr{U} \times \mathbb{R}^{d+1}$ onto $\mathscr{V}^0$, $\mathscr{V}^+$ and $\mathscr{V}^-$, respectively. Then there are constants $c_1, c_2 \geq 1$ and $\alpha_1, \alpha_2 > 0$ such that

$$c_1^{-1} e^{\alpha_1 t} |\varphi(t, \omega) P^-(\omega)x| \leq |\varphi(t, \omega) P^0(\omega)x| \leq c_2 e^{-\alpha_2 t} |\varphi(t, \omega) P^+(\omega)x|$$

$$(7.30)$$

for all $t \geq 0$ and $(\omega, x) \in \mathscr{U} \times \mathbb{R}^{d+1}$ with $|P^-(\omega)x| = |P^0(\omega)x| = |P^+(\omega)x|$. Now we can define the desired subbundles of (7.29) by setting

$$E^0_{\omega, \mathbb{P}x} := \mathrm{d}_x \pi \, \mathscr{V}^0_\omega, \quad E^\pm_{\omega, \mathbb{P}x} := \mathrm{d}_x \pi \, \mathscr{V}^\pm_\omega \quad \text{for all } (\omega, \mathbb{P}x) \in \mathcal{Q}. \qquad (7.31)$$

Proposition 7.8. *The following assertions hold:*

(i) $E^0_{\omega, \mathbb{P}x}$ *and* $E^\pm_{\omega, \mathbb{P}x}$ *are well-defined linear subspaces of* $T_{\mathbb{P}x}\mathbb{P}^d$. *Their dimensions are constant on* $\mathcal{Q}$ *with*

$$\dim E^0_{\omega, \mathbb{P}x} = \mathrm{rk}\, \mathscr{V}^0 - 1 \quad \text{and} \quad \dim E^\pm_{\omega, \mathbb{P}x} = \mathrm{rk}\, \mathscr{V}^\pm.$$

(ii) We have a decomposition

$$T_{\mathbb{P}x}\mathbb{P}^d = E^0_{\omega,\mathbb{P}x} \oplus E^+_{\omega,\mathbb{P}x} \oplus E^-_{\omega,\mathbb{P}x} \quad \text{for all } (\omega, \mathbb{P}x) \in \mathcal{Q}.$$

(iii) The spaces $E^0_{\omega,\mathbb{P}x}$ and $E^{\pm}_{\omega,\mathbb{P}x}$ are the fibers of subbundles $E^0 \to \mathcal{Q}$ and $E^{\pm} \to \mathcal{Q}$ of (7.29) which are invariant under the differential $d\psi_{(\cdot,\cdot)}$.

Proof. (i) It follows from (7.22) that the definitions (7.31) are independent of the choice of x in the corresponding equivalence class. The assertions on the dimensions follow from the definitions of $E^0_{\omega,\mathbb{P}x}$ and $E^{\pm}_{\omega,\mathbb{P}x}$ and the fact that $\ker d_x \pi \subset \mathcal{V}^0_\omega$ whenever $(\omega, \mathbb{P}x) \in \mathcal{Q}$. The latter holds, since $\ker d_x \pi = \mathbb{R}x$ and $\mathcal{V}^0_\omega = \{0\} \cup \{x \in \mathbb{R}^{d+1} \setminus \{0\} : (\omega, \mathbb{P}x) \in \mathcal{Q}\}$.

(ii) Since $\mathbb{R}^{d+1} = \mathcal{V}^0_\omega \oplus \mathcal{V}^+_\omega \oplus \mathcal{V}^-_\omega$ for all ω and $d_x \pi$ is surjective for all x, it follows that $T_{\mathbb{P}x}\mathbb{P}^d = E^0_{\omega,\mathbb{P}x} + E^+_{\omega,\mathbb{P}x} + E^-_{\omega,\mathbb{P}x}$. Looking at the dimensions of the three subspaces, it follows that this must be a direct sum.

(iii) To prove that the spaces $E^0_{\omega,\mathbb{P}x}$ define a subbundle, we only have to show that $E^0 = \bigcup_{\mathcal{Q}} \{\omega\} \times E^0_{\omega,\mathbb{P}x}$ is closed in $F := \bigcup \{\omega\} \times T_{\mathbb{P}x}\mathbb{P}^d$. To this end, consider a sequence $(\omega_n, v_n) \in E^0$ converging to some $(\omega, v) \in F$. Then $v_n \in E^0_{\omega_n,\mathbb{P}x_n} = d_{x_n} \pi \mathcal{V}^0_{\omega_n}$ for a sequence x_n with $|x_n| = 1$ and $(\omega_n, \mathbb{P}x_n) \in \mathcal{Q}$. We may assume that $x_n \to x$. By compactness of $\mathcal{Q}$ we have $(\omega, \mathbb{P}x) \in \mathcal{Q}$. Then it remains to prove that $v \in E^0_{\omega,\mathbb{P}x}$. To this end, let $w_n \in \mathcal{V}^0_{\omega_n}$ such that $d_{x_n} \pi(w_n) = v_n$. We can decompose $w_n = \lambda_n x_n + \tilde{w}_n$ with $\langle \tilde{w}_n, x_n \rangle = 0$. It follows that $\tilde{w}_n \in T_{x_n}S^d \cap \mathcal{V}^0_{\omega_n}$, since $T_{x_n}S^d = x_n^\perp$ and both x_n and w_n are in $\mathcal{V}^0_{\omega_n}$. Moreover, $d_{x_n}\pi(\tilde{w}_n) = v_n$. The vector $\tilde{w}_n$ is unique and independent of the choice of w_n. Hence, $\tilde{w}_n = (d_{x_n} \pi|_{T_{x_n}S^d})^{-1}(v_n)$ with the isomorphism $d_{x_n}\pi : T_{x_n}S^d \to T_{\mathbb{P}x_n}\mathbb{P}^d$. This implies that $(\tilde{w}_n)$ is bounded. Hence, we may assume that $\tilde{w}_n \to w$ which implies $w \in \mathcal{V}^0_\omega$ and hence

$$v = \lim_{n\to\infty} v_n = \lim_{n\to\infty} d_{x_n}\pi(\tilde{w}_n) = d_x\pi(w) \in E^0_{\omega,\mathbb{P}x}.$$

For the subbundles $E^{\pm}$ the proof is similar. To show invariance, we use (7.23). Differentiating this identity and using invariance of the subbundle $\mathcal{V}^0$, we find

$$d_{\pi(x)}\psi_{t,\omega} E^0_{\omega,\mathbb{P}x} = d_{\pi(x)}\psi_{t,\omega} d_x \pi \mathcal{V}^0_\omega$$

$$= d_x \pi(\varphi(t,\omega)x)\varphi(t,\omega)\mathcal{V}^0_\omega$$

$$= d_x \pi(\varphi(t,\omega)x)\mathcal{V}^0_{\Theta_t\omega} = E^0_{\Theta_t\omega,\psi_{t,\omega}(\pi(x))}.$$

The same arguments also apply to the other two subbundles. $\square$

If the subbundle $E^0 \to \mathcal{Q}$ has rank zero, we have a decomposition $E^+ \oplus E^-$. We want to show that this gives the chain control set Q the structure of a uniformly hyperbolic set in the sense of Definition 6.2.

Proposition 7.9. *The restriction of the differential* $d\psi_{(\cdot,\cdot)}$ *to the bundle* E^+ *is uniformly expanding and its restriction to* E^- *is uniformly contracting. In particular, if* rk $\mathcal{V}^0 = 1$, *the chain control set* Q *is uniformly hyperbolic.*

Proof. We may work with system $\Sigma_{\mathbb{S}}$ instead of $\Sigma_{\mathbb{P}}$. Then the projection is given by $\pi(x) = x/|x|$. For given $v \in E^+_{\omega,x} \subset T_x \mathbb{S}^d$, we have $v = d_x \pi(\overline{v})$, where $\overline{v} \in \mathcal{V}^+_\omega$. We find

$$\left| d_x \tilde{\psi}_{t,\omega}(v) \right| = \left| d_x \tilde{\psi}_{t,\omega} d_x \pi(\overline{v}) \right| = \left| d_{\varphi(t,\omega)x} \pi \varphi(t,\omega)\overline{v} \right|.$$

Using that $d_x \pi \equiv \alpha d_{\alpha x} \pi$ for $\alpha > 0$ and $d_x \pi = I - x x^T$ for $x \in \mathbb{S}^d$, we obtain

$$
\left| d_x \tilde{\psi}_{t,\omega}(v) \right|^2 = \frac{\left| \varphi(t,\omega)\overline{v} - \frac{\langle \varphi(t,\omega)\overline{v},\varphi(t,\omega)x \rangle}{|\varphi(t,\omega)x|^2} \varphi(t,\omega)x \right|^2}{|\varphi(t,\omega)x|^2}
$$

$$
= \frac{|\varphi(t,\omega)\overline{v}|^2}{|\varphi(t,\omega)x|^2} - \frac{\langle \varphi(t,\omega)\overline{v}, \varphi(t,\omega)x \rangle^2}{|\varphi(t,\omega)x|^4}
$$

$$
= \frac{|\varphi(t,\omega)\overline{v}|^2}{|\varphi(t,\omega)x|^2} \left(1 - \cos^2 \angle(\varphi(t,\omega)x, \varphi(t,\omega)\overline{v}) \right).
$$

Hence, we find

$$
\left| d_x \tilde{\psi}_{t,\omega}(v) \right| = \frac{|\varphi(t,\omega)\overline{v}|}{|\varphi(t,\omega)x|} \left| \sin \angle(\varphi(t,\omega)x, \varphi(t,\omega)\overline{v}) \right|. \tag{7.32}
$$

Since $x \in \mathcal{V}^0_\omega$ and $\overline{v} \in \mathcal{V}^+_\omega$, we conclude from (7.30) that

$$
\left| d_x \tilde{\psi}_{t,\omega}(v) \right| \geq c_2^{-1} e^{\alpha_2 t} \frac{|\varphi(t,\omega)x| |\overline{v}||}{|\varphi(t,\omega)x|} \left| \sin \angle (\varphi(t,\omega)x, \varphi(t,\omega)\overline{v}) \right|
$$

$$
= c_2^{-1} e^{\alpha_2 t} |\overline{v}| \left| \sin \angle (\varphi(t,\omega)x, \varphi(t,\omega)\overline{v}) \right|.
$$

Setting $t = 0$ in (7.32), we find $|v| = |\overline{v}| |\sin \angle(x, \overline{v})|$. Hence,

$$
\left| d_x \tilde{\psi}_{t,\omega}(v) \right| \geq c_2^{-1} e^{\alpha_2 t} |v| \frac{|\sin \angle (\varphi(t,\omega)x, \varphi(t,\omega)\overline{v})|}{|\sin \angle(x, \overline{v})|}.
$$

Since x and $\overline{v}$ are in different invariant bundles, the angle between $\varphi(t,\omega)x$ and $\varphi(t,\omega)\overline{v}$ is bounded away from zero uniformly for all $x, \overline{v}, \omega$ and $t \geq 0$ (as a consequence of compactness of $\mathcal{U}$). This proves uniform expansiveness on E^+. Now let $v \in E^-_{\omega,x}$ and $v = d_x \pi(\overline{v})$, $\overline{v} \in \mathcal{V}^-_\omega$. As above, the identity (7.32) follows. Now using the first inequality in (7.30), we arrive at

$$
\left| d_x \tilde{\psi}_{t,\omega}(v) \right| \leq c_1 e^{-\alpha_1 t} |v| \frac{|\sin \angle(\varphi(t,\omega)x, \varphi(t,\omega)\overline{v})|}{|\sin \angle(x, \overline{v})|}.
$$

This concludes the proof. $\square$

From Proposition 1.24 it follows that every main control set is contained in one of the chain control sets. Hence, the preceding proposition implies that the assumptions of Theorem 7.8 are satisfied for the main control set contained in Q_1, provided that rk $\mathcal{W}^1 = 1$, since then the contracting subbundle E^- vanishes, which implies that all Lyapunov exponents are positive (cf. the proof of Theorem 7.4).

It is tempting to conjecture that a similar formula as the one in Theorem 7.8 holds for other hyperbolic main control sets, now only involving the volume growth rate on the unstable bundle E^+. Indeed, hyperbolicity guarantees that we can apply both the upper estimates of Sect. 5.2, since the number of positive Lyapunov exponents is constant, and the lower estimates of Chap. 6, since we have an unstable subbundle. However, there are still the following severe problems to solve in order to merge the upper and lower estimates into a formula:

1. In the upper estimate, one has to bridge the gap between the Lyapunov exponents of trajectories which stay in compact subsets of the interior and those of arbitrary trajectories in the closure of the control set;
2. In the lower estimate, one must show that $h_{esc}(K, Q) \leq 0$ or $\hat{h}_{esc}(K, Q) = 0$, respectively.

By inspecting the proof of Theorem 7.8 one sees that problem (1) in general cannot be solved in the same way as under the assumption of positive Lyapunov exponents only. This is related to the fact that the sum of the positive Lyapunov exponents can be described by a subadditive cocycle, but in general not by an additive one which satisfies the essential assumption (5.18). As noticed by Luiz San Martin, problem (2) probably can be solved by using methods from topological entropy with regard to the fact that projective flows (classical projective systems) are easily seen to have vanishing topological entropy, since they act as isometries on their nonwandering sets (this is a special case of Ferraiol, Patrão, and Seco [43, Theorem 5.4]). Since $\hat{h}_{esc}(K, Q)$ is by definition close to topological entropy, one might use similar techniques to show that this quantity is zero.

The following proposition could be a first step to solve problem (1).

Proposition 7.10. *Assume that $\Sigma_{\mathbb{P}}$ is locally accessible, let Q be a chain control set of multiplicity one, and D the corresponding main control set with cl $D = Q$. Then for every compact $K \subset D$ it holds that*

$$h_{inv}(K, Q) \leq \inf_{(\omega, \mathbb{P}x) \in \mathcal{Q}_{Per}} \limsup_{\tau \to \infty} \frac{1}{\tau} \log \left| \det d_{\mathbb{P}x} \psi_{\tau, \omega} \big|_{E^+_{\omega, \mathbb{P}x}} \right|,$$

where $\mathcal{Q}_{Per}$ denotes the subset of $\mathcal{Q}$ consisting of the periodic points.

Proof. We define

$$\mathcal{Q}_{Per} := \{(\omega, \mathbb{P}x) \in \mathcal{Q} : (\omega, \mathbb{P}x) \text{ is periodic}\},$$
$$\mathcal{Q}_{Per,Int} := \{(\omega, \mathbb{P}x) \in \mathcal{Q}_{Per} : (\omega, \mathbb{P}x) \in \text{int } \mathcal{U} \times \text{int } D\}.$$

We further write

$$a_t(\omega, \mathbb{P}x) := \log \left| \det d_{\mathbb{P}x} \psi_{\tau, \omega} \big|_{E^+_{\omega, \mathbb{P}x}} \right|$$

and note that a is a continuous additive cocycle over the control flow on $\mathcal{U} \times \mathbb{P}^d$ (cf. the proof of Theorem 6.2). From Proposition 5.9 the estimate

$$h_{\text{inv}}(K, Q) \leq \inf_{(\omega, \mathbb{P}x) \in \mathcal{Q}_{\text{Per,Int}}} \lim_{t \to \infty} \frac{1}{t} a_t(\omega, \mathbb{P}x) \tag{7.33}$$

follows (using Floquet theory), since there are exactly $k := \operatorname{rk} E^+$ positive Lyapunov exponents for every trajectory in $\mathcal{Q}$.

Now let $(\omega, \mathbb{P}x) \in \mathcal{Q}_{\text{Per}}$ be a periodic point and let $\tau > 0$ denote its period. From Proposition 1.6 it follows that we can find a sequence ω_n of τ-periodic piecewise constant control functions in $\mathcal{U}$ such that $g_n := \varphi(t, \omega_n) \to \varphi(t, \omega) =: g$ for $n \to \infty$ uniformly in $t \in [0, \tau]$. Since $\mathbb{P}x$ is periodic, x is an eigenvector of $\varphi(\tau, \omega)$ corresponding to a real eigenvalue λ. Moreover, the algebraic multiplicity of λ is one which follows from the assumption that Q has multiplicity one. Each g_n has an eigenvalue λ_n such that $\lambda_n \to \lambda$. Moreover, for n sufficiently large, λ_n must be real and its multiplicity must be one as well (cf. [25, Lemma 7.3.9 (ii)]). It follows that there is an eigenvalue x_n of g_n corresponding to λ_n with $x_n \to x$. Hence, the points $(\omega_n, \mathbb{P}x_n)$ are periodic points of the control flow on $\mathcal{U} \times \mathbb{P}^d$. By Lemma 7.3 each g_n can be approximated by elements of $\mathscr{S}_\tau(\Sigma) \cap \operatorname{int} \mathscr{S}_{\leq \tau+1}(\Sigma)$ and hence we may assume as well that $g_n \in \mathscr{S}_\tau(\Sigma) \cap \operatorname{int} \mathscr{S}_{\leq \tau+1}(\Sigma)$, which by Theorem 7.6 (iii) implies that $\mathbb{P}x_n \in \operatorname{int} D$. By a slight perturbation (which is not destroying the convergence $g_n \to g$), we may further assume that $\omega_n \in \operatorname{int} \mathcal{U}$. We claim that this implies

$$\lim_{t \to \infty} \frac{1}{t} a_t(\omega_n, \mathbb{P}x_n) = \frac{1}{\tau} a_\tau(\omega_n, \mathbb{P}x_n) \to \frac{1}{\tau} a_\tau(\omega, \mathbb{P}x) = \lim_{t \to \infty} \frac{1}{t} a_t(\omega, \mathbb{P}x)$$

for $n \to \infty$. Existence of the limits is an easy consequence of periodicity of the points and additivity of the cocycle. To show the convergence statement, consider the natural action of $GL(d+1, \mathbb{R})$ on $\mathbb{P}^d$, $(h, \mathbb{P}x) \mapsto \mathbb{P}(hx)$, and denote the diffeomorphism of $\mathbb{P}^d$ induced by some $h \in GL(d+1, \mathbb{R})$ by $\overline{h}$. We restrict this action to the set

$$A := \{(h, \mathbb{P}x) : x \text{ is an eigenvector of } h\}.$$

For every $(h, \mathbb{P}x) \in A$ we have $\overline{h}(\mathbb{P}x) = \mathbb{P}x$ and hence, $d_{\mathbb{P}x}\overline{h} : T_{\mathbb{P}x}\mathbb{P}^d \to T_{\mathbb{P}x}\mathbb{P}^d$. Let $E^u(h, \mathbb{P}x) \subset T_{\mathbb{P}x}\mathbb{P}^d$ denote the unstable subspace of this linear map, and consider the continuous map

$$\alpha : A \to \mathbb{R}, \quad (h, \mathbb{P}x) \mapsto \left| \det d_{\mathbb{P}x}\overline{h} \big|_{E^u(h, \mathbb{P}x)} \right|.$$

Then it is easy to see that

$$\frac{1}{\tau} a_\tau(\omega_n, \mathbb{P}x_n) = \frac{1}{\tau} \log \alpha(g_n, \mathbb{P}x_n) \quad \text{and} \quad \frac{1}{\tau} a_\tau(\omega, \mathbb{P}x) = \frac{1}{\tau} \log \alpha(g, \mathbb{P}x).$$

This implies the claim. Together with (7.33) we find

$$h_{\mathrm{inv}}(K, Q) \leq \inf_{(\omega, \mathbb{P}x) \in \mathcal{Q}_{\mathrm{Per}}} \lim_{t \to \infty} \frac{1}{t} a_t(\omega, \mathbb{P}x), \tag{7.34}$$

which concludes the proof.   □

One idea how to bridge the gap between periodic and arbitrary trajectories of points in $\mathcal{Q}$ would be via the Morse spectrum of the cocycle a over $\mathcal{Q}$. It is well-known that the infimum of the Morse spectrum coincides with the infimum of the Lyapunov spectrum. On the other hand, the full Morse spectrum can be obtained via periodic chains. The natural idea how to get from periodic chains to periodic trajectories would be the use of a shadowing lemma. However, it seems that an appropriate shadowing lemma for this situation is not available. The skew product shadowing lemma by Meyer and Sell [81], for instance, is not sufficient, since it does not allow jumps in the base space.

Looking beyond the examples of this section, one might expect that the general case of a hyperbolic control set can be handled by extending the methods developed in Chaps. 5 and 6, using further concepts from the classical hyperbolic theory of dynamical systems. However, for non-hyperbolic control sets the situation might be much more complicated and formulas for the invariance entropy might look quite different than the ones obtained in this chapter. In general, for example, there is no reason to expect that the escape entropy vanishes. But if this is not the case, it is unlikely that the upper estimates of Chap. 5 will match with the lower ones of Chap. 6. The upper estimates were based on the idea of keeping the system in the vicinity of a periodic trajectory in order to stay in the given control set. However, if the control set contains more complicated subsets (consisting of many trajectories) at which the system can be stabilized, the corresponding data rates (or entropies) might be smaller than the ones that can be obtained by stabilizing at single trajectories. In this way, similar terms as the escape entropy could enter the upper estimates. But at the current state of the theory this is pure fantasy, and so we stop the discussion here.

7.5 Comments and Bibliographical Notes

The first main result of Sect. 7.1, Theorem 7.1, is new and has not appeared before in the literature. Theorem 7.2 can be found in Kawan [62–64] with different proofs. For instance, in Kawan [62], it is proved by establishing a topological conjugacy to another control-affine system whose control vector field is constant. Applying the results of Chap. 4 to this system and using Corollary 5.1, one immediately arrives at formula (7.2). Another possibility, which can be found in Kawan [63], is based on the use of an adapted Riemannian metric and the upper estimate of Corollary 4.1. Moreover, this result provides another example where $h_{\mathrm{inv,out}}(K, Q)$ and $h_{\mathrm{inv}}(K, Q)$ coincide. Except for Theorem 7.4, the results about uniformly expanding systems have appeared before in [62, 65]. The result of Sect. 7.3 can be found in [65]. Theorem 7.8 is new, but parts of the discussion of projective systems can be found in [62, 63, 65]. The proof of Theorem 7.8 is basically an adaptation of Colonius and Kliemann [25, Theorem 7.3.25]. Projective systems with $\mathrm{tr}\, A_i = 0, i = 0, 1, \ldots, m$, can be considered as a special class of right-invariant systems on flag manifolds of semisimple Lie groups (in this case, the Lie group is $\mathrm{SL}(d + 1, \mathbb{R})$). It is part of the current research to compute the invariance entropy of the main control sets of general right-invariant systems on flag manifolds of semisimple Lie groups.

Appendix A
General Concepts

In the first part of the appendix, we give an overview of some of the basic mathematical tools used in this book.

A.1 Linear and Multilinear Algebra

In this section, we present some of the (multi-)linear algebra used in this book which may not be as well known as other linear algebraic concepts.

Singular Values of a Linear Operator

Let $(E, \langle \cdot, \cdot \rangle_E)$ and $(F, \langle \cdot, \cdot \rangle_F)$ be d-dimensional Euclidean spaces and $L : E \to F$ a linear operator. Then the *adjoint* of L is the unique linear operator $L^* : F \to E$ such that $\langle Lx, y \rangle_F = \langle x, L^*y \rangle_E$ for all $x \in E$ and $y \in F$. The *singular values* of L are the nonnegative square roots of the eigenvalues of the positive semi-definite self-adjoint operator $L^*L : E \to E$. We denote them by $\sigma_1(L) \geq \cdots \geq \sigma_d(L) \geq 0$. The number of the positive singular values equals the rank of L. Using the singular values we can define the *absolute determinant* of L by

$$|\det L| := |\det L^*L|^{1/2} = \sigma_1(L) \cdot \ldots \cdot \sigma_d(L). \tag{A.1}$$

If L is an isomorphism, the singular values of L are all positive and $\sigma_i(L)^{-1}$, $i = 1, \ldots, d$, are the singular values of L^{-1}. The geometric meaning of the singular values becomes clear in the following proposition which can be found in Boichenko et al. [9, Chap. I, Propositions 1.2.2 and 7.2.1] or Temam [108, Sect. V.1.3].

C. Kawan, *Invariance Entropy for Deterministic Control Systems*, Lecture Notes in Mathematics 2089, DOI 10.1007/978-3-319-01288-9,
© Springer International Publishing Switzerland 2013

Proposition A.1. *Let $L : E \to F$ be a linear operator between d-dimensional Euclidean spaces. If B is a closed (or open) ball in E of radius r, then LB is a closed (or open) ellipsoid in $\operatorname{im} L \subset F$ with semi-axes of lengths $r\sigma_i(L)$, $\sigma_i(L) > 0$.*

Furthermore, we define the *singular value function of order k* $(0 \le k \le d)$ of a linear operator $L : E \to F$ by

$$\alpha_k(L) := \begin{cases} \sigma_1(L)\sigma_2(L) \cdots \sigma_k(L) & \text{for } k > 0, \\ 1 & \text{for } k = 0. \end{cases}$$

Note that singular value functions can also be defined for non-integer values of k (cf. Boichenko et al. [9]). If $T : E \to F$ and $S : F \to G$ are linear operators between d-dimensional Euclidean spaces, then *Horn's inequality*

$$\alpha_k(ST) \le \alpha_k(S)\alpha_k(T) \quad \text{for } k = 0, 1, \dots, d$$

holds. In the case that $k = d$ the singular value functions coincide with the absolute determinants and therefore equality holds (cf. [9, Chap. I, Proposition 7.4.3]).

Tensors on a Vector Space

Let V be a d-dimensional vector space over $\mathbb{R}$. As usual, V^* denotes the dual space of V, the space of *covectors* or real-valued linear functionals on V. A *covariant k-tensor* on V is a multilinear map

$$F : \underbrace{V \times \cdots \times V}_{k \text{ copies}} \to \mathbb{R}.$$

Similarly, a *contravariant l-tensor* is a multilinear map

$$F : \underbrace{V^* \times \cdots \times V^*}_{l \text{ copies}} \to \mathbb{R}.$$

A *tensor of type (k, l)* or (k, l)-*tensor* is a multilinear map

$$F : \underbrace{V^* \times \cdots \times V^*}_{l \text{ copies}} \times \underbrace{V \times \cdots \times V}_{k \text{ copies}} \to \mathbb{R}.$$

By convention, a tensor of type $(0, 0)$ is just a real number.

The space of all covariant k-tensors is denoted by $T^k(V)$, the space of contravariant l-tensors by $T_l(V)$, and the space of (k, l)-tensors by $T_l^k(V)$. There are some natural identifications: $T_0^k(V) = T^k(V)$, $T_l^0(V) = T_l(V)$, $T^1(V) = V^*$, and $T_1(V) = V^{**} = V$.

The *tensor product* $F \otimes G$ of two tensors $F \in T_l^k(V)$ and $G \in T_q^p(V)$ is the $(k+p, l+q)$-tensor defined by

$$
F \otimes G(\omega^1, \ldots, \omega^{l+q}, X_1, \ldots, X_{k+p})
$$
$$
:= F(\omega^1, \ldots, \omega^l, X_1, \ldots, X_k) G(\omega^{l+1}, \ldots, \omega^{l+q}, X_{k+1}, \ldots, X_{k+p}).
$$

If $(E_1, \ldots, E_d)$ is a basis of V, we denote by $(\varphi^1, \ldots, \varphi^d)$ the corresponding dual basis of V^*, defined by $\varphi^i(E_j) = \delta_j^i$. A basis of $T_l^k(V)$ is given by the set of tensors of the form

$$
E_{j_1} \otimes \cdots \otimes E_{j_l} \otimes \varphi^{i_1} \otimes \cdots \otimes \varphi^{i_k},
$$

where the indices i_p, j_q range from 1 to d. Hence, the dimension of $T_l^k(V)$ is d^{k+l}. Every tensor of type (k, l) can be written in terms of this basis as

$$
F = F_{i_1 \ldots i_k}^{j_1 \ldots j_l} E_{j_1} \otimes \cdots \otimes E_{j_l} \otimes \varphi^{i_1} \otimes \cdots \otimes \varphi^{i_k}.
$$

Here we use the *Einstein summation convention*, that is, if in any term the same index appears twice, once as a lower and once as an upper index, that term is assumed to be summed over the possible values of this index (which is usually from 1 to the dimension of the space).

An important class of tensors are the *alternating tensors*, those which change sign whenever two arguments are interchanged. We denote by $\bigwedge^k V^*$ the space of all covariant alternating k-tensors on V, also called *(exterior) k-forms*. The space $\bigwedge^k V^*$ is called the *k-th exterior power* of V. By $\bigwedge^k V$ we denote the space of all contravariant alternating k-tensors, also called *k-multivectors*.[1] By convention, $\bigwedge^0 V = \bigwedge^0 V^* = \mathbb{R}$.

There is a natural bilinear associative product on forms called the *wedge product*, defined on one-forms by setting

$$
\omega^1 \wedge \ldots \wedge \omega^k (X_1, \ldots, X_k) := \det \left(\omega^i(X_j) \right),
$$

and extending by linearity. If $(\omega_1, \ldots, \omega_d)$ is a basis of V^*, an associated basis of $\bigwedge^k V^*$ is given by the tensors of the form

$$
\omega_{i_1} \wedge \ldots \wedge \omega_{i_k},
$$

where $1 \le i_1 < i_1 < \cdots < i_k \le d$. Consequently, the dimension of $\bigwedge^k V^*$ is $\binom{d}{k} = d!/(k!(d-k)!)$.

---

[1] A word of caution: Some authors write $\bigwedge^k V$ for the alternating covariant tensors and $\bigwedge^k V^*$ for the alternating contravariant tensors. A discussion of the reason why can be found in Lee [75, Chap. 12].

The wedge product can also be defined on k-multivectors by setting

$$\xi^1 \wedge \ldots \wedge \xi^k(\omega_1, \ldots, \omega_k) := \det\left(\omega_i(\xi^j)\right)$$

for one-multivectors (which are the elements of $V^{**} = V$), and extending by linearity.

Taking the direct sum of the spaces $\bigwedge^k V$ (or alternatively $\bigwedge^k V^*$) for $0 \leq k \leq d$, one obtains another vector space

$$\bigwedge V := \bigoplus_{k=0}^{d} \bigwedge^k V,$$

of dimension 2^d, which is called the *exterior algebra* of V. If $L : V \to W$ is a linear operator between real vector spaces of dimension d_1 and d_2, respectively, then the k-*th exterior power* of L is the linear operator defined by

$$L^{\wedge k} : \bigwedge^k V \to \bigwedge^k W, \quad \xi^1 \wedge \ldots \wedge \xi^k \mapsto L\xi^1 \wedge \ldots \wedge L\xi^k$$

for all $\xi^1, \ldots, \xi^k \in V$ and extending by linearity. This definition naturally gives an induced operator $L^\wedge : \bigwedge V \to \bigwedge W$, called the *exterior power* of L.

Now assume that V is endowed with an inner product $\langle \cdot, \cdot \rangle$. Then an associated inner product on $\bigwedge^k V$ is defined by

$$\langle v_1 \wedge \ldots \wedge v_k, w_1 \wedge \ldots \wedge w_k \rangle_{\bigwedge^k V} := \det\left(\langle v_i, w_j \rangle\right),$$

and extending by linearity in each argument. The associated norm is denoted by $|\cdot|_{\bigwedge^k V}$. If V and W are Euclidean spaces of the same dimension and $L : V \to W$ is a linear operator, the operator norm of $L^{\wedge k} : \bigwedge^k V \to \bigwedge^k W$ with respect to the norms $|\cdot|_{\bigwedge^k V}$ and $|\cdot|_{\bigwedge^k W}$ is the product of the k greatest singular values of L, $\|L^{\wedge k}\| = \sigma_1(L) \cdots \sigma_k(L)$ (cf. Arnold [4, Proposition 3.2.7]). The operator norm of the exterior power $L^\wedge : \bigwedge V \to \bigwedge W$ then is $\|L^\wedge\| = \max_{0 \leq k \leq d}(\sigma_1(L) \cdots \sigma_k(L))$.

An operator $L \in \mathscr{L}(V, V)$ induces an operator $L_k \in \mathscr{L}(\bigwedge^k V, \bigwedge^k V)$ by

$$L_k(v_1 \wedge \ldots \wedge v_k) := Lv_1 \wedge \ldots \wedge v_k + v_1 \wedge Lv_2 \wedge \ldots \wedge v_k + \ldots$$

$$+ v_1 \wedge \ldots \wedge Lv_k,$$

called the k-*th derivation operator of L*. The eigenvalues of this operator are the sums $\lambda_{i_1} + \cdots + \lambda_{i_k}$, where $1 \leq i_1 < \cdots < i_k \leq d$ and $\lambda_1, \ldots, \lambda_d$ are the eigenvalues of L. Moreover, we have the relation $(e^{tL})^{\wedge k} = e^{tL_k}$ (cf. Arnold [4, Lemma 3.2.6]).

The following lemma can be used to prove generalizations of the Liouville formula for ordinary differential equations. It can be found in Temam [108, Chap. V, Lemma 1.2].

Lemma A.1. *For all $k \in \{1, \ldots, d\}$ and all $v_1, \ldots, v_k \in V$ it holds that*

$$\langle L_k(v_1 \wedge \ldots \wedge v_k), v_1 \wedge \ldots \wedge v_k \rangle_{\bigwedge^k V} = |v_1 \wedge \ldots \wedge v_k|^2_{\bigwedge^k V} \, \mathrm{tr}(L \circ Q),$$

where $Q = Q(v_1, \ldots, v_k)$ denotes the orthogonal projection in V onto the linear subspace spanned by $v_1, \ldots, v_k$.

A.2 Differentiable Manifolds

The natural state space of a control system given by ordinary differential equations is a differentiable manifold. Usually, this is a submanifold of some Euclidean space $\mathbb{R}^n$. But for the analysis of bilinear systems, for instance, also systems on more abstractly defined manifolds like projective spaces play an important role. In this section, we provide the necessary background on differentiable manifolds which is needed for the treatment of smooth systems in this book. In particular, for the understanding of Chaps. 4–6, the reader should be familiar with the material presented here. Good references are, for instance, the books Gallot et al. [48], Lee [75] or Bullo and Lewis [15]. However, we note that in the last reference (which is a control theory book) no proofs for the differential-geometric results can be found, but the exposition is very clear, many examples are given, and, in contrast to almost all books on differential geometry, the theory is exposed under minimal differentiability assumptions.

Definition of a Manifold

Let M be a second-countable Hausdorff space.[2] A family $\mathscr{A} = \{(\phi_\alpha, U_\alpha)\}_{\alpha \in A}$ is called a $\mathscr{C}^k$-*atlas* on M for some $k \in \mathbb{Z}_+ \cup \{\infty\} \cup \{\omega\}$ if the following axioms are satisfied:

(i) $\{U_\alpha\}_{\alpha \in A}$ is an open cover of M;
(ii) For each $\alpha \in A$, $\phi_\alpha : U_\alpha \to V_\alpha$ is a homeomorphism onto an open subset V_α of $\mathbb{R}^d$ for some $d \in \mathbb{N}$;

---

[2]We recall that a topological space X is called *Hausdorff* if any two distinct points $x, y \in X$ have disjoint open neighborhoods. The space X is called *second-countable* if its topology has a countable basis, that is, there is a countable family of open sets such that every open set can be written as the union of sets in this family.

(iii) For all $\alpha, \beta \in A$, the transition function

$$\phi_\beta \circ \phi_\alpha^{-1} : \phi_\alpha(U_\alpha \cap U_\beta) \to \phi_\beta(U_\alpha \cap U_\beta)$$

is of class $\mathscr{C}^k$ (if $U_\alpha \cap U_\beta = \emptyset$, this is trivially satisfied).

If $k = 0$, the transition functions are only assumed to be continuous, which is already satisfied by Axiom (ii). In the case $k = \omega$, the transition functions are assumed to be real-analytic. Every $\mathscr{C}^k$-atlas $\mathscr{A}$ is contained in a unique maximal $\mathscr{C}^k$-atlas $\mathscr{A}_{\max}$, that is, a $\mathscr{C}^k$-atlas with the property that no further charts can be added without violating Axiom (iii). The pair $(M, \mathscr{A}_{\max})$ is then called a $\mathscr{C}^k$-*manifold* and if $k \geq 1$, $\mathscr{A}_{\max}$ is called a *differentiable structure* on M. In the case that $k = 0$ we also speak of a *topological manifold*, in the case $k \geq 1$ of a *differentiable manifold of class* $\mathscr{C}^k$, in the case $k = \infty$ of a *smooth manifold*, and in the case $k = \omega$ of a *real-analytic manifold*. In the rest of this section we restrict ourselves to the case $k \geq 1$. Usually, when we speak of a $\mathscr{C}^k$-manifold, we do not explicitly mention the atlas, that is, we only write M instead of $(M, \mathscr{A})$ or $(M, \mathscr{A}_{\max})$.

The elements (ϕ_α, U_α) of a $\mathscr{C}^k$-atlas $\mathscr{A}$ are called *charts* and the inverse maps $\phi_\alpha^{-1} : V_\alpha \to U_\alpha$ *local coordinate systems* of M. A chart (ϕ_α, U_α) is said to be a *chart around* $p \in M$ if $p \in U_\alpha$. If the natural number d (the dimension of the Euclidean space where ϕ_α takes its values) is the same for all charts, we call this number the *dimension* of M and write $d = \dim M$. If M is connected, this is automatically satisfied. Throughout this book we assume that all manifolds have a well-defined dimension.

The definition of a $\mathscr{C}^k$-manifold implies several topological properties of the underlying topological space M. In particular, M is locally compact, locally path-connected, and metrizable. If M is connected, it is automatically path-connected. In general, the connected components of M coincide with the path-connected components.

When speaking of a d-dimensional real vector space V as a differentiable (real-analytic) manifold, we mean V endowed with the maximal $\mathscr{C}^\omega$-atlas which contains a chart (ϕ, V), where $\phi : V \to \mathbb{R}^d$ is a linear isomorphism.

Every open subset N of a d-dimensional $\mathscr{C}^k$-manifold $(M, \mathscr{A})$ is itself a d-dimensional $\mathscr{C}^k$-manifold with atlas $\{(\phi|_{U \cap N}, U \cap N) : (\phi, U) \in \mathscr{A}\}$. Given two $\mathscr{C}^k$-manifolds $(M, \mathscr{A})$ and $(N, \mathscr{B})$ of dimensions k and l, respectively, their Cartesian product $M \times N$ (endowed with the product topology) becomes a $(k + l)$-dimensional $\mathscr{C}^k$-manifold with the maximal $\mathscr{C}^k$-atlas which contains the *product atlas*

$$\{(\phi \times \psi, U \times V) : (\phi, U) \in \mathscr{A}, (\psi, V) \in \mathscr{B}\}.$$

A manifold of this type is called a *product manifold*. Inductively, the product of any finite number of $\mathscr{C}^k$-manifolds can be defined.

Tangent Spaces and Derivatives

In order to develop a differential calculus on $\mathscr{C}^k$-manifolds ($k \geq 1$), the notions of tangent vectors and tangent spaces have to be introduced. In this and the following sections, when we speak of a $\mathscr{C}^{r-1}$-manifold or $\mathscr{C}^{r-1}$-map, we use the convention that $r - 1 = \infty$ if $r = \infty$ and $r - 1 = \omega$ if $r = \omega$.

Let M be a d-dimensional $\mathscr{C}^k$-manifold. On the set of all triples (p, ϕ, ξ), where $p \in M$, (ϕ, U) is a chart around p, and $\xi \in \mathbb{R}^d$, we introduce an equivalence relation by

$$(p, \phi, \xi) \sim (p, \psi, \eta) \quad :\Leftrightarrow \quad \eta = \mathrm{D}(\psi \circ \phi^{-1})(\phi(p))\xi.$$

The equivalence class $[p, \phi, \xi]$ of a triple (p, ϕ, ξ) is called a *tangent vector at p*. The *tangent space at p*, denoted by $T_p M$, is defined as the set of all tangent vectors at p, and is endowed with the structure of a real vector space, given by

- $[p, \phi, \xi] + [p, \phi, \eta] := [p, \phi, \xi + \eta]$ for all $\xi, \eta \in \mathbb{R}^d$;
- $\lambda[p, \phi, \xi] := [p, \phi, \lambda\xi]$ for all $\lambda \in \mathbb{R}, \xi \in \mathbb{R}^d$.

It can easily be shown that these operations are well-defined and give $T_p M$ the structure of a vector space isomorphic to $\mathbb{R}^d$. The zero vector $[p, \phi, 0] \in T_p M$ is denoted by 0_p.

A map $f : M \to N$ between $\mathscr{C}^k$-manifolds M and N is said to be *differentiable at* $p \in M$ if there are charts (ϕ, U) of M around p and (ψ, V) of N around $f(p)$ such that $f(U) \subset V$ and the *local representation*

$$\psi \circ f \circ \phi^{-1} : \phi(U) \to \psi(V)$$

of f is differentiable at $\phi(p)$ in the usual sense. If $\psi \circ f \circ \phi^{-1}$ is of class $\mathscr{C}^r$, $r \in \{1, \ldots, k\}$, in a neighborhood of $\phi(p)$, then f is said to be $\mathscr{C}^r$-*differentiable at* p. It follows from Axiom (iii) in the definition of $\mathscr{C}^k$-manifolds that this definition is independent of the chosen charts. If f is $\mathscr{C}^r$-differentiable at every $p \in M$, then f is called a $\mathscr{C}^r$-*map*. If f is additionally invertible and also $f^{-1} : N \to M$ is a $\mathscr{C}^r$-map, then f is called a $\mathscr{C}^r$-*diffeomorphism*. It is easy to see that every $\mathscr{C}^r$-map is continuous and hence every $\mathscr{C}^r$-diffeomorphism is a homeomorphism. For the set of all $\mathscr{C}^r$-maps $f : M \to N$ we use the notation $\mathscr{C}^r(M, N)$.

Given a $\mathscr{C}^1$-map $f : M \to N$ between $\mathscr{C}^k$-manifolds M and N, the *derivative of f at $p \in M$* is the linear map $\mathrm{d}_p f : T_p M \to T_{f(p)} N$, defined by

$$\mathrm{d}_p f[p, \phi, \xi] := [f(p), \psi, \mathrm{D}(\psi \circ f \circ \phi^{-1})(\phi(p))\xi],$$

where (ϕ, U) and (ψ, V) are charts of M and N around p and $f(p)$, respectively. One easily shows that this definition is independent of the choice of the charts.

A $\mathscr{C}^r$-*curve* is a continuous map $c : I \to M$ defined on an interval $I \subset \mathbb{R}$ with values in a $\mathscr{C}^k$-manifold M such that the restriction of c to the interior of I is a

$\mathscr{C}^r$-map. Given a $\mathscr{C}^1$-curve $c : I \to M$ and $t \in \text{int } I$, the *tangent vector to c at t* is an element of $T_{c(t)}M$, given by

$$\frac{\mathrm{d}}{\mathrm{d}t}c(t) = \dot{c}(t) := \left[c(t), \phi, \frac{\mathrm{d}}{\mathrm{d}t}(\phi \circ c)(t) \right],$$

where (ϕ, U) is any chart around $c(t)$. Every tangent vector can be obtained as the tangent vector to a $\mathscr{C}^1$-curve and hence

$$T_p M = \{ \dot{c}(0) \mid c : (-\varepsilon, \varepsilon) \to M \text{ is a } \mathscr{C}^1\text{-curve with } c(0) = p \}.$$

The derivative satisfies the following properties:

- The *chain rule* holds:

$$\mathrm{d}_p(f \circ g) = \mathrm{d}_{g(p)} f \circ \mathrm{d}_p g$$

  for all $\mathscr{C}^1$-maps $g : M \to N$ and $f : N \to P$.
- If $f : M \to N$ is a $\mathscr{C}^1$-map and $c : (-\varepsilon, \varepsilon) \to M$ is a $\mathscr{C}^1$-curve with $c(0) = p$, then

$$\mathrm{d}_p f(\dot{c}(0)) = \frac{\mathrm{d}}{\mathrm{d}t}(f \circ c)(0).$$

- If $f : M \to N$ is a $\mathscr{C}^1$-diffeomorphism, then $\mathrm{d}_p f : T_p M \to T_{f(p)} N$ is an isomorphism for all $p \in M$.
- The *inverse function theorem* holds: If $f : M \to N$ is a $\mathscr{C}^r$-map, $r \in \mathbb{N} \cup \{\infty\} \cup \{\omega\}$, and $\mathrm{d}_p f$ is invertible for some $p \in M$, then there are open neighborhoods U of p and V of $f(p)$ such that $V = f(U)$ and the restriction $f|_U : U \to V$ is a $\mathscr{C}^r$-diffeomorphism with

$$\mathrm{d}_{f(p)} f^{-1} = (\mathrm{d}_p f)^{-1}.$$

To every chart (ϕ, U) around a point $p \in M$ we can associate an isomorphism $T_p M \to \mathbb{R}^d$ by $\alpha_{p,\phi} : [p, \phi, \xi] \mapsto \xi$. The preimages of the standard basis vectors $e_1, \ldots, e_d \in \mathbb{R}^d$ under $\alpha_{p,\phi}$ form a basis of $T_p M$. They are denoted by

$$\partial_i \phi(p) := \alpha_{p,\phi}^{-1} e_i = [p, \phi, e_i], \quad i = 1, \ldots, d.$$

The reason for this notation stems from the fact that every tangent vector $[p, \phi, \xi]$ can be identified canonically with a directional derivative acting on differentiable functions on M via

$$[p, \phi, \xi] f := \mathrm{D}(f \circ \phi^{-1})(\phi(p))\xi \quad \text{for all } f \in \mathscr{C}^1(M, \mathbb{R}). \tag{A.2}$$

Hence, $\partial_i \phi(p)$ corresponds to the i-th partial derivative. To make formulas better readable, we introduce another (more common) notation for the expression $\partial_i \phi(p) f$. Namely, we write

$$\frac{\partial f}{\partial \phi^i}(p) := \partial_i \phi(p) f = \frac{\partial (f \circ \phi^{-1})}{\partial x_i}(\phi(p)).$$

If V is a d-dimensional real vector space with its standard $\mathscr{C}^\omega$-atlas, the tangent space $T_p V$ at any point $p \in V$ can be identified canonically with V itself via

$$T_p V \ni [p, \phi, \xi] \mapsto \phi^{-1} \xi \in V$$

for every chart (ϕ, V) such that $\phi : V \to \mathbb{R}^d$ is a linear isomorphism.

The *tangent bundle* TM of the d-dimensional $\mathscr{C}^k$-manifold M is defined as the disjoint union of all tangent spaces $T_p M$, $p \in M$. It can be endowed with an atlas in a canonical way such that it becomes a $2d$-dimensional $\mathscr{C}^{k-1}$-manifold. The charts of this atlas are defined as follows: If (ϕ, U) is a chart of M, every tangent vector $v \in T_p M$ with $p \in U$ can be written uniquely as $v = v^i \partial_i \phi(p)$. Then a chart of TM is given by (ψ, TU) with

$$\psi(v) = (\phi(p), v^1, \dots, v^d) \quad \text{for } v \in T_p M, \ p \in U.$$

For every $p \in M$ we denote by $T_p^* M$ the dual space of $T_p M$, that is, $T_p^* M := (T_p M)^*$. The disjoint union $T^* M$ of all these dual spaces is called the *cotangent bundle* of M, and it can also be endowed with a canonical $\mathscr{C}^{k-1}$-atlas.

If (ϕ, U) is a chart of M around p, then a basis $d\phi^1(p), \dots, d\phi^d(p)$ of $T_p^* M$ is given by

$$d\phi^i(p)[p, \phi, \xi] := \xi_i \quad \text{for } \xi = (\xi_1, \dots, \xi_d) \in \mathbb{R}^d.$$

This basis is the dual basis of $\partial_1 \phi(p), \dots, \partial_d \phi(p)$, that is, $d\phi^i(p)\partial_j \phi(p) = \delta_j^i$.

Given a product manifold $M_1 \times M_2$ such that $k = \dim M_1$ and $l = \dim M_2$, the tangent space $T_{(p,q)}(M_1 \times M_2)$ at some point $(p, q) \in M_1 \times M_2$ can be identified canonically with $T_p M_1 \times T_q M_2$ by

$$[(p, q), \phi \times \psi, \xi] \mapsto ([p, \phi, (\xi_1, \dots, \xi_k)], [q, \psi, (\xi_{k+1}, \dots, \xi_{k+l})]) ,$$

where $\xi_1, \dots, \xi_{k+l}$ are the coordinates of ξ in the standard basis of $\mathbb{R}^{k+l}$. Using this identification, the derivative of a $\mathscr{C}^1$-map $f : M_1 \times M_2 \to P$ can be computed as

$$d_{(p,q)} f(v, w) = d_p f(\cdot, q)v + d_q f(p, \cdot)w \quad \text{for all } (v, w) \in T_p M_1 \times T_q M_2$$

with the partial maps $f(\cdot, q) : M_1 \to P$ and $f(p, \cdot) : M_2 \to P$. For the partial derivatives we also use the common notation $\partial f / \partial x_1, \partial f / \partial x_2$.

Tensor Fields

All the objects defined on $\mathscr{C}^k$-manifolds which are interesting for us, in particular
vector fields and Riemannian metrics, can be regarded as special tensor fields.
Tensor fields are the natural extensions of tensors on a vector space (see Sect. A.1)
to the tangent bundle of a manifold.

Given a $\mathscr{C}^k$-manifold M with $k \geq 2$, the *bundle of* (k, l)-*tensors* $T_l^k M$ is defined
as the disjoint union of the spaces $T_l^k(T_p M)$, $p \in M$. Analogously, the *bundle of
k-forms* $\bigwedge^k M$ is the disjoint union of the spaces $\bigwedge^k T_p M$, $p \in M$. There are the
usual identifications $T_1^0 M = TM$ and $T_0^1 M = \bigwedge^1 M = T^* M$. Each of these
spaces can be endowed with a canonical $\mathscr{C}^{k-1}$-atlas (and also with the structure of
a differentiable vector bundle).

A *tensor field* on M of type (k, l) is a map $t : M \to T_l^k M$, $p \mapsto t_p$, such that
$t_p \in T_l^k(T_p M)$ for all $p \in M$. Given a chart (ϕ, U) of M, we can express each t_p
in terms of the bases of $T_p M$ and $T_p^* M$ introduced above, that is

$$t_p = t_{i_1 \dots i_k}^{j_1 \dots j_l}(p) \partial_{j_1} \phi(p) \otimes \cdots \otimes \partial_{j_l} \phi(p) \otimes d\phi^{i_1}(p) \otimes \cdots \otimes d\phi^{i_k}(p).$$

If the so-defined coordinate functions $t_{i_1 \dots i_k}^{j_1 \dots j_l} : U \to \mathbb{R}$ are of class $\mathscr{C}^r$ ($r \in
\{0, \dots, k-1\}$) for every chart (ϕ, U), we say that the tensor field t is of class
$\mathscr{C}^r$. A *differential k-form* is a tensor field $\omega : M \to \bigwedge^k M$ of class $\mathscr{C}^r$, $r \geq 1$.

Obviously, the tensor product and the wedge product for tensors on vector
spaces extend to tensor fields by performing these operations pointwise. For the
corresponding operations on tensor fields we use the same notation as we do for
tensors. For instance, the wedge product of two differential forms ω and μ is denoted
by $\omega \wedge \mu$ and defined by $(\omega \wedge \mu)_p := \omega_p \wedge \mu_p$ for all $p \in M$.

A function $f : M \to \mathbb{R}$ can be regarded as a tensor field of type $(0, 0)$ (since
tensors of type $(0, 0)$ are by convention just real numbers). We introduce the notation
$\mathscr{C}^r(M)$ for the space of all $\mathscr{C}^r$-functions from M to $\mathbb{R}$. For the tensor product $f \otimes t$
of a function $f \in \mathscr{C}^r(M)$ and an arbitrary $\mathscr{C}^r$-tensor field t we simply write ft.

Vector Fields

To define ordinary differential equations on $\mathscr{C}^k$-manifolds, we need the notion of a
vector field. Given a $\mathscr{C}^k$-manifold M with $k \geq 2$, a tensor field X of type $(0, 1)$ and
class $\mathscr{C}^r$, $r \in \{0, \dots, k-1\}$, is called a $\mathscr{C}^r$-*vector field* on M. Such X assigns
to each $p \in M$ a tangent vector $X_p \in T_p M$ (using the natural identification
$T_1 M = TM$). For the real vector space of all $\mathscr{C}^r$-vector fields on M we introduce
the notation $\mathscr{X}^r(M)$.

Each vector field $X \in \mathscr{X}^r(M)$ defines an ordinary differential equation

$$\frac{dx}{dt} = X(x).$$

Assuming that $r \geq 1$, for each $x \in M$ there exists a unique maximal solution $\lambda : I \to M$ with initial condition $\lambda(0) = x$ whose domain I is an open interval containing 0. That is, the curve λ is of class $\mathscr{C}^1$ and satisfies $\dot{\lambda}(t) = X(\lambda(t))$ for all $t \in I$. The solutions to all initial conditions $(0, x)$, $x \in M$, can be condensed into one map $(t, x) \mapsto X_t(x)$, that is, $X_0(x) = x$ and $(\partial/\partial t)X_t(x) \equiv X(X_t(x))$. For fixed $t \in \mathbb{R}$, the map $X_t : x \mapsto X_t(x)$ is a local $\mathscr{C}^r$-diffeomorphism of M in the sense that the domain $\operatorname{dom} X_t$ of X_t is an open set in M and $X_t : \operatorname{dom} X_t \to X_t(\operatorname{dom} X_t)$ is a $\mathscr{C}^r$-diffeomorphism. The domain $\operatorname{dom} X_t$ is the set of all elements of M whose maximal solutions extend up to time t, that is, their interval of definition (α, ω) contains t. The vector field X is called *complete* if $\operatorname{dom} X_t = M$ for all $t \in \mathbb{R}$. Equivalently, X is complete if all maximal solutions are defined on $\mathbb{R}$. If the vector field X has the property that the image of every maximal solution of the associated differential equation is relatively compact, X is automatically complete. In particular, this is the case if M is compact.

The map $(t, x) \mapsto X_t(x)$ is called the *flow* of the vector field. Restricted to the domains, the flow satisfies the homomorphism property: $X_{t+s} = X_t \circ X_s$, that is, if $X_s(x)$ and $X_t(X_s(x))$ are defined, then $X_{t+s}(x)$ is defined and the equality $X_{t+s}(x) = X_t(X_s(x))$ holds. It is clear that $\operatorname{dom} X_{t+s} = X_t(\operatorname{dom} X_s) \cap \operatorname{dom} X_s$. In particular, the elements of the flow commute with each other: $X_t \circ X_s = X_s \circ X_t$ and $X_{-t} = (X_t)^{-1}$.

Given two vector fields $X, Y \in \mathscr{X}^r(M)$, $r \geq 1$, another vector field of class $\mathscr{C}^{r-1}$, called the *Lie bracket of X and Y* and denoted by $[X, Y]$, is defined via

$$[X, Y](p) = \frac{d}{dt}\bigg|_{t=0} \left(d_{X_t(p)} X_{-t}\right) Y(X_t(p)).$$

The Lie bracket satisfies the following properties:

- Bilinearity over $\mathbb{R}$: If X, Y, W and Z are vector fields and $a, b \in \mathbb{R}$, then

$$[aX + Y, bZ + W] = ab[X, Z] + a[X, W] + b[Y, Z] + [Y, W].$$

- Anti-symmetry: $[X, Y] = -[Y, X]$.
- Jacobi-identity: $[X, [Y, Z]] + [Y, [Z, X]] + [Z, [X, Y]] = 0$.

In particular, for a smooth manifold M the vector space $\mathscr{X}^\infty(M)$ becomes a Lie algebra when endowed with the Lie bracket of vector fields.

Using the interpretation of tangent vectors as derivations (A.2), a vector field $X \in \mathscr{X}^r(M)$ can be applied to a function $f \in \mathscr{C}^r(M)$, $r \geq 1$:

$$(Xf)(p) := X_p f \quad \text{for all } p \in M.$$

The resulting function $Xf : M \rightarrow \mathbb{R}$ is of class $\mathscr{C}^{r-1}$. If $X, Y \in \mathscr{X}^r(M)$ and $f \in \mathscr{C}^r(M)$, then $[X, Y]f = X(Yf) - Y(Xf)$.

Riemannian Metrics

Every $\mathscr{C}^k$-manifold is metrizable, but there is no canonical way to measure distances. However, there exists a large class of nice "smooth" metrics. These are defined as follows. Let M be a connected $\mathscr{C}^k$-manifold with $k \geq 2$. Given a symmetric $\mathscr{C}^r$-tensor field g of type $(2, 0)$ on M, $r \in \{0, \ldots, k-1\}$, every tangent space $T_p M$ becomes endowed with a bilinear symmetric form[3]

$$g_p : T_p M \times T_p M \rightarrow \mathbb{R}.$$

We assume that g_p is positive definite for every $p \in M$, that is, g_p is an inner product on $T_p M$. The induced norms $| \cdot |_p$ on $T_p M$, $p \in M$, allow to measure the lengths of tangent vectors and therefore also the lengths of differentiable curves on M. Precisely, let $c : [a, b] \rightarrow M$ be a piecewise $\mathscr{C}^1$-curve, that is, there exists a partition $a = t_0 < t_1 < \cdots < t_n = b$ such that each of the restrictions $c|_{[t_i, t_{i+1}]}$, $i = 0, \ldots, n-1$, is a $\mathscr{C}^1$-curve. Then the *length* of c is defined as

$$\mathscr{L}(c) := \int_a^b |\dot{c}(t)|_{c(t)} \, \mathrm{d}t.$$

From the assumption that M is connected it follows that each two points $p, q \in M$ can be joined by a piecewise $\mathscr{C}^1$-curve. In this case, a metric which is compatible with the topology of M is given by

$$\varrho(p, q) := \inf \{ \mathscr{L}(c) \mid c : [a, b] \rightarrow M \text{ piecewise } \mathscr{C}^1 \text{ with } c(a) = p, \ c(b) = q \}. \tag{A.3}$$

The tensor field g is called a *Riemannian metric* on M and the metric ϱ the *Riemannian distance* associated with g. If M is a $\mathscr{C}^k$-manifold and g a $\mathscr{C}^{k-1}$-Riemannian metric on M, then (M, g) is called a *Riemannian manifold of class $\mathscr{C}^k$* or a *Riemannian $\mathscr{C}^k$-manifold*. Using partitions of unity, one can construct a Riemannian metric of class $\mathscr{C}^{k-1}$ on every $\mathscr{C}^k$-manifold, $2 \leq k \leq \infty$.

With respect to a chart (ϕ, U) of M, the Riemannian metric g has a local expression

$$g_p = g_{ij}(p) \mathrm{d}\phi^i(p) \otimes \mathrm{d}\phi^j(p),$$

---

[3]The tensor field g is called symmetric if it is symmetric at every point, that is, $g_p(v, w) = g_p(w, v)$ for all $v, w \in T_p M$ and $p \in M$.

where the real numbers $g_{ij}(p)$, $1 \leq i, j \leq d$, define a positive definite symmetric matrix $(g_{ij}(p))$. The entries of the inverse of this matrix are denoted by $g^{ij}(p)$, that is, $g_{ik}g^{kj} = \delta_i^j$.

For two points $p, q \in M$ the infimum in (A.3) need not be attained, that is, a curve of minimal length joining p and q not necessarily exists. However, locally (in sufficiently small neighborhoods of points) such shortest curves do exist. A curve which locally realizes the shortest distance between two points in its image is called a *geodesic*. However, it is not convenient to define geodesics via the property of realizing shortest distances, but rather by the property of being "straight lines" in M, that is, being as straight as possible. In $\mathbb{R}^d$, a straight line, given as a curve $t \mapsto a + tv$, is characterized by the property that its second derivative vanishes. To adapt this criterion to Riemannian manifolds, the notion of a *connection* needs to be introduced.

Let (M, g) be a d-dimensional Riemannian manifold of class $\mathscr{C}^k$ with $k \geq 3$. To each chart (ϕ, U) of M one can associate d^3 $\mathscr{C}^{k-2}$-functions by

$$\Gamma_{ij}^k := \frac{g_{kl}}{2}\left(\frac{\partial g_{il}}{\partial \phi^j} + \frac{\partial g_{jl}}{\partial \phi^i} - \frac{\partial g_{ij}}{\partial \phi^l}\right), \quad \Gamma_{ij}^k : U \to \mathbb{R}.$$

These functions are called the *Christoffel symbols* of (M, g) with respect to the chart (ϕ, U). They have the property that $\Gamma_{ij}^k = \Gamma_{ji}^k$, that is, they are symmetric in the lower two indices.

Using the Christoffel symbols, one can define the *Levi–Civita connection* associated with (M, g), which is an operator assigning to a pair (X, Y) of $\mathscr{C}^r$-vector fields with $r \in \{1, \ldots, k-2\}$ a $\mathscr{C}^{r-1}$-vector field $\nabla_X Y$. Locally, we can write any vector fields X and Y as $X = X^i \partial_i \phi$ and $Y = Y^j \partial_j \phi$. Then $\nabla_X Y$ is defined by

$$(\nabla_X Y)(p) := X^i(p)\left(Y^j(p)\Gamma_{ij}^k(p)\partial_k\phi(p) + \frac{\partial Y^j}{\partial \phi^i}(p)\partial_j\phi(p)\right).$$

It can be checked easily that this definition is independent of the chosen charts. The Levi–Civita connection satisfies the following identities for all $X, X_1, X_2, Y, Y_1, Y_2 \in \mathscr{X}^r(M)$ and $f \in \mathscr{C}^r(M)$:

- $\nabla_{X_1 + X_2} Y = \nabla_{X_1} Y + \nabla_{X_2} Y$;
- $\nabla_{fX} Y = f\nabla_X Y$;
- $\nabla_X(Y_1 + Y_2) = \nabla_X Y_1 + \nabla_X Y_2$;
- $\nabla_X(fY) = f\nabla_X Y + (Xf)Y$;
- $[X, Y] = \nabla_X Y - \nabla_Y X$;
- $Z(g(X, Y)) = g(\nabla_Z X, Y) + g(X, \nabla_Z Y)$.

We can interpret $\nabla_X Y$ as the vector field obtained by computing (pointwise) the directional derivative of Y in direction X. In fact, $(\nabla_X Y)_p$ depends only on $X(p)$ and the values of Y in an arbitrarily small neighborhood of p. Hence, to every

$\mathscr{C}^r$-vector field we can assign its *covariant derivative* at $p \in M$ by $\nabla X(p)v :=$ $(\nabla_v X)(p)$, $\nabla X(p) : T_p M \to T_p M$. The *symmetrized covariant derivative* of X at p is defined by

$$S\nabla X(p) := \frac{1}{2}\left(\nabla X(p) + \nabla X(p)^*\right),$$

where $\nabla X(p)^*$ denotes the adjoint operator (with respect to the Riemannian metric). With respect to a chart (ϕ, V) and the associated basis $\partial_1 \phi(p), \dots, \partial_d \phi(p)$, we have a matrix expression for $S\nabla X(p)$, which is given by

$$s_\nu^\mu = \frac{1}{2}\left(\frac{\partial X^\mu}{\partial \phi^\nu} + \frac{\partial X^\kappa}{\partial \phi^\theta} g^{\mu\theta} g_{\kappa\nu} + X^i g^{\mu l} \frac{\partial g_{\nu l}}{\partial \phi^i}\right). \tag{A.4}$$

In order to define geodesics as the "straight lines" in M, we need a way to compute the second derivative of a curve. To this end, a further concept derived from the Levi–Civita connection has to be introduced. Given a $\mathscr{C}^r$-curve $c : I \to M$, a *vector field along c* is a map $X : I \to TM$ such that $X(t) \in T_{c(t)}M$ for all $t \in I$. The vector field X is of class $\mathscr{C}^r$ if the coordinate functions $X^1, \dots, X^d : I \to \mathbb{R}$, defined by writing $X(t) = X^i(t)\partial_i \phi(c(t))$ with respect to a chart (ϕ, U), are of class $\mathscr{C}^r$ for every chart around some point $c(t_0)$, $t_0 \in I$. The real vector space of all $\mathscr{C}^r$-vector fields along a $\mathscr{C}^r$-curve c is denoted by $\mathscr{X}_c^r$. Now we can differentiate vector fields along c by the local formula

$$\frac{\mathrm{D}X}{\mathrm{d}t}(t) := (X^i)'(t)\partial_i \phi(c(t)) + X^i(t)\left(\nabla_{\dot{c}(\cdot)}\partial_i \phi\right)(c(t)).$$

This defines an operator $\mathrm{D}/\mathrm{d}t$, called the *covariant derivative along c*, which assigns to a $\mathscr{C}^r$-vector field along c a $\mathscr{C}^{r-1}$-vector field, and has the following properties:

- For all $X_1, X_2 \in \mathscr{X}_c^r$,

$$\frac{\mathrm{D}(X_1 + X_2)}{\mathrm{d}t} = \frac{\mathrm{D}X_1}{\mathrm{d}t} + \frac{\mathrm{D}X_2}{\mathrm{d}t};$$

- For each $X \in \mathscr{X}_c^r$ and $f \in \mathscr{C}^r(M)$,

$$\frac{\mathrm{D}(fX)}{\mathrm{d}t} = f'X + f\frac{\mathrm{D}X}{\mathrm{d}t};$$

- For every $Y \in \mathscr{X}^r(M)$,

$$\frac{\mathrm{D}(Y \circ c)}{\mathrm{d}t} = \left(\nabla_{\dot{c}(\cdot)}Y\right) \circ c;$$

- For all $X, Y \in \mathscr{X}_c^r$,

$$\frac{\mathrm{d}}{\mathrm{d}t} g_{c(t)}(X(t), Y(t)) = g_{c(t)}\left(\frac{\mathrm{D}X}{\mathrm{d}t}(t), Y(t)\right) + g_{c(t)}\left(X(t), \frac{\mathrm{D}Y}{\mathrm{d}t}(t)\right). \quad \text{(A.5)}$$

Finally, we can define a *geodesic* as a $\mathscr{C}^2$-curve $c : I \to M$ such that $(\mathrm{D}\dot{c}/\mathrm{d}t) \equiv 0$, that is, the covariant derivative of the tangent vector field $t \mapsto \dot{c}(t)$ along c vanishes. In local coordinates, this reads

$$\ddot{c}^k(t) + \dot{c}^i(t)\dot{c}^j(t)\Gamma_{ij}^k(c(t)) = 0 \quad \text{for } k = 1, \dots, d,$$

where $(c^1(t), \dots, c^d(t)) = \phi(c(t))$. This is a second-order ordinary differential equation and the Picard–Lindelöf theorem guarantees existence and uniqueness of solutions. Therefore, for every $p \in M$ and $v \in T_pM$ there exists a unique maximal geodesic $c_v : I \to M$ with $c(0) = p$ and $\dot{c}(0) = v$. Geodesics have the following desired properties: Every $\mathscr{C}^1$-curve $c : [a, b] \to M$ with $\mathscr{L}(c) \leq \mathscr{L}(\tilde{c})$ for all $\mathscr{C}^1$-curves $\tilde{c} : [a, b] \to M$ with $\tilde{c}(a) = c(a)$ and $\tilde{c}(b) = c(b)$, is a geodesic. On the other hand, for every $p \in M$ there exists $\varepsilon > 0$ such that for all $\delta \in [0, \varepsilon)$ and for every $v \in T_pM$ with $|v|_p = 1$ the geodesic $c_v : [0, \delta] \to M$ is the shortest curve between its endpoints. Furthermore, it can be seen easily that every geodesic is parametrized proportionally to its arclength, that is, $|\dot{c}(t)|_{c(t)}$ is constant.

The subset of T_pM, where $c_v(1)$ is defined, contains an open neighborhood U_p of 0_p, such that the map

$$\exp_p : U_p \to M, \quad \exp_p(v) := c_v(1),$$

is a $\mathscr{C}^{k-2}$-diffeomorphism onto its image. In particular, it holds that

$$\mathrm{d}_{0_p} \exp_p = \mathrm{id}_{T_pM}.$$

The map $\exp_p$ is called the *Riemannian exponential map* at $p \in M$.

By the *theorem of Hopf–Rinow* (cf., for instance, Gallot et al. [48, Theorem 2.103]), the following assertions are equivalent for a Riemannian manifold:

(a) All maximal geodesics are defined on $\mathbb{R}$;
(b) There exists a point $p_0 \in M$ such that all geodesics starting at p_0 are defined on $\mathbb{R}$;
(c) Every bounded and closed subset of M is compact;
(d) Endowed with the Riemannian distance, M is a complete metric space.

On a Riemannian manifold (M, g) one can define the absolute determinant $|\det \mathrm{d}_p f|$ for a $\mathscr{C}^r$-map $f : M \to M$ by using the definition (A.1) via the singular values. Then $|\det \mathrm{d}_{(\cdot)} f| : M \to \mathbb{R}_+$ is a $\mathscr{C}^{r-1}$-function. Moreover, one can define the divergence of a vector field $X \in \mathscr{X}^r(M)$ by

$$\operatorname{div} X(p) := \operatorname{tr}\left(\nabla X(p) : T_pM \to T_pM\right).$$

Then $\operatorname{div} X : M \to \mathbb{R}$ is a $\mathscr{C}^{r-1}$-function.

On every Riemannian $\mathscr{C}^k$-manifold (M, g), $k \geq 2$, one can define a canonical Borel measure $\mathrm{vol} = \mathrm{vol}_g$, called the *Riemannian volume*, as follows. For a Borel set $A \subset M$ which is contained in the domain of a chart (ϕ, U), we set

$$\mathrm{vol}(A) := \int_{\phi(A)} \sqrt{\det[g_{ij}(\phi^{-1}(x))]} \mathrm{d}x,$$

where the integral is the usual Lebesgue integral on $\mathbb{R}^d$, $d = \dim M$. Using the transformation theorem for Lebesgue integration, one shows that this definition is independent of the chosen chart. Then vol is extended naturally to all Borel subsets of M. Using these definitions, one finds that the transformation rule

$$\int_{f(A)} \varphi \, \mathrm{dvol} = \int_A \varphi \circ f \, |\det \mathrm{d}f| \, \mathrm{dvol}$$

holds for every $\mathscr{C}^1$-diffeomorphism $f : M \to M$ and every integrable function $\varphi : M \to \mathbb{R}$.

A.3   Carathéodory Differential Equations

Continuous-time control systems are usually given by ordinary differential equations of the form $\dot{x}(t) = F(x(t), \omega(t))$ with measurable control functions ω. The standard results about ordinary differential equations such as the Picard–Lindelöf theorem about existence and uniqueness of solutions assume that the right-hand side of the equation is continuous in t. Differential equations whose right-hand side is only measurable are called *Carathéodory differential equations* and most of the theory for equations with continuous right-hand side is also valid for those (with minor modifications which are mostly obvious). In this section, we present the results about Carathéodory equations needed for the treatment of control systems in this book.

The Carathéodory Flow Box Theorem

Recall that an *absolutely continuous curve* is a map $\gamma : I \to \mathbb{R}^d$, defined on some interval $I \subset \mathbb{R}$, with the property that for every $\varepsilon > 0$ there exists $\delta > 0$ such that for every finite system $\{[\alpha_1, \beta_1], \ldots, [\alpha_n, \beta_n]\}$ of disjoint subintervals of I the implication

$$\sum_{i=1}^n (\beta_i - \alpha_i) < \delta \quad \Rightarrow \quad \sum_{i=1}^n |\gamma(\beta_i) - \gamma(\alpha_i)| < \varepsilon$$

holds (for some norm $|\cdot|$ on $\mathbb{R}^d$). Equivalently, we can require that every coordinate function $\gamma_i : I \to \mathbb{R}, i = 1, \ldots, d$, has this property. Any absolutely continuous curve is continuous and differentiable Lebesgue almost everywhere. A function $\gamma : I \to \mathbb{R}$ is absolutely continuous if and only if its derivative exists almost everywhere and defines a Lebesgue integrable function $\dot{\gamma} : I \to \mathbb{R}$ such that $\gamma(t) \equiv \gamma(t_0) + \int_{t_0}^t \dot{\gamma}(s)\mathrm{d}s$ for some $t_0 \in I$. A curve $\gamma : I \to \mathbb{R}^d$ is called *locally absolutely continuous* if the restriction of γ to every compact subinterval $J \subset I$ is absolutely continuous. A locally absolutely continuous curve on a differentiable manifold is defined as follows.

Definition A.1. Let M be a $\mathscr{C}^k$-manifold, $k \in \mathbb{N} \cup \{\infty\} \cup \{\omega\}$. A map $\gamma : I \to M$, defined on some interval $I \subset \mathbb{R}$, is called a *locally absolutely continuous curve* if $\varphi \circ \gamma : I \to \mathbb{R}$ is locally absolutely continuous for every $\varphi \in \mathscr{C}^k(M)$.

To describe the properties of the right-hand sides of Carathéodory differential equations, we need to introduce the following notions.

Definition A.2. Let M be a $\mathscr{C}^k$-manifold, $k \in \mathbb{N} \cup \{\infty\} \cup \{\omega\}$, and $I \subset \mathbb{R}$ a nonempty interval. A *Carathéodory function* on M is a map $\varphi : I \times M \to \mathbb{R}$ with the property that $\varphi_t : x \mapsto \varphi(t, x)$ is continuous for each $t \in I$ and $\varphi_x : t \mapsto \varphi(t, x)$ is Lebesgue measurable for each $x \in M$. A Carathéodory function is *locally integrally bounded* if, for each compact subset $K \subset M$, there exists a positive locally integrable function $\psi_K : I \to \mathbb{R}$ such that $|\varphi(t, x)| \leq \psi_K(t)$ for all $t \in I$ and $x \in K$. A Carathéodory function $\varphi : I \times M \to \mathbb{R}$ is of *class $\mathscr{C}^k$* if $\varphi_t = \varphi(t, \cdot) : M \to \mathbb{R}$ is of class $\mathscr{C}^k$ for each $t \in I$. If $r \in \mathbb{N}$, then a Carathéodory function φ is *locally integrally of class $\mathscr{C}^r$* if it is of class $\mathscr{C}^r$, and $X_1 \cdots X_r(\varphi_t)^4$ is locally integrally bounded for all $t \in I$ and $X_1, \ldots, X_r \in \mathscr{X}^r(M)$. If φ is locally integrally of class $\mathscr{C}^r$ for every $r \in \mathbb{N}$, then it is *locally integrally of class $\mathscr{C}^\infty$*.

For $k = \omega$ locally integrally class $\mathscr{C}^k$-functions are defined in a different way. We refer to Bullo and Lewis [15, Sect. A.2.1] for further details.

Definition A.3. Let M be a $\mathscr{C}^{k+1}$-manifold, $k \in \mathbb{N} \cup \{\infty\} \cup \{\omega\}$, and $I \subset \mathbb{R}$ a nonempty interval. A map $f : I \times M \to TM$ with the property that $f(t, x) \in T_x M$ for all $(t, x) \in I \times M$, is called a *Carathéodory vector field* on M if for every continuous one-form $\alpha : M \to T^*M$ the function $\alpha \cdot f : I \times M \to \mathbb{R}$, $(t, x) \mapsto \alpha(x) f(t, x)$, is a Carathéodory function. A Carathéodory vector field f is *locally integrally of class $\mathscr{C}^k$* if $\alpha \cdot f$ is locally integrally of class $\mathscr{C}^k$ for every $\mathscr{C}^k$-one-form α. If $f : I \times M \to TM$ is a Carathéodory vector field on M, the equation

$$\dot{x}(t) = f(t, x(t)) \tag{A.6}$$

---

[4]Here we mean the application of the vector fields $X_1, \ldots, X_r$ as differential operators acting on functions $M \to \mathbb{R}$.

is called a *Carathéodory differential equation* or a *differential equation of Carathéodory type*. A *solution* of (A.6) is a locally absolutely continuous curve $\gamma : J \to M$, defined on some open subinterval $J \subset I$, such that $\dot{\gamma}(t) = f(t, \gamma(t))$ for Lebesgue almost all $t \in J$.

The following result about existence and uniqueness of solutions for Carathéodory differential equations can be found in Bullo and Lewis [15, Theorem A.11] as the *time-dependent flow box theorem*.

Theorem A.1 (Flow Box Theorem). *Let $f : I \times M \to TM$ be a locally integrally class $\mathscr{C}^k$-vector field, $k \in \mathbb{N} \cup \{\infty\} \cup \{\omega\}$, and let $(t_0, x_0) \in I \times M$. Then there exists a triple (U, T, Φ) (called a* flow box *of f at (t_0, x_0)) with the following properties:*

  (i) *U is an open subset of M containing x_0;*
  (ii) *$T > 0$ or $T = \infty$;*
  (iii) *$\Phi : (t_0 - T, t_0 + T) \times U \to M$ is a map having the following properties:*

   (a) *the map $t \mapsto \Phi(t, x)$ is locally absolutely continuous for each $x \in U$;*
   (b) *the map $x \mapsto \Phi(t, x)$ is of class $\mathscr{C}^k$ for each $t \in (t_0 - T, t_0 + T)$;*
   (c) *$t \mapsto \Phi(t, x)$ is the unique solution of $\dot{x}(t) = f(t, x(t))$ with $\Phi(t_0, x) = x$;*

  (iv) *for all $t \in (t_0 - T, t_0 + T)$, $\Phi_t : U \to M$ is a $\mathscr{C}^k$-diffeomorphism onto its image, where $\Phi_t(x) = \Phi(t, x)$.*

Furthermore, if $(\tilde{U}, \tilde{T}, \tilde{\Phi})$ is another such triple, then Φ and $\tilde{\Phi}$ agree when restricted to $((t_0 - T, t_0 + T) \cap (t_0 - \tilde{T}, t_0 + \tilde{T})) \times (U \cap \tilde{U})$.

For linear equations of Carathéodory type the usual *variation-of-constants formula* holds (see Aulbach and Wanner [5, Theorem 2.10]).

Proposition A.2. *Let $I \subset \mathbb{R}$ be a nonempty interval and $A : I \to \mathbb{R}^{d \times d}$, $b : I \to \mathbb{R}^d$, locally integrable mappings. Then the equation*

$$\dot{x}(t) = A(t)x(t) + b(t) \tag{A.7}$$

is a Carathéodory differential equation. The solution $\Phi(t; t_0, x_0)$ of the corresponding initial value problem (A.7), $x(t_0) = x_0$, exists and is unique with

$$\Phi(t; t_0, x_0) = \Psi(t, t_0)x_0 + \int_{t_0}^{t} \Psi(t, s)b(s)\mathrm{d}s$$

for all $(t, t_0, x_0) \in I \times I \times \mathbb{R}^d$, where $t \mapsto \Psi(t, t_0) \in \mathrm{GL}(\mathbb{R}^d)$ is the unique solution of the initial value problem

$$\dot{X}(t) = A(t)X(t), \quad X(t_0) = I \in \mathbb{R}^{d \times d}.$$

The Variational Equation and Applications

For a Carathéodory differential equation on a Riemannian manifold (M, g) the variational equation can be written in a covariant way. See the following proposition whose proof is standard and will be omitted.

Proposition A.3. *Let (M, g) be a Riemannian $\mathscr{C}^k$-manifold, $k \geq 2$. Consider a locally integrally class $\mathscr{C}^{k-1}$-vector field $f : I \times M \to TM$ and the corresponding differential equation*

$$\dot{x}(t) = f(t, x(t))$$

with flow box (U, T, Φ) at $(t_0, x_0) \in I \times M$. Then for any $v \in T_{x_0}M$ the curve

$$c(t) := \mathrm{d}_{x_0}\Phi_t(v), \quad c : (t_0 - T, t_0 + T) \to TM,$$

is locally absolutely continuous and satisfies the Riemannian variational equation

$$\frac{\mathrm{D}z}{\mathrm{d}t}(t) = \nabla f_t(\Phi_t(x_0))z(t) \tag{A.8}$$

almost everywhere, where $\mathrm{D}/\mathrm{d}t$ denotes the covariant derivative along the solution $\Phi(\cdot, x_0)$.[5]

The preceding proposition has two important applications, the *Wazewski inequality* and the *(generalized) Liouville formula*. The Wazewski inequality gives an estimate for the operator norm of the derivative $\mathrm{d}_x\Phi_t$ (given a flow box (U, T, Φ) of a Carathéodory vector field).

Proposition A.4 (Wazewski Inequality). *Let (M, g) be a Riemannian $\mathscr{C}^k$-manifold, $k \geq 3$. Consider a locally integrally class $\mathscr{C}^{k-1}$-vector field $f : I \times M \to TM$ and the corresponding differential equation*

$$\dot{x}(t) = f(t, x(t))$$

with flow box (U, T, Φ) at $(t_0, x_0) \in I \times M$. Then it holds that

$$\|\mathrm{d}_{x_0}\Phi_t\| \leq \exp\left(\int_{t_0}^{t} \lambda_{\max}(S\nabla f_s(\Phi_s(x_0)))\mathrm{d}s\right)$$

for all $t \in [t_0, t_0 + T)$, where $\lambda_{\max}(\cdot)$ denotes the maximal eigenvalue and $S\nabla(\cdot)$ the symmetrized covariant derivative.

---

[5]Although we have only defined the covariant derivative of a $\mathscr{C}^r$-vector field along a $\mathscr{C}^r$-curve, this notion also makes sense if both the curve and the vector field are only locally absolutely continuous.

Proof. Let $x_t :\equiv \Phi_t(x_0)$ and $\lambda(t) :\equiv \lambda_{\max}(S\nabla f_t(x_t))$. Let $z : J \to TM$, $t_0 \in J \subset I$, be a locally absolutely continuous vector field along x_t which solves the variational equation (A.8). Then for almost all $t \in J$ we obtain

$$\frac{\mathrm{d}}{\mathrm{d}t}|z(t)|^2 = \frac{\mathrm{d}}{\mathrm{d}t}g_{x_t}(z(t),z(t)) \stackrel{(A.5)}{=} g_{x_t}\left(\frac{\mathrm{D}z}{\mathrm{d}t}(t),z(t)\right) + g_{x_t}\left(z(t),\frac{\mathrm{D}z}{\mathrm{d}t}(t)\right)$$

$$= g_{x_t}\left(\nabla f_t(x_t)z(t),z(t)\right) + g_{x_t}\left(z(t),\nabla f_t(x_t)z(t)\right)$$

$$= g_{x_t}\left(\nabla f_t(x_t)z(t),z(t)\right) + g_{x_t}\left(\nabla f_t(x_t)^*z(t),z(t)\right)$$

$$= 2g_{x_t}\left(\frac{1}{2}\left[\nabla f_t(x_t) + \nabla f_t(x_t)^*\right]z(t),z(t)\right)$$

$$\leq 2\lambda(t)|z(t)|^2.$$

Now we assume that $z(t) \neq 0$ for all $t \in J \cap [t_0,\infty)$. For almost all t, this implies

$$\frac{\frac{\mathrm{d}}{\mathrm{d}t}|z(t)|^2}{|z(t)|^2} \leq 2\lambda(t) \Rightarrow \int_{t_0}^t \frac{\frac{\mathrm{d}}{\mathrm{d}s}|z(s)|^2}{|z(s)|^2}\mathrm{d}s \leq 2\int_{t_0}^t \lambda(s)\mathrm{d}s$$

$$\Rightarrow \log\left(|z(t)|^2\right) - \log\left(|z(t_0)|^2\right) \leq 2\int_{t_0}^t \lambda(s)\mathrm{d}s$$

$$\Rightarrow \log|z(t)| - \log|z(t_0)| \leq \int_{t_0}^t \lambda(s)\mathrm{d}s$$

$$\Rightarrow |z(t)| \leq |z(t_0)|\exp\left(\int_{t_0}^t \lambda(s)\mathrm{d}s\right).$$

It is easy to see that the integral over λ exists. Since for each nonzero $v \in T_{x_0}M$ the map $z(t) = \mathrm{d}_{x_0}\Phi_t(v)$ is a solution of (A.8) with $z(t) \neq 0$ for all $t \in (t_0 - T, t_0 + T)$, we obtain

$$\|\mathrm{d}_{x_0}\Phi_t\| = \max_{|v|=1}\|\mathrm{d}_{x_0}\Phi_t(v)\|$$

$$\leq \max_{|v|=1}\underbrace{\|\mathrm{d}_{x_0}\Phi_{t_0}(v)\|}_{=\mathrm{id}}\exp\left(\int_{t_0}^t \lambda(s)\mathrm{d}s\right) = \exp\left(\int_{t_0}^t \lambda(s)\mathrm{d}s\right),$$

which finishes the proof.   □

The classical Liouville formula expresses the absolute determinant $|\det \mathrm{d}_x\Phi_t|$ in terms of the integral over the divergence of f_t along the solution $\Phi_t(x)$. There exist several generalizations of this formula. We use the following one which involves an invariant subbundle of the tangent bundle.

Proposition A.5 (Generalized Liouville Formula). *Let (M,g) be a d-dimensional Riemannian $\mathscr{C}^k$-manifold, $k \geq 2$. Consider a locally integrally class $\mathscr{C}^{k-1}$-vector* *field $f : I \times M \to TM$ and the corresponding differential equation*

$$\dot{x}(t) = f(t, x(t))$$

with flow box (U, T, Φ) at $(t_0, x_0) \in I \times M$. Let $E \to M$ be a subbundle of $I \times TM \to M$, $(t, v) \mapsto \pi_{TM}(v)$ (with the base point projection $\pi_{TM} : TM \to M$), of rank $n \in \{1, \dots, d\}$, which is invariant under $\mathrm{d}\Phi$ in the sense that

$$\mathrm{d}_x \Phi_t(E_{t_0, x}) = E_{t, \Phi_t(x)}$$

holds for all $x \in U$ and $t \in (t_0 - T, t_0 + T)$. Then

$$\left| \det \mathrm{d}_{x_0} \Phi_t |_{E_{t_0, x_0}} \right| = \exp\left(\int_{t_0}^{t} \mathrm{tr}\left[\nabla f_s(\Phi_s(x)) \circ Q(s, \Phi_s(x)) \right] \mathrm{d}s \right),$$

where $Q(t, x) : T_x M \to E_{t, x}$ denotes the orthogonal projection.

Proof. For every $t \in (t_0 - T, t_0 + T)$ we write

$$L(t) := \mathrm{d}_{x_0} \Phi_t |_{E_{t_0, x_0}} : E_{t_0, x_0} \to E_{t, \Phi_t(x_0)}.$$

Let $(v_1, \dots, v_n)$ be an orthonormal basis of E_{t_0, x_0}. Then

$$| \det L(t) |^2 = \det(L(t)^* L(t)) = \det\left(\langle L(t)^* L(t) v_i, v_j \rangle \right)_{i,j=1}^{n}$$

$$= \det\left(\langle L(t) v_i, L(t) v_j \rangle \right)_{i,j=1}^{n}.$$

Using that $v_i(t) := L(t) v_i$ solves the Riemannian variational equation for each $i \in \{1, \dots, n\}$, we obtain for almost all $t \in (t_0 - T, t_0 + T)$ that

$$\frac{1}{2} \frac{\mathrm{d}}{\mathrm{d}t} |\det L(t)|^2 = \frac{1}{2} \frac{\mathrm{d}}{\mathrm{d}t} \langle v_1(t) \wedge \dots \wedge v_n(t), v_1(t) \wedge \dots \wedge v_n(t) \rangle_{\bigwedge^n T_{\Phi_t(x)} M}$$

$$= \left\langle \frac{\mathrm{D}v_1}{\mathrm{d}t}(t) \wedge \dots \wedge v_n(t), v_1(t) \wedge \dots \wedge v_n(t) \right\rangle_{\bigwedge^n T_{\Phi_t(x)} M}$$

$$+ \dots +$$

$$\left\langle v_1(t) \wedge \dots \wedge \frac{\mathrm{D}v_n}{\mathrm{d}t}(t), v_1(t) \wedge \dots \wedge v_n(t) \right\rangle_{\bigwedge^n T_{\Phi_t(x)} M}$$

$$= \langle \nabla f_t(\Phi_t(x)) v_1(t) \wedge \dots \wedge v_n(t), v_1(t) \wedge \dots \wedge v_n(t) \rangle_{\bigwedge^n T_{\Phi_t(x)} M}$$

$$+ \dots +$$

$$\langle v_1(t) \wedge \dots \wedge \nabla f_t(\Phi_t(x)) v_n(t), v_1(t) \wedge \dots \wedge v_n(t) \rangle_{\bigwedge^n T_{\Phi_t(x)} M}.$$

From Lemma A.1 and invariance of E it thus follows that $|\det L(t)|$ satisfies the scalar linear Carathéodory differential equation

$$\frac{\mathrm{d}}{\mathrm{d}t}\,|\det L(t)| = \frac{\frac{\mathrm{d}}{\mathrm{d}t}\,|\det L(t)|^2}{2\,|\det L(t)|}$$

$$= \mathrm{tr}\,[\nabla f_t(\Phi_t(x)) \circ Q(t, \Phi_t(x))]\,|\det L(t)|.$$

Hence, the variation-of-constants formula gives

$$|\det L(t)| = \exp\left(\int_{t_0}^t \mathrm{tr}\,[\nabla f_s(\Phi_s(x)) \circ Q(s, \Phi_s(x))]\,\mathrm{d}s\right),$$

since $|\det L(t_0)| = |\det \mathrm{id}_{E_{t_0,x_0}}| = 1$.   □

Cut-Off Functions

Every $\mathscr{C}^k$-manifold, $k \in \mathbb{N} \cup \{\infty\}$, admits partitions of unity of class $\mathscr{C}^k$.[6] As for instance shown in Lee [75, Proposition 2.26], one can construct cut-off functions from such partitions which yields the following proposition.

Proposition A.6. *Let M be a $\mathscr{C}^k$-manifold, $k \in \mathbb{N} \cup \{\infty\}$. For any closed set $A \subset M$ and any open set U containing A there exists a cut-off function $\theta : M \to \mathbb{R}$ of class $\mathscr{C}^k$, that is, $\theta(x) \in [0, 1]$ for all $x \in M$, $\theta(x) \equiv 1$ on A, and $\theta(x) \equiv 0$ on U^c.*

Given an arbitrary $\mathscr{C}^k$-vector field f on a manifold M and a class $\mathscr{C}^k$ cut-off function $\theta : M \to [0, 1]$ with compact support, one obtains a complete $\mathscr{C}^k$-vector field θf whose integral curves coincide with those of f on the set where $\theta(x) \equiv 1$.

A.4 Metric Spaces

In this short section we prove two simple lemmas about metric spaces. To this end, we first introduce some notation: Let (X, ϱ) be a metric space and $K \subset X$ a subset. Then for every $\varepsilon > 0$ the ε-neighborhood of K is defined by

$$N_\varepsilon(K) := \{x \in X \mid \exists y \in K : \varrho(x, y) < \varepsilon\}.$$

---

[6]A *partition of unity* is a family of nonnegative functions $f_\alpha : M \to \mathbb{R}$ such that for every $x \in M$ only finitely many of the values $f_\alpha(x)$ are different from zero and $\sum_\alpha f_\alpha(x) = 1$.

That is, $N_\varepsilon(K)$ is the union of the open balls $B(x, \varepsilon)$, $x \in K$, and thus an open neighborhood of K. For a point $x \in X$ and a nonempty set $A \subset X$ the distance from x to A is defined by

$$\mathrm{dist}(x, A) := \inf_{a \in A} \varrho(x, a).$$

Lemma A.2. *Let (X, ϱ) be a metric space and $A \subset X$ nonempty. Then the function*

$$x \mapsto \mathrm{dist}(x, A), \quad X \to \mathbb{R}_+,$$

is continuous.

Proof. For all $x, y \in X$ and $a \in A$ we have

$$\mathrm{dist}(x, A) \leq \varrho(x, a) \leq \varrho(x, y) + \varrho(a, y).$$

Hence, $\mathrm{dist}(x, A) - \varrho(x, y) \leq \varrho(a, y)$, which implies

$$\mathrm{dist}(x, A) - \varrho(x, y) \leq \inf_{a \in A} \varrho(y, a) = \mathrm{dist}(y, A).$$

Therefore, $\mathrm{dist}(x, A) - \mathrm{dist}(y, A) \leq \varrho(x, y)$. By changing the roles of x and y we obtain

$$|\mathrm{dist}(x, A) - \mathrm{dist}(y, A)| \leq \varrho(x, y),$$

which proves the assertion.   □

Recall that a topological space X is called *locally compact* if every neighborhood of a point $x \in X$ contains a compact neighborhood of x.

Lemma A.3. *Let (X, ϱ) be a locally compact metric space. Then for every compact set $K \subset X$ there exists some $\varepsilon > 0$ such that $\mathrm{cl}\, N_\varepsilon(K)$ is compact.*

Proof. Since X is locally compact, every $x \in K$ has an open neighborhood K_x with compact closure. Since K is compact, there are $x_1, \ldots, x_n \in K$ with $K \subset \bigcup_{i=1}^{n} K_{x_i}$. Let $W := \bigcup_{i=1}^{n} \mathrm{cl}\, K_{x_i}$. Then, as a finite union of compact sets, W is a compact neighborhood of K. Assume to the contrary that for each $\varepsilon > 0$ there is some $x \in X$ with $\mathrm{dist}(x, K) < \varepsilon$ and $x \notin W$. Then there are sequences $(y_n)_{n \in \mathbb{N}}$ and $(z_n)_{n \in \mathbb{N}}$ with $y_n \in X \backslash W$, $z_n \in K$, and $\varrho(y_n, z_n) < 1/n$ for all $n \in \mathbb{N}$. By compactness of K we may assume that $z_n \to z \in K$ for $n \to \infty$. Consequently, also $y_n \to z$. Let $i \in \{1, \ldots, n\}$ such that $z \in K_{x_i}$. Then, for sufficiently large n we obtain $y_n \in K_{x_i} \subset W$ in contradiction to $y_n \in X \backslash W$. Hence, there exists some $\varepsilon > 0$ with $N_\varepsilon(K) \subset W$ which implies that $\mathrm{cl}\, N_\varepsilon(K) \subset W$ is compact.   □

Appendix B
Dynamical Systems

In this part of the appendix, we recall basic concepts from the theory of dynamical systems. By a (classical) dynamical system we understand a mapping $\Phi : \mathbb{T} \times X \to X$, where $\mathbb{T} \in \{\mathbb{Z}_+, \mathbb{Z}, \mathbb{R}_+, \mathbb{R}\}$, which satisfies the axioms $\Phi(0, x) = x$ and $\Phi(t + s, x) = \Phi(t, \Phi(s, x))$ for all $x \in X$ and $t, s \in \mathbb{T}$. In other words, a dynamical system is a group or semigroup action of $\mathbb{T}$ on a set X. The set $\mathbb{T}$ is also called the time set of the dynamical system. In the case $\mathbb{T} = \mathbb{R}$ we also speak of a flow, or in the case $\mathbb{T} = \mathbb{R}_+$ of a semiflow. Alternatively, we speak of a continuous-time dynamical system if $\mathbb{T} \in \{\mathbb{R}_+, \mathbb{R}\}$ and of a discrete-time dynamical system if $\mathbb{T} \in \{\mathbb{Z}_+, \mathbb{Z}\}$. Often, we additionally assume that X is a topological or metric space and Φ is continuous. For fixed $t \in \mathbb{T}$, the map $\Phi_t : X \to X$, $x \mapsto \Phi(t, x)$, is called the time-t-map of the dynamical system. If $\mathbb{T} \in \{\mathbb{Z}, \mathbb{R}\}$, this map is invertible with inverse Φ_{-t}. The orbit through a point $x \in X$ is the set $\mathcal{O}(x) = \{\Phi(t, x) : t \in \mathbb{T}\}$.

B.1 Chain Recurrence and Chain Transitivity

In this section, we collect some definitions and elementary results about continuous-time dynamical systems on compact metric spaces. Throughout we assume that $\Phi : \mathbb{R} \times X \to X$ is a continuous flow on a compact metric space (X, ϱ). All of the following definitions and results (together with proofs) can be found in Colonius and Kliemann [25, Appendix B]. Further references are Conley [29] and Katok and Hasselblatt [61].

Definition B.1. The *ω-limit set* of a subset $Y \subset X$ is defined as

$$\omega(Y) := \bigcap_{t>0} \mathrm{cl} \bigcup_{s \geq t} \Phi(s, Y).$$

The *α-limit set* of Y is

C. Kawan, *Invariance Entropy for Deterministic Control Systems*, Lecture Notes in Mathematics 2089, DOI 10.1007/978-3-319-01288-9,
© Springer International Publishing Switzerland 2013

$$\alpha(Y) := \bigcap_{t>0} \mathrm{cl} \bigcup_{s \le -t} \Phi(s, Y).$$

If Y consists of only one element y, we write $\omega(y) := \omega(\{y\})$ and $\alpha(y) := \alpha(\{y\})$.

Definition B.2. A compact set $K \subset X$ is called *invariant* if $\Phi_t(K) \subset K$ for all $t \in \mathbb{R}$. It is called *isolated invariant* if it is invariant and there is a neighborhood N of K such that $\Phi(\mathbb{R}, x) \subset N$ implies $x \in K$.

Definition B.3. A *Morse decomposition* of Φ is a finite collection $\{\mathcal{M}_i\}_{i=1}^n$ of nonempty, pairwise disjoint, and isolated compact invariant sets such that:

(i) For all $x \in X$ one has $\alpha(x), \omega(x) \subset \bigcup_{i=1}^n \mathcal{M}_i$.
(ii) If there are $\mathcal{M}_{j_0}, \mathcal{M}_{j_1}, \dots, \mathcal{M}_{j_l}$ and $x_1, \dots, x_l \in X \setminus \bigcup_{i=1}^n \mathcal{M}_i$ with $\alpha(x_i) \subset \mathcal{M}_{j_{i-1}}$ and $\omega(x_i) \subset \mathcal{M}_{j_i}$ for $i = 1, \dots, l$, then $\mathcal{M}_{j_0} \ne \mathcal{M}_{j_l}$.

The elements of a Morse decomposition are called *Morse sets*. A Morse decomposition is *finer* than another one if the elements of the first one are contained in those of the second one.

Definition B.4. For $x, y \in X$ and $\varepsilon, \tau > 0$, an (ε, τ)-*chain* from x to y is given by a natural number $n \in \mathbb{N}$ together with points

$$x_0 = x, x_1, \dots, x_n = y \quad \text{and times} \quad \tau_0, \dots, \tau_{n-1} \ge \tau,$$

such that $\varrho(\Phi(\tau_i, x_i), x_{i+1}) < \varepsilon$ for $i = 0, 1, \dots, n - 1$.

Definition B.5. A subset $Y \subset X$ is called *chain transitive* if for all $x, y \in Y$ and $\varepsilon, \tau > 0$ there exists an (ε, τ)-chain from x to y. A point $x \in X$ is *chain recurrent* if for all $\varepsilon, \tau > 0$ there exists an (ε, τ)-chain from x to x. The *chain recurrent set* $\mathcal{R}$ of Φ is the set of all chain recurrent points.

Proposition B.1. *The following assertions hold:*

(i)   *The set $\mathcal{R}$ is closed and invariant. The flow Φ restricted to a maximal (with respect to set inclusion) chain transitive subset of the chain recurrent set $\mathcal{R}$ is chain transitive. In particular, the flow restricted to $\mathcal{R}$ is chain recurrent.*
(ii) *A closed subset Y of X is chain transitive if it is chain recurrent and connected. Conversely, if Φ is chain transitive on X, then X is connected.*
(iii) *The connected components of the chain recurrent set $\mathcal{R}$ coincide with the maximal chain transitive subsets of $\mathcal{R}$.*

Proposition B.2. *There exists a finest Morse decomposition $\{\mathcal{M}_1, \dots, \mathcal{M}_l\}$ if and only if the chain recurrent set $\mathcal{R}$ has only finitely many connected components. In this case, the Morse sets coincide with the chain recurrent components of $\mathcal{R}$ and the flow restricted to each Morse set is chain transitive and chain recurrent.*

B.2   Vector Bundles and Linear Flows

In this section, we collect some definitions and results about real finite-dimensional vector bundles and linear flows. We start with the definition of a vector bundle following Lee [75, Chap. 5].

Definition B.6. Let B be a topological space. A *(real) vector bundle of rank k over* B is a topological space E together with a continuous surjective map $\pi : E \to B$ satisfying:

 (i) For each $b \in B$ the set $E_b := \pi^{-1}(b) \subset E$ (called the *fiber* of E over b) is endowed with the structure of a k-dimensional real vector space;
 (ii) For each $b \in B$ there exist a neighborhood U of b in B and a homeomorphism $\Phi : \pi^{-1}(U) \to U \times \mathbb{R}^k$ (called a *local trivialization* of E over U) such that the following diagram commutes:

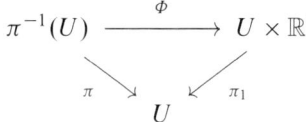

Here π_1 is the projection onto the first factor. Furthermore, for each $c \in U$ the restriction of Φ to E_c is a linear isomorphism from E_c to $\{c\} \times \mathbb{R}^k \cong \mathbb{R}^k$.

The space E is called the *total space* of the vector bundle, B is called the *base space*, and π the *projection*. Often we simply write $\pi : E \to B$, $E \to B$, or E for a vector bundle. The subset of E consisting of all the zero vectors of the fibers E_b, $b \in B$, is called the *zero section* of the vector bundle. A vector bundle $\pi : E \to B$ is called *trivial* if there exists a local trivialization over all of B (called a *global trivialization* of E). In this case, E itself is homeomorphic to the product $B \times \mathbb{R}^k$. A *subbundle* of a vector bundle $\pi : E \to B$ is a vector bundle $\pi' : E' \to B$ such that $E' \subset E$ is a closed subset of E which intersects each fiber E_b, $b \in B$, in a linear subspace, and such that $\pi' = \pi|_{E'}$ (E' is endowed with the subspace topology).

Definition B.7. Let $E \to B$ be a vector bundle and $E^1, E^2 \subset E$ subbundles with $E_b^1 \cap E_b^2 = \{0\}$ for each $b \in B$. The *Whitney sum* of E^1 and E^2 is the vector bundle $E' = E^1 \oplus E^2 \subset E$ with fibers

$$E_b' = E_b^1 \oplus E_b^2 = \left\{ e_1 + e_2 \ : \ e_i \in E_b^i \right\}.$$

Then $E' \to B$ is a subbundle of $E \to B$.

Lemma B.1. *Consider a trivial vector bundle $\pi : B \times X \to B$, $\pi(b, x) = b$, where* (B, ϱ) *is a compact metric space and* $(X, \langle \cdot, \cdot \rangle)$ *a d-dimensional Euclidean space. Suppose that there is a decomposition*

$$B \times X = \mathcal{V} \oplus \mathcal{W}$$

into a Whitney sum of subbundles $\mathcal{V}$ and $\mathcal{W}$. For each $b \in B$ let P_b denote the projection onto $\mathcal{V}_b$ along $\mathcal{W}_b$. Then the mapping

$$b \mapsto P_b, \quad B \to \mathcal{L}(X, X),$$

is continuous.

Proof. Let $\pi_{\mathcal{V}} : \mathcal{V} \to B$ denote the projection of $\mathcal{V}$ (that is, $\pi_{\mathcal{V}} = \pi|_{\mathcal{V}}$), let k be the rank of $\mathcal{V}$, and fix $b_0 \in B$. Then, by definition, there exists an open neighborhood $U \subset B$ of b_0 and a homeomorphism $\varphi : \pi_{\mathcal{V}}^{-1}(U) \to U \times \mathbb{R}^k$ of the form

$$\varphi(b, x) = (b, \hat{\varphi}(b, x)).$$

Hence, for every $(b, y) \in U \times \mathbb{R}^k$, there exists a unique $x \in \mathcal{V}_b$ with $\hat{\varphi}(b, x) = y$. In particular, the map $\hat{\varphi}_b : \mathcal{V}_b \to \mathbb{R}^k$, $x \mapsto \hat{\varphi}(b, x)$, is a linear isomorphism, and it holds that

$$\mathcal{V}_b = \hat{\varphi}_b^{-1}(\mathbb{R}^k) \quad \text{for every } b \in U.$$

Now let $\{e_1(b_0), \ldots, e_k(b_0)\}$ be a fixed basis of $\mathcal{V}_{b_0}$ and define $e_1(b), \ldots, e_k(b) \in \mathcal{V}_b$ by

$$e_j(b) := \hat{\varphi}_b^{-1}(\hat{\varphi}_{b_0}(e_j(b_0)))$$

for all $b \in U$. It follows that $\{e_1(b), \ldots, e_k(b)\}$ is a basis of $\mathcal{V}_b$ for all $b \in U$. Analogously, we find such a basis $\{e_{k+1}(b), \ldots, e_d(b)\}$ for $\mathcal{W}_b$, depending continuously on b, and we can assume that both bases are defined on the same neighborhood U of b_0. Then for each $(b, x) \in U \times X$ there are unique $\alpha_1(b, x), \ldots, \alpha_d(b, x) \in \mathbb{R}$ such that

$$x = \underbrace{\sum_{i=1}^{k} \alpha_i(b, x) e_i(b)}_{= P_b x} + \sum_{i=k+1}^{d} \alpha_i(b, x) e_i(b).$$

Let $a_{ij}(b) := \langle e_i(b), e_j(b) \rangle$ for all $b \in U$ and $i, j = 1, \ldots, d$. Then $A(b) := (a_{ij}(b))_{1 \le i, j \le d}$ is a symmetric positive definite matrix and for $j = 1, \ldots, d$, $x \in X$, it holds that

$$x_j(b) := \langle x, e_j(b) \rangle = \sum_{i=1}^{d} a_{ij}(b) \alpha_i(b, x).$$

Hence, the vectors $\hat{x} := (x_1(b), \dots, x_d(b))$ and $\alpha(b, x) := (\alpha_1(b, x), \dots, \alpha_d(b, x))$ satisfy $\hat{x} = A(b)\alpha(b, x)$ which implies $\alpha(b, x) = A(b)^{-1}\hat{x}$. Therefore, in particular $\alpha_1(b, x), \dots, \alpha_k(b, x)$ depend continuously on (b, x) and thus also $P_b x$. Continuity of P_b is then shown as follows. We have

$$\| P_b - P_{b_0} \| = \max_{|x|=1} |P_b x - P_{b_0} x|$$

$$= \max_{|x|=1} \underbrace{\left| \sum_{i=1}^{k} \alpha_i(b, x)e_i(b) - \sum_{i=1}^{k} \alpha_i(b_0, x)e_i(b_0) \right|}_{=:f(b,x)}.$$

Since f is uniformly continuous on the compact set $W \times S(X)$, where $W \subset U$ is a compact neighborhood of b_0 and $S(X) = \{x \in X : |x| = 1\}$, for every $\varepsilon > 0$ we find $\delta > 0$ such that $\varrho(b, b_0) < \delta$ implies $|f(b, x) - f(b_0, x)| < \varepsilon$ for all $x \in S(X)$. This implies continuity of $b \mapsto P_b$ at b_0. $\square$

Definition B.8. A *(discrete- or continuous-time) linear flow* on a vector bundle $\pi : E \to B$ is a continuous flow $\Phi : \mathbb{T} \times E \to E$, $\mathbb{T} \in \{\mathbb{Z}, \mathbb{R}\}$, such that for each $t \in \mathbb{T}$ the time-t-map $\Phi_t : E \to E$ maps fibers into fibers, that is, $\pi(\Phi(t, e_1)) = \pi(\Phi(t, e_2))$ if $\pi(e_1) = \pi(e_2)$, and the restrictions $\Phi_t|_{E_b} : E_b \to E_{\pi(\Phi_t(e))}$ are linear maps. Every linear flow induces a flow Θ on the base space B by $\Theta(t, b) := \pi(\Phi(t, e))$ for $b \in B$ and $e \in E_b$. Analogously, a *linear semiflow* on a vector bundle is defined by replacing $\mathbb{T}$ with $\mathbb{T}_+$ in the above definition.

If the base space B of the vector bundle $\pi : E \to B$ is trivial, that is, B consists of only one point, the space E is a finite-dimensional real vector space and each continuous-time linear semiflow on E has the form $(t, x) \mapsto \mathrm{e}^{At}x$ for some $A \in \mathscr{L}(E, E)$. This is proved in the following proposition. The arguments of the proof are borrowed from the theory of strongly continuous semigroups on Banach spaces (see, for instance, Pazy [89]).

Proposition B.3. *Let X be a finite-dimensional real vector space and $T : \mathbb{R}_+ \times X \to X$, $(t, x) \mapsto T(t)x$, a linear semiflow on X. Then the mapping $t \mapsto T(t)$, $\mathbb{R}_+ \to \mathscr{L}(X, X)$, is continuous and there exists a unique linear operator $A \in \mathscr{L}(X, X)$ such that $T(t) = \mathrm{e}^{At}$ for all $t \geq 0$, which for all $x \in X$ is given by*

$$Ax = \lim_{t \searrow 0} \frac{T(t)x - x}{t}. \tag{B.1}$$

Proof. Let X be endowed with some norm $|\cdot|$. Then continuity of $t \mapsto T(t)$ follows from uniform continuity of $(t, x) \mapsto T(t)x$ on compact sets of the form $[a, b] \times S(X)$, where $[a, b] \subset \mathbb{R}_+$ and $S(X) := \{x \in X : |x| = 1\}$. Now let $D(A) \subset X$ be the set of all $x \in X$ such that the limit in (B.1) exists and define $A : D(A) \to X$ according to (B.1). In the following, we show that $D(A) = X$, that is, that the definition of A is correct. It is easy to see that $D(A)$ is a linear subspace of X and therefore a closed set. Hence, it suffices to prove that $D(A)$ is dense in X. From continuity of $T(\cdot)$ we can conclude that for every $x \in X$ it holds that

$$\frac{1}{t} \int_0^t T(s)x \mathrm{d}s \to x \quad \text{for } t \searrow 0. \tag{B.2}$$

For every $x \in X$ we have

$$\frac{T(s) - I}{s} \int_0^t T(r)x \mathrm{d}r = \frac{1}{s} \int_0^t T(s + r)x \mathrm{d}r - \frac{1}{s} \int_0^t T(r)x \mathrm{d}r.$$

Substituting $\rho = s + r$ in the second integral gives

$$\frac{T(s) - I}{s} \int_0^t T(r)x \mathrm{d}r = \frac{1}{s} \int_s^{t+s} T(\rho)x \mathrm{d}\rho - \frac{1}{s} \int_0^t T(r)x \mathrm{d}r$$

$$= \frac{1}{s} \left(\int_t^{t+s} T(\rho)x \mathrm{d}\rho + \int_s^t T(\rho)x \mathrm{d}\rho \right.$$

$$\left. - \int_s^t T(r)x \mathrm{d}r - \int_0^s T(r)x \mathrm{d}r \right)$$

$$= \frac{1}{s} \left(\int_0^s (T(t + r) - T(r))x \mathrm{d}r \right)$$

$$= \frac{1}{s} \int_0^s T(r)(T(t) - I)x \mathrm{d}r.$$

From (B.2) it follows that the right-hand side tends to $(T(t) - I)x$ as $s \searrow 0$. Hence,

$$\int_0^t T(r)x \mathrm{d}r \in D(A) \quad \text{and} \quad A \int_0^t T(r)x \mathrm{d}r = (T(t) - I)x.$$

Consequently, (B.2) implies that for any $x \in X$ there exists a sequence (x_n) in $D(A)$ such that $x_n \to x$, which proves that $D(A) = \operatorname{cl} D(A) = X$. It is clear that A is a linear operator. Now for $s > 0$ consider the equalities

$$\frac{T(t + s)x - T(t)x}{s} = T(t)\frac{(T(s) - I)x}{s} = \frac{T(s) - I}{s}T(t)x.$$

The limit for $s \searrow 0$ of the second term exists and is equal to $T(t)Ax$. Hence, also the other limits exist and the right derivative of $t \mapsto T(t)x$ equals $AT(t)x$. For $t > 0$ and $s > 0$ sufficiently small we have

$$\frac{T(t - s)x - T(t)x}{-s} = T(t - s)\frac{(T(s) - I)x}{s}.$$

Therefore, also the left derivative exists and equals $T(t)Ax$. We have thus proven that $(d/dt)T(t)x = AT(t)x$ for all $x \in X$ and $t > 0$, which implies $T(t) = e^{At}$. Uniqueness of A is obvious.   □

The following lemma gives an estimate for the growth of linear flows on Euclidean space.

Lemma B.2. *Let $A \in \mathbb{R}^{d \times d}$ and denote by $\alpha(A)$ the maximum of the real parts of all eigenvalues of A. Then it holds that*

$$\forall \delta > 0 \, \exists c \geq 1 \, \forall t \geq 0 : \left\| e^{At} \right\| \leq c e^{(\alpha(A) + \delta)t},$$

where $\| \cdot \|$ denotes the operator norm induced by an arbitrary vector norm on $\mathbb{R}^d$.

Proof. Given $\delta > 0$, define $B_\delta := A - (\alpha(A) + \delta)I$. Then all eigenvalues of B_δ have negative real parts, and hence, by Robinson [93, Chap. IV, Theorem 5.1], there exist constants $a > 0$ and $c \geq 1$ such that

$$\left\| e^{B_\delta t} \right\| \leq c e^{-at} \quad \text{for all } t \geq 0.$$

Since $e^{B_\delta t} = e^{-(\alpha(A) + \delta)t} e^{At}$, this implies

$$\left\| e^{At} \right\| \leq c e^{-at} e^{(\alpha(A) + \delta)t} \leq c e^{(\alpha(A) + \delta)t},$$

which proves the assertion.   □

Finally, we cite Selgrade's theorem (see, for instance, Colonius and Kliemann [25, Theorem 5.2.5]).

Theorem B.1 (Selgrade). *Consider a continuous-time linear flow Φ on a vector bundle $\pi : \mathcal{V} \to B$ of rank d with connected and compact metric base space B. Suppose that the induced flow on B is chain transitive. Then there exists a unique finest Morse decomposition $\{\mathcal{M}_1, \ldots, \mathcal{M}_r\}$ of the induced flow $\mathbb{P}\Phi$ on the projective bundle[1] $\mathbb{P}\mathcal{V}$ with $1 \leq r \leq d$. Every chain recurrent component $\mathcal{M}_i$ defines an invariant subbundle of $\mathcal{V}$ via*

$$\mathcal{V}^i = \mathbb{P}^{-1}(\mathcal{M}_i) = \{v \in \mathcal{V} : v \notin Z \Rightarrow \mathbb{P}v \in \mathcal{M}_i\},$$

where Z denotes the zero section of $\mathcal{V}$, and the following decomposition into a Whitney sum holds:

$$\mathcal{V} = \mathcal{V}^1 \oplus \cdots \oplus \mathcal{V}^r.$$

---

[1] The *projective bundle* $\mathbb{P}\mathcal{V} \to B$ of a vector bundle $\pi : \mathcal{V} \to B$ is the fiber bundle whose fibers are the projective spaces of the fibers $\pi^{-1}(b)$, $b \in B$, defined as the quotient space $\mathbb{P}\mathcal{V} := (\mathcal{V} \setminus Z)/\sim$ under the equivalence relation $\sim$ whose equivalence classes are the lines through the origins of the fibers $\pi^{-1}(b)$.

B.3   Dimension Theory and Topological Entropy

The Subadditivity Lemma

The following lemma is a well-known result in analysis frequently used in connection with exponential growth rates, in particular with entropy.[2] For the sake of completeness, we give a proof.

Lemma B.3 (Subadditivity Lemma). *Let* $\mathbb{T} \in \{\mathbb{Z}, \mathbb{R}\}$ *and let* $f : \mathbb{T}_+ \to \mathbb{R}$ *be a subadditive function, that is,*

$$f(t + s) \le f(t) + f(s) \quad \text{for all } t, s \in \mathbb{T}_+.$$

Suppose further that f *is bounded from above on an interval of the form* $\mathbb{T} \cap [0, t_0]$ *with* $t_0 > 0$. *Then the limit* $\lim_{t \to \infty} f(t)/t$ *exists and equals* $\inf_{t > 0} f(t)/t$.

Proof. From boundedness of f on $\mathbb{T} \cap [0, t_0]$ and subadditivity it follows that f is bounded from above on any bounded interval. Let $\gamma := \inf_{t > 0} f(t)/t$. Fix a positive number $T \in \mathbb{T}$ and write each $t \in \mathbb{T}$, $t > 0$, as $t = k(t)T + r(t)$ with $k(t) \in \mathbb{Z}_+$ and $r(t) \in \mathbb{T} \cap [0, T)$. Then $k(t)/t \to 1/T$ for $t \to \infty$. By subadditivity, for any $t, T > 0$ it holds that

$$\gamma \le \frac{f(t)}{t} \le \frac{1}{t} \left[k(t) f(T) + f(r(t)) \right].$$

Hence, for every $\varepsilon > 0$ there exists $T_0 = T_0(\varepsilon, T)$ such that for all $t > T_0$

$$\gamma \le \frac{f(t)}{t} \le \frac{f(T)}{T} + \varepsilon,$$

where we used boundedness of f on $\mathbb{T} \cap [0, T]$. Since ε and T are arbitrary, the result follows. □

Remark B.1. The lemma also applies to subadditive functions $f : \mathbb{T} \cap (0, \infty) \to \mathbb{R}$, since one can extend such a function to $\mathbb{T}_+$ by setting $f(0) := 0$ without destroying subadditivity.

Hausdorff and Capacitive Dimension

There exist several notions of dimension for topological and metric spaces, generalizing the dimension concept in vector spaces. In the following, we introduce the notions of *Hausdorff* and *capacitive dimension* for metric spaces both used in several entropy estimates in this book.

---

[2]This result is also known as *Fekete's Lemma* due to Michael Fekete.

Let (X, ϱ) be a metric space, $Z \subset X$, and $d \geq 0$, $\varepsilon > 0$. Define

$$\mu_H(Z, d, \varepsilon) = \mu_H(Z, d, \varepsilon; \varrho) := \inf \left\{ \sum_{j \geq 1} r_j^d \; : \; r_j \leq \varepsilon, \; Z \subset \bigcup_{j \geq 1} B(x_j, r_j) \right\},$$

where the infimum is taken over all countable covers of Z by metric balls $B(x_j, r_j)$ of radii $r_j \leq \varepsilon$ and centers $x_j \in X$.[3] It is easy to see that the function $\mu_H(\cdot, d, \varepsilon)$ is an outer measure on X. For fixed Z and d, the function $\mu_H(Z, d, \cdot)$ does not decrease with decreasing ε and hence the limit

$$\mu_H(Z, d) = \mu_H(Z, d; \varrho) := \lim_{\varepsilon \searrow 0} \mu_H(Z, d, \varepsilon) = \sup_{\varepsilon > 0} \mu_H(Z, d, \varepsilon)$$

exists (it may be ∞). The number $\mu_H(Z, d)$ is called the d-*dimensional outer Hausdorff measure of* Z. The function $\mu_H(\cdot, d)$ is a metric outer measure on X, that is, the restriction of $\mu_H(\cdot, d)$ to the Borel-σ-algebra of X is a measure. For every $Z \subset X$ there exists a critical value $d_{\mathrm{crit}}(Z)$ such that

$$\mu_H(Z, d) = \begin{cases} 0 & \text{for } d > d_{\mathrm{crit}}(Z), \\ \infty & \text{for } d < d_{\mathrm{crit}}(Z). \end{cases}$$

This unique value is called the *Hausdorff dimension of* Z and is denoted by $\dim_H(Z)$.[4] For a totally bounded set $Z \subset X$ (that is, for every $\varepsilon > 0$ finitely many ε-balls are sufficient to cover Z), $d \geq 0$ and $\varepsilon > 0$ we also introduce the quantity

$$\mu_C(Z, d, \varepsilon) = \mu_C(Z, d, \varepsilon; \varrho) := \varepsilon^d n(\varepsilon, Z),$$

where $n(\varepsilon, Z)$ is the minimal number of ε-balls necessary to cover Z:

$$n(\varepsilon, Z) := \min \left\{ \#\mathscr{C} \; : \; \mathscr{C} = \{B(x_j, \varepsilon)\}_j, \; Z \subset \bigcup_j B(x_j, \varepsilon) \right\}.$$

It is easy to see that $\mu_H(Z, d, \varepsilon) \leq \mu_C(Z, d, \varepsilon)$. We define the d-*dimensional upper capacitive measure of* Z by

$$\mu_C(Z, d) = \mu_C(Z, d; \varrho) := \limsup_{\varepsilon \searrow 0} \mu_C(Z, d, \varepsilon).$$

The properties of $\mu_C(Z, d)$ are similar to those of $\mu_H(Z, d)$. In particular, $\mu_C(\cdot, d, \varepsilon)$ and $\mu_C(\cdot, d)$ are outer measures if X is totally bounded.

---

[3] Taking balls centered at elements of Z makes no essential difference, that is, it does not change the value of the Hausdorff dimension of Z.

[4] Equivalently, one can introduce the Hausdorff measures and the Hausdorff dimension by replacing the covers of Z by metric balls with radii $\leq \varepsilon$ with covers consisting of arbitrary sets with diameters $\leq \varepsilon$.

The *upper capacitive dimension of Z* is defined by

$$\overline{\dim}_C(Z) := \limsup_{\varepsilon \searrow 0} \frac{\log n(\varepsilon, Z)}{\log(1/\varepsilon)}.$$

Analogously, one defines the *lower capacitive dimension* by replacing the limit superior by a limit inferior. In the literature, one finds several other names for this notion of dimension such as *(upper and lower) box dimension* or *fractal dimension*. Alternatively, one can introduce the upper capacitive dimension in the same way as the Hausdorff dimension as a critical value for the upper capacitive measure. The following proposition shows that the upper capacitive dimension of a totally bounded set Z does not depend on the space it is embedded in.

Proposition B.4. *Let (X, ϱ) be a metric space and $Z \subset X$ a totally bounded set. Let $\overline{\dim}_C(Z; X)$ denote the upper capacitive dimension of Z as a subspace of (X, ϱ), and $\overline{\dim}_C(Z; Z)$ the upper capacitive dimension of Z as a subspace of (Z, ϱ). Then $\overline{\dim}_C(Z; X) = \overline{\dim}_C(Z; Z)$.*

Proof. By $n(\varepsilon, Z; X)$ ($n(\varepsilon, Z; Z)$) we denote the minimal cardinality of a cover of Z with ε-balls in X (in Z). For given $\varepsilon > 0$, let $\mathcal{B} = \{B(x_1, \varepsilon), \ldots, B(x_n, \varepsilon)\}$, $x_i \in X$, be a minimal cover of Z with ε-balls in X (in particular, $n = n(\varepsilon, Z; X)$). Then for every $i \in \{1, \ldots, n\}$ there exists some $z_i \in B(x_i, \varepsilon) \cap Z$, since otherwise $\mathcal{B}$ would not be minimal. Let $\tilde{\mathcal{B}} := \{B(z_1, 2\varepsilon), \ldots, B(z_n, 2\varepsilon)\}$. Now take an arbitrary point $z \in Z$. Then there exists $i \in \{1, \ldots, n\}$ with $\varrho(z, x_i) < \varepsilon$. It follows that

$$\varrho(z, z_i) \leq \varrho(z, x_i) + \varrho(x_i, z_i) < \varepsilon + \varepsilon = 2\varepsilon.$$

Hence, $\tilde{\mathcal{B}}$ is a cover of Z consisting of n balls in Z of radius 2ε. This implies

$$n(2\varepsilon, Z; X) \leq n(2\varepsilon, Z; Z) \leq n(\varepsilon, Z; X).$$

Therefore, for all $\varepsilon \in (0, 1)$ it holds that

$$\frac{\log n(2\varepsilon, Z; X)}{\log(1/\varepsilon)} \leq \frac{\log n(2\varepsilon, Z; Z)}{\log(1/\varepsilon)} \leq \frac{\log n(\varepsilon, Z; X)}{\log(1/\varepsilon)}.$$

Using that $\log(1/\varepsilon) = \log(2) + \log(1/(2\varepsilon))$, we obtain

$$\limsup_{\varepsilon \searrow 0} \frac{\log n(2\varepsilon, Z; X)}{\log(2) + \log(1/(2\varepsilon))} \leq \limsup_{\varepsilon \searrow 0} \frac{\log n(2\varepsilon, Z; Z)}{\log(2) + \log(1/(2\varepsilon))} \leq \overline{\dim}_C(Z; X).$$

This implies $\overline{\dim}_C(Z; X) \leq \overline{\dim}_C(Z; Z) \leq \overline{\dim}_C(Z; X)$. □

Some more properties of the Hausdorff and upper capacitive dimensions are summarized in the following proposition. For proofs we refer to Boichenko et al. [9, Chap. III].

Proposition B.5. *Let (X, ϱ) be a metric space. Then the following assertions hold:*

(i) $0 \leq \dim_H(Z) \leq \overline{\dim}_C(Z)$ *for any totally bounded set $Z \subset X$.*

(ii) $\dim_H(\emptyset) = 0$ *and* $\overline{\dim}_C(\emptyset) = 0$.

(iii) $\dim_H(Z_1) \leq \dim_H(Z_2)$ *if* $Z_1 \subset Z_2 \subset X$.

(iv) $\dim_H(\bigcup_{j \geq 1} Z_j) = \sup_{j \geq 1} \dim_H(Z_j)$ *for any sequence* $Z_j \subset X$.

(v) $\overline{\dim}_C(Z_1) \leq \overline{\dim}_C(Z_2)$ *if* $Z_1 \subset Z_2 \subset X$ *are totally bounded sets.*

(vi) $\overline{\dim}_C(\bigcup_{j \geq 1} Z_j) \geq \sup_{j \geq 1} \overline{\dim}_C(Z_j)$ *for a sequence* $Z_j \subset X$ *of totally bounded sets whose union is totally bounded. For finite unions equality holds.*

(vii) *If $Z \subset X$ is a totally bounded set, then* $\overline{\dim}_C(Z) = \overline{\dim}_C(\mathrm{cl}\, Z)$.

(viii) *If X is a d-dimensional Riemannian manifold, then* $\dim_H(X) = d$. *If, additionally, X is compact, then* $\overline{\dim}_C(X) = d$.

Topological Entropy

The concept of topological entropy for discrete-time dynamical systems on compact topological spaces was first introduced by Adler et al. [1] as a topological analog to the measure-theoretic entropy of Kolmogorov [69] and Sinai [99]. Topological entropy can be regarded as a measure of the global exponential complexity of the orbit structure, and it has proved to be an important topological invariant. Later, equivalent definitions were given by Dinaburg [37] and Bowen [10] for maps on metric spaces. In Chap. 3, we use Bowen's definition of topological entropy and his result on the entropy of a linear dynamical system. In the following, we give the necessary background for understanding the concepts involved.

Let $f : X \to X$ be a uniformly continuous map on a metric space (X, ϱ). The iterates of f are defined inductively by $f^0 := \mathrm{id}_X$ and $f^{n+1} := f \circ f^n$ for all $n \in \mathbb{Z}_+$.[5] It is easy to see that for each integer $n \geq 1$ the following function defines a metric on X which induces the same topology as ϱ:

$$\varrho_{n,f}(x, y) := \max_{0 \leq i \leq n} \varrho(f^i(x), f^i(y)).$$

Usually, a metric of this form is called a *Bowen-metric* or a *Bowen–Dinaburg-metric*.[6] The metric balls with respect to $\varrho_{n,f}$ are also called *Bowen-balls of order n*. A set $E \subset X$ is called (n, ε, f)-*separated* if the distance of any two distinct points $x, y \in E$ measured by the metric $\varrho_{n,f}$ is at least ε. A set $F \subset X$ (n, ε, f)-*spans* another set $K \subset X$ if for every $x \in K$ there exists $y \in F$ such that $\varrho_{n,f}(x, y) < \varepsilon$.

---

[5]This defines a discrete-time dynamical system by $\Phi : \mathbb{Z}_+ \times X \to X$, $\Phi(n, x) := f^n(x)$.

[6]Usually, the maximum in the definition of $\varrho_{n,f}$ is only taken over $i \in \{0, \ldots, n-1\}$. However, this makes no essential difference, and we use the slightly different definition only for formal reasons.

Equivalently, K is covered by the ε-balls in the metric $\varrho_{n,f}$ centered at the elements of F.[7]

It is clear that an (n, ε, f)-separated subset of a compact set $K \subset X$ is finite and that there is an upper bound for its cardinality, since otherwise one could place infinitely many disjoint Bowen-balls of radius $\varepsilon/2$ and order n in K contradicting compactness. The maximal cardinality of an (n, ε, f)-separated subset of K is denoted by $r_{\mathrm{sep}}(n, \varepsilon, K, f)$. For the minimal cardinality of a set which (n, ε, f)-spans K we write $r_{\mathrm{span}}(n, \varepsilon, K, f)$. A maximal (n, ε, f)-separated subset E of K automatically (n, ε, f)-spans K. Otherwise there would exist a point $x \in K$ which has distance at least ε to every element of E, and $E \cup \{x\}$ would also be (n, ε, f)-separated. On the other hand, given an (n, ε, f)-separated subset E of K and a set F which $(n, \varepsilon/2, f)$-spans K, one finds that two distinct elements of E cannot be contained in the same Bowen-ball of radius $\varepsilon/2$ and order n around an element of F. This defines an injective map from E to F which shows that $r_{\mathrm{sep}}(n, \varepsilon, K, f) \leq r_{\mathrm{span}}(n, \varepsilon/2, K, f)$. Altogether,

$$r_{\mathrm{span}}\left(n, \varepsilon, K, f\right) \leq r_{\mathrm{sep}}\left(n, \varepsilon, K, f\right) \leq r_{\mathrm{span}}\left(n, \frac{\varepsilon}{2}, K, f\right) < \infty.$$

Moreover, these quantities are non-decreasing with decreasing ε. Therefore, the following definitions make sense:

$$h_{\mathrm{sep},\varrho}(\varepsilon, K, f) := \limsup_{n \to \infty} \frac{1}{n} \log r_{\mathrm{sep}}(n, \varepsilon, K, f),$$

$$h_{\mathrm{span},\varrho}(\varepsilon, K, f) := \limsup_{n \to \infty} \frac{1}{n} \log r_{\mathrm{span}}(n, \varepsilon, K, f),$$

$$h_{\mathrm{top},\varrho}(K, f) := \lim_{\varepsilon \searrow 0} h_{\mathrm{sep},\varrho}(\varepsilon, K, f) = \lim_{\varepsilon \searrow 0} h_{\mathrm{span},\varrho}(\varepsilon, K, f).$$

One defines the *topological entropy* of f as

$$h_{\mathrm{top},\varrho}(f) := \sup_{K \subset X} h_{\mathrm{top},\varrho}(K, f),$$

where the supremum is taken over all nonempty compact subsets of X. In general, this quantity depends on the metric ϱ. If two metrics ϱ_1 and ϱ_2 are uniformly equivalent, that is, if the identity maps id : $(X, \varrho_1) \to (X, \varrho_2)$ and id : $(X, \varrho_2) \to (X, \varrho_1)$ are uniformly continuous, then the corresponding entropies coincide. In particular, this is the case if X is compact. Then the topological entropy can also be defined in a purely topological way using open covers of the space X as done

---

[7]In the definitions of separated and spanning sets, Bowen requires that $\varrho_{n,f}(x, y) > \varepsilon$ and $\varrho_{n,f}(x, y) \leq \varepsilon$, respectively. For our purposes however it is more convenient to relax the strict inequality and vice versa. For the value of topological entropy this makes no difference.

in [1]. One elementary property of topological entropy which we use in Chap. 3 is the following power rule (see also Bowen [10, Proposition 4]).

Lemma B.4. *Let* $f : X \to X$ *be a uniformly continuous map on a metric space* (X, ϱ) *and* $K \subset X$ *a compact set. Then for each integer* $m \geq 1$ *it holds that*

$$h_{\text{top},\varrho}(K, f^m) = m h_{\text{top},\varrho}(K, f).$$

Proof. It is clear that $r_{\text{span}}(n, \varepsilon, K, f^m) \leq r_{\text{span}}(mn, \varepsilon, K, f)$ which implies

$$h_{\text{span},\varrho}(\varepsilon, K, f^m) = \limsup_{n \to \infty} \frac{1}{n} \log r_{\text{span}}(n, \varepsilon, K, f^m)$$

$$\leq m \limsup_{n \to \infty} \frac{1}{mn} \log r_{\text{span}}(mn, \varepsilon, K, f)$$

$$\leq m \limsup_{n \to \infty} \frac{1}{n} \log r_{\text{span}}(n, \varepsilon, K, f) = m h_{\text{span},\varrho}(\varepsilon, K, f).$$

This gives $h_{\text{top},\varrho}(K, f^m) \leq m h_{\text{top},\varrho}(K, f)$. For the converse inequality, fix $m \geq 1$ and $\varepsilon > 0$. Choose $\delta = \delta(\varepsilon)$ such that $\varrho(x, y) < \delta$ implies $\varrho(f^j(x), f^j(y)) < \varepsilon$ for $0 \leq j \leq m$, which is possible by uniform continuity of f. Then an (n, δ, K, f^m)-spanning set is automatically (mn, ε, K, f)-spanning, which implies $r_{\text{span}}(mn, \varepsilon, K, f) \leq r_{\text{span}}(n, \delta, K, f^m)$. For each $k \geq 1$ let $n_k \geq 1$ be such that $m(n_k - 1) < k \leq m n_k$. Then we obtain

$$h_{\text{span},\varrho}(\varepsilon, K, f) = \limsup_{k \to \infty} \frac{1}{k} \log r_{\text{span}}(k, \varepsilon, K, f)$$

$$\leq \limsup_{k \to \infty} \frac{1}{k} \log r_{\text{span}}(m n_k, \varepsilon, K, f)$$

$$\leq \limsup_{k \to \infty} \frac{n_k}{k} \frac{1}{n_k} r_{\text{span}}(n_k, \delta, K, f^m).$$

Since $n_k / k \to 1/m$ for $k \to \infty$, it follows that

$$h_{\text{span},\varrho}(\varepsilon, K, f) \leq \frac{1}{m} h_{\text{span},\varrho}(\delta, K, f^m),$$

which implies the desired inequality.   □

The following result can be found in Bowen [10, Theorem 15]. An elementary proof can also be found in Matveev and Savkin [79, Theorem 2.4.2].

Proposition B.6. *If* $T : \mathbb{R}^d \to \mathbb{R}^d$ *is a linear map, then*

$$h_{\text{top},\varrho}(T) = \sum_{\lambda \in \sigma(T)} \max\{0, n_\lambda \log |\lambda|\},$$

where ϱ denotes a metric on $\mathbb{R}^d$ induced by a norm and n_λ is the algebraic multiplicity of the eigenvalue λ.

In the same manner as for maps, topological entropy can be defined for continuous-time dynamical systems on metric spaces: Let $\Phi : \mathbb{R}_+ \times X \to X$ be a semiflow which is uniformly continuous in the sense of [10, Sect. 5], that is, for all $t_0 > 0$ it holds that

$$\forall \varepsilon > 0 : \exists \delta > 0 : \forall t \in [0, t_0] : \forall x, y \in X :$$

$$\varrho(x, y) < \delta \quad \Rightarrow \quad \varrho(\Phi_t(x), \Phi_t(y)) < \varepsilon. \qquad (B.3)$$

As we did for maps, we define the Bowen-metrics

$$\varrho_{\tau,\Phi}(x, y) := \max_{t \in [0,\tau]} \varrho(\Phi_t(x), \Phi_t(y)), \quad \tau > 0.$$

For any real number $\tau > 0$ a set $E \subset X$ is called $(\tau, \varepsilon, \Phi)$-separated if $\varrho_{\tau,\Phi}(x, y) \geq \varepsilon$ for any two distinct points $x, y \in E$. A set $F \subset X$ $(\tau, \varepsilon, \Phi)$-spans another set K if for each $x \in K$ there is $y \in F$ with $\varrho_{\tau,\Phi}(x, y) < \varepsilon$. Then $r_{\text{sep}}(\tau, \varepsilon, K, \Phi)$ and $r_{\text{span}}(\tau, \varepsilon, K, \Phi)$ are the maximal and minimal cardinalities of (n, ε, Φ)-separated and (n, ε, Φ)-spanning sets, respectively. The topological entropy is again defined by

$$h_{\text{top},\varrho}(K, \Phi) := \lim_{\varepsilon \searrow 0} \limsup_{\tau \to \infty} \frac{1}{\tau} \log r_{\text{sep}}(\tau, \varepsilon, K, \Phi)$$

$$= \lim_{\varepsilon \searrow 0} \limsup_{\tau \to \infty} \frac{1}{\tau} \log r_{\text{span}}(\tau, \varepsilon, K, \Phi),$$

$$h_{\text{top},\varrho}(\Phi) := \sup_{K \subset X \text{ compact}} h_{\text{top},\varrho}(K, \Phi).$$

The following proposition shows that the topological entropy of a semiflow coincides with the entropy of its time-one-map. In particular, together with Proposition B.6, this shows that the entropy of a linear flow $(t, x) \mapsto e^{At}x$ on a Euclidean space is given by the sum of the positive real parts of the eigenvalues of A (counting multiplicities):

$$h_{\text{top},\varrho}(\{e^{At}\}) = \sum_{\lambda \in \sigma(A)} \max\{0, n_\lambda \operatorname{Re}(\lambda)\}.$$

As in Proposition B.6, ϱ denotes a metric induced by a norm.

Proposition B.7. *The topological entropy of a uniformly continuous semiflow Φ on a metric space (X, ϱ) equals the topological entropy of its time-one-map:* $h_{\text{top},\varrho}(\Phi) = h_{\text{top},\varrho}(\Phi_1)$.

Proof. Fix a compact set $K \subset X$ and real numbers $\tau, \varepsilon > 0$. Let $F \subset X$ be a set which $(\tau, \varepsilon, \Phi)$-spans K and define $n(\tau) \in \mathbb{Z}_+$ to be the greatest integer such that $n(\tau) \le \tau$. Then for every $x \in K$ there is some $y \in F$ with $\max_{t \in [0,\tau]} \varrho(\Phi_t(x), \Phi_t(y)) < \varepsilon$. Since $\Phi_j = (\Phi_1)^j$ for all $j \in \mathbb{Z}_+$, this implies

$$\varrho_{n(\tau), \Phi_1}(x, y) = \max_{0 \le j \le n(\tau)} \varrho((\Phi_1)^j(x), (\Phi_1)^j(y)) \le \max_{t \in [0,\tau]} \varrho(\Phi_t(x), \Phi_t(y)) < \varepsilon.$$

Thus, F $(n(\tau), \varepsilon, \Phi_1)$-spans the set K, which implies $r_{\mathrm{span}}(n(\tau), \varepsilon, K, \Phi_1) \le r_{\mathrm{span}}(\tau, \varepsilon, K, \Phi)$. It follows that

$$
\begin{aligned}
h_{\mathrm{span}, \varrho}(\varepsilon, K, \Phi_1) &= \limsup_{n \to \infty} \frac{1}{n} \log r_{\mathrm{span}}(n, \varepsilon, K, \Phi_1) \\
&\le \limsup_{n \to \infty} \frac{1}{n} \log r_{\mathrm{span}}(n, \varepsilon, K, \Phi) \\
&\le \limsup_{\tau \to \infty} \frac{1}{\tau} \log r_{\mathrm{span}}(\tau, \varepsilon, K, \Phi) \\
&= h_{\mathrm{span}, \varrho}(\varepsilon, K, \Phi).
\end{aligned}
$$

Consequently, $h_{\mathrm{top}, \varrho}(\Phi_1) \le h_{\mathrm{top}, \varrho}(\Phi)$.

In order to show the converse inequality, let $\tau, \varepsilon > 0$ and choose $\delta = \delta(\varepsilon)$ according to (B.3) with $t_0 = 1$. Let $n(\tau) \in \mathbb{Z}_+$ be the smallest integer such that $\tau \le n(\tau)$ and let $F \subset X$ be a set which $(n(\tau), \delta, \Phi_1)$-spans K. Then for every $x \in K$ there is some $y \in F$ such that $\varrho_{n(\tau), \Phi_1}(x, y) < \delta$. For every $t \in [0, \tau]$ there are unique $j \in \{0, 1, \ldots, n(\tau)\}$ and $s \in [0, 1)$ such that $t = j + s$, which implies

$$
\begin{aligned}
\varrho(\Phi_t(x), \Phi_t(y)) &= \varrho(\Phi_s(\Phi_j(x)), \Phi_s(\Phi_j(y))) \\
&= \varrho(\Phi_s((\Phi_1)^j(x)), \Phi_s((\Phi_1)^j(y))) < \varepsilon.
\end{aligned}
$$

Consequently, F also $(\tau, \varepsilon, \Phi)$-spans the set K. Finally, we obtain

$$
\begin{aligned}
h_{\mathrm{span}, \varrho}(\varepsilon, K, \Phi) &= \limsup_{\tau \to \infty} \frac{1}{\tau} \log r_{\mathrm{span}}(\tau, \varepsilon, K, \Phi) \\
&\le \limsup_{\tau \to \infty} \frac{1}{\tau} \log r_{\mathrm{span}}(n(\tau), \delta, K, \Phi_1) \\
&\le \limsup_{n \to \infty} \frac{1}{n-1} \log r_{\mathrm{span}}(n, \delta, K, \Phi_1) \\
&= \limsup_{n \to \infty} \frac{1}{n} \log r_{\mathrm{span}}(n, \delta, K, \Phi_1) = h_{\mathrm{span}, \varrho}(\delta, K, \Phi_1).
\end{aligned}
$$

Thus, $h_{\mathrm{top}, \varrho}(K, \Phi) \le h_{\mathrm{top}, \varrho}(K, \Phi_1)$ and $h_{\mathrm{top}, \varrho}(\Phi) \le h_{\mathrm{top}, \varrho}(\Phi_1)$. $\square$

Finally, we prove a simple estimate for the topological entropy of a Lipschitz map. The proof is taken from Katok and Hasselblatt [61, Theorem 3.2.9].

Proposition B.8. *Let* $f : X \to X$ *be a map on a metric space* (X, ϱ), *which satisfies a global Lipschitz condition with Lipschitz constant* $L(f)$. *Assume further that* $K \subset X$ *is a compact set of finite upper capacitive dimension. Then*

$$h_{\text{top},\varrho}(K, f) \leq \max\{0, \log L(f)\} \cdot \overline{\dim}_C(K) < \infty.$$

Proof. Let $L := \max\{1, L(f)\}$, $n \geq 1$ and $\varepsilon > 0$. Pick $x, y \in X$ with $\varrho(x, y) < L^{-n}\varepsilon$. Then for any $0 \leq i \leq n$ we have

$$\varrho(f^i(x), f^i(y)) \leq L^i \varrho(x, y) < L^{i-n}\varepsilon \leq \varepsilon.$$

Hence,

$$\varrho_{n,f}(x, y) = \max_{0 \leq i \leq n} \varrho(f^i(x), f^i(y)) < \varepsilon.$$

If $F \subset X$ is a minimal set which (n, ε, f)-spans K, then K is covered by the Bowen-balls of radius ε and order n, centered at the elements of F. Each of these balls contains an $(L^{-n}\varepsilon)$-ball (with respect to ϱ), as we have proved. We thus obtain

$$r_{\text{span}}(n, \varepsilon, K, f) \leq n(L^{-n}\varepsilon, K).$$

For $L^{-n}\varepsilon < 1$ we have $|\log(L^{-n}\varepsilon)| = |-n \log L + \log \varepsilon| = n \log L - \log \varepsilon$, and therefore

$$n = \frac{|\log(L^{-n}\varepsilon)| + \log \varepsilon}{\log L} = \frac{|\log(L^{-n}\varepsilon)|}{\log L} \left(1 + \frac{\log \varepsilon}{|\log(L^{-n}\varepsilon)|}\right).$$

We may assume that $L > 1$ and hence

$$\lim_{n \to \infty} \left(1 + \frac{\log \varepsilon}{|\log(L^{-n}\varepsilon)|}\right) = 1.$$

This implies

$$h_{\text{span},\varrho}(\varepsilon, K, f) = \limsup_{n \to \infty} \frac{\log r_{\text{span}}(n, \varepsilon, K, f)}{n} \leq \limsup_{n \to \infty} \frac{\log n(L^{-n}\varepsilon, K)}{n}$$

$$= \log L \cdot \limsup_{n \to \infty} \frac{\log n(L^{-n}\varepsilon, K)}{|\log(L^{-n}\varepsilon)|} \leq \log L \cdot \overline{\dim}_C(K).$$

It follows that $h_{\text{top},\varrho}(K, f) \leq \log L \cdot \overline{\dim}_C(K)$, as claimed. $\square$

B.4   Additive and Subadditive Cocycles

Let $\Phi : \mathbb{T} \times X \rightarrow X$ be a dynamical system on a set X with time set $\mathbb{T} \in \{\mathbb{Z}_+, \mathbb{Z}, \mathbb{R}_+, \mathbb{R}\}$. By an *additive cocycle over* Φ we understand a function

$$a : \mathbb{T} \times X \rightarrow \mathbb{R}$$

which satisfies the equality

$$a(t + s, x) = a(t, x) + a(s, \Phi(t, x)) \quad \text{for all } t, s \in \mathbb{T} \text{ and } x \in X.$$

In general, we do not impose any continuity assumptions on Φ and a. However, in a topological context, we have the following result proved in [66, Corollary 2] via investigation of the *uniform growth spectrum* introduced by Grüne [53].

Theorem B.2. *Let* $\Phi : \mathbb{T} \times X \rightarrow X$ *be a continuous dynamical system on a Hausdorff space* X *and* $a : \mathbb{T} \times X \rightarrow \mathbb{R}$ *a continuous additive cocycle over* Φ. *Then, given a compact* Φ-*invariant set* $M \subset X$, *that is,* $\Phi_t(M) \subset M$ *for all* $t \in \mathbb{T}$, *we have*

$$\inf_{x \in M} \limsup_{t \to \infty} \frac{1}{t} a(t, x) = \inf_{x \in M} \liminf_{t \to \infty} \frac{1}{t} a(t, x)$$

$$= \lim_{t \to \infty} \inf_{x \in M} \frac{1}{t} a(t, x) = \sup_{t > 0} \inf_{x \in M} \frac{1}{t} a(t, x) \qquad (\text{B.4})$$

and

$$\sup_{x \in M} \limsup_{t \to \infty} \frac{1}{t} a(t, x) = \sup_{x \in M} \liminf_{t \to \infty} \frac{1}{t} a(t, x)$$

$$= \lim_{t \to \infty} \sup_{x \in M} \frac{1}{t} a(t, x) = \inf_{t > 0} \sup_{x \in M} \frac{1}{t} a(t, x). \qquad (\text{B.5})$$

Furthermore, there are $x_*, x^* \in M$ *such that*

$$\inf_{x \in M} \limsup_{t \to \infty} \frac{1}{t} a(t, x) = \lim_{t \to \infty} \frac{1}{t} a(t, x_*),$$

$$\sup_{x \in M} \limsup_{t \to \infty} \frac{1}{t} a(t, x) = \lim_{t \to \infty} \frac{1}{t} a(t, x^*).$$

A *subadditive cocycle* over the dynamical system Φ is a function

$$a : \mathbb{T} \times X \rightarrow \mathbb{R}$$

which satisfies the inequality

$$a(t + s, x) \leq a(t, x) + a(s, \Phi(t, x)) \quad \text{for all } t, s \in \mathbb{T} \text{ and } x \in X. \tag{B.6}$$

In the case where X is a compact metric space and both Φ and a are continuous, Schreiber [97, Theorem 1] shows that

$$\sup_{x \in X} \limsup_{t \to \infty} \frac{1}{t} a(t, x) = \lim_{t \to \infty} \sup_{x \in X} \frac{1}{t} a(t, x) = \inf_{t > 0} \sup_{x \in X} \frac{1}{t} a(t, x),$$

using methods from ergodic theory, in particular Kingman's subadditive ergodic theorem. For a superadditive cocycle a (where the inequality in (B.6) is reversed), one has the analogous result with suprema replaced by infima and vice versa, and limsup replaced by liminf.

References

1. Adler, R.L., Konheim, A.G., McAndrew, M.H.: Topological entropy. Trans. Am. Math. Soc. **114**(2), 309–319 (1965)
2. Albertini, F., Sontag, E.D.: Some connections between chaotic dynamical systems and control systems. Report SYCON-90-13, Rutgers Center for Systems and Control, New Brunswick (1990)
3. Albertini, F., Sontag, E.D.: Discrete-time transitivity and accessibility: Analytic systems. SIAM J. Control Optim. **31**(6), 1599–1622 (1993)
4. Arnold, L.: Random Dynamical Systems. Springer Monographs in Mathematics. Springer, Berlin (1998)
5. Aulbach, B., Wanner, T.: Integral manifolds for Carathéodory type differential equations in Banach spaces. In: Aulbach, B., Colonius, F. (eds.) Six Lectures on Dynamical Systems (Augsburg, 1994), pp. 45–119. World Scientific, River Edge (1996)
6. Bachman, G., Narici, L.: Functional Analysis. Academic, New York (1966)
7. Baillieul, J.: Feedback designs in information-based control. In: Stochastic Theory and Control (Lawrence, KS, 2001). Lecture Notes in Control and Information Scienes, vol. 280, pp. 35–57. Springer, Berlin (2002)
8. Boichenko, V.A., Leonov, G.A.: The direct Lyapunov method in estimates for topological entropy. J. Math. Sci. (New York) **91**(6), 3370–3379 (1998). Translation from Zap. Nauchn. Sem. S.-Peterburg. Otdel. Mat. Inst. Steklov. (POMI) **231** (1995), Issled. po Topol. **8**, 62–75, 323 (1996)
9. Boichenko, V.A., Leonov, G.A., Reitmann, V.: Dimension Theory for Ordinary Differential Equations. Teubner-Texte zur Mathematik [Teubner Texts in Mathematics], vol. 141. B.G. Teubner, Stuttgart (2005)
10. Bowen, R.: Entropy for group endomorphisms and homogeneous spaces. Trans. Am. Math. Soc. **153**, 401–414 (1971); Errata, Trans. Am. Math. Soc. **181**, 509–510 (1973)
11. Bowen, R.: Entropy-expansive maps. Trans. Am. Math. Soc. **164**, 323–331 (1972)
12. Bowen, R.: Topological entropy for noncompact sets. Trans. Am. Math. Soc. **184**, 125–136 (1973)
13. Bowen, R.: Equilibrium States and the Ergodic Theory of Anosov Diffeomorphisms, 2nd revised edn. Lecture Notes in Mathematics, vol. 470. Springer, Berlin (2008)
14. Bowen, R., Ruelle, D.: The ergodic theory of axiom A flows. Invent. Math. **29**(3), 181–202 (1975)
15. Bullo, F., Lewis, A.D.: Geometric Control of Mechanical Systems. Texts in Applied Mathematics, vol. 49. Springer, New York (2005)
16. Catalan, T., Tahzibi, A.: A lower bound for topological entropy of generic non Anosov symplectic diffeomorphisms. Preprint, arXiv:1011.2441 [math.DS] (2010)

C. Kawan, *Invariance Entropy for Deterministic Control Systems*, Lecture Notes in Mathematics 2089, DOI 10.1007/978-3-319-01288-9,
© Springer International Publishing Switzerland 2013

17. Chicone, C.: Ordinary Differential Equations with Applications, 2nd edn. Texts in Applied Mathematics, vol. 34. Springer, New York (2006)
18. Cohn, D.: Measure Theory. Birkhäuser, Boston (1980)
19. Colonius, F.: Minimal data rates and invariance entropy. In: Electronic Proceedings of the Conference on Mathematical Theory of Networks and Systems (MTNS), Budapest, 5–9 July 2010
20. Colonius, F.: Minimal bit rates and entropy for stabilization. SIAM J. Control Optim. **50**, 2988–3010 (2012)
21. Colonius, F., Du, W.: Hyperbolic control sets and chain control sets. J. Dyn. Control Syst. **7**(1), 49–59 (2001)
22. Colonius, F., Helmke, U.: Entropy of controlled-invariant subspaces. Z. Angew. Math. Mech. (2013) (to appear)
23. Colonius, F., Kawan, C.: Invariance entropy for control systems. SIAM J. Control Optim. **48**(3), 1701–1721 (2009)
24. Colonius, F., Kawan, C.: Invariance entropy for outputs. Math. Control Signals Syst. **22**(3), 203–227 (2011)
25. Colonius, F., Kliemann, W.: The Dynamics of Control. Birkhäuser, Boston (2000)
26. Colonius, F., Spadini, M.: Uniqueness of local control sets. J. Dyn. Control Syst. **9**(4), 513–530 (2003)
27. Colonius, F., Fukuoka, R., Santana, A.: Invariance entropy for topological semigroup actions. Proc. Am. Math. Soc. (2013) (in press)
28. Colonius, F., Kawan, C., Nair, G.N.: A note on topological feedback entropy and invariance entropy. Syst. Control Lett. **62**, 377–381 (2013)
29. Conley, C.: Isolated Invariant Sets and the Morse Index. Regional Conference Series in Mathematics, vol. 38. American Mathematical Society, Providence (1978)
30. Coron, J.-M.: Linearized control systems and applications to smooth stabilization. SIAM J. Control Optim. **32**(2), 358–386 (1994)
31. Dai, X., Zhou, Z., Geng, X.: Some relations between Hausdorff-dimensions and entropies. Sci. China Ser. A **41**(10), 1068–1075 (1998)
32. Da Silva, A.: Invariance entropy for random control systems. Math. Control Signals Syst. (2013). doi:10.1007/s00498-013-0111-9
33. Delchamps, D.: Stabilizing a linear system with quantized state feedback. IEEE Trans. Autom. Control **35**(8), 916–924 (1990)
34. Delvenne, J.-C.: An optimal quantized feedback strategy for scalar linear systems. IEEE Trans. Autom. Control **51**(2), 298–303 (2006)
35. Demers, M.F., Young, L.-S.: Escape rates and conditionally invariant measures. Nonlinearity **19**, 377–379 (2006)
36. De Persis, C.: n-bit stabilization of n-dimensional nonlinear systems in feedforward form. IEEE Trans. Autom. Control **50**(3), 299–311 (2005)
37. Dinaburg, E.I.: A connection between various entropy characteristics of dynamical systems (Russian). Izv. Akad. Nauk SSSR Ser. Mat. **35**, 324–366 (1971)
38. Douady, A., Oesterlé, J.: Dimension de Hausdorff des attracteurs (French). C. R. Acad. Sci. Paris Ser. A–B **290**(24), A1135–A1138 (1980)
39. Downarowicz, T.: Entropy in Dynamical Systems. New Mathematical Monographs, vol. 18. Cambridge University Press, Cambridge (2011)
40. Dunford, N., Schwartz, J.T.: Linear Operators, Part I: General Theory. With the assistance of William G. Bade and Robert G. Bartle. Wiley, New York (1988). Reprint of the 1958 original. Wiley Classics Library. A Wiley-Interscience Publication
41. Fagnani, F., Zampieri, S.A.: A symbolic approach to performance analysis of quantized feedback systems: The scalar case. SIAM J. Control Optim. **44**(3), 816–866 (2005)
42. Federer, H.: Geometric Measure Theory. Die Grundlehren der mathematischen Wissenschaften, Band 153. Springer, New York (1969)
43. Ferraiol, T., Patrão, M., Seco, L.: Jordan decomposition and dynamics on flag manifolds. Discrete Contin. Dyn. Syst. **26**(3), 923–947 (2010)

44. Franz, A.: Hausdorff dimension estimates for invariant sets with an equivariant tangent bundle splitting. Nonlinearity **11**(4), 1063–1074 (1998)
45. Fried, D., Shub, M.: Entropy, linearity and chain-recurrence. Inst. Hautes Études Sci. Publ. Math. No. **50**, 203–214 (1979)
46. Froyland, G., Stancevic, O.: Escape rates and Perron-Frobenius operators: Open and closed dynamical systems. Discrete Contin. Dyn. Syst. Ser. B **14**, 457–472 (2010)
47. Froyland, G., Junge, O., Ochs, G.: Rigorous computation of topological entropy with respect to a finite partition. Phys. D **154**(1–2), 68–84 (2001)
48. Gallot, S., Hulin, D., Lafontaine, J.: Riemannian Geometry. Universitext. Springer, Berlin (1987)
49. Gelfert, K.: Abschätzungen der kapazitiven Dimension und der topologischen Entropie für partiell volumenexpandierende sowie volumenkontrahierende Systeme auf Mannig-faltigkeiten. Dissertation, Technical University of Dresden (2001)
50. Gelfert, K.: Lower bounds for the topological entropy. Discrete Contin. Dyn. Syst. **12**(3), 555–565 (2005)
51. Goodwyn, L.W.: The product theorem for topological entropy. Trans. Am. Math. Soc. **158**, 445–452 (1971)
52. Grasse, K.A., Sussmann, H.J.: Global controllability by nice controls. In: Nonlinear Controllability and Optimal Control. Monographs and Textbooks in Pure and Applied Mathematics, vol. 133, pp. 33–79. Dekker, New York (1990)
53. Grüne, L.: A uniform exponential spectrum for linear flows on vector bundles. J. Dyn. Differ. Equ. **12**(2), 435–448 (2000)
54. Gu, X.: An upper bound for the Hausdorff dimension of a hyperbolic set. Nonlinearity **4**(3), 927–934 (1991)
55. Gundlach, V.M., Kifer, Y.: Random hyperbolic systems. In: Stochastic Dynamics (Bremen, 1997), pp. 117–145. Springer, New York (1999)
56. Hagihara, R., Nair, G.N.: Two extensions of topological feedback entropy. Math. Control Signals Syst. (2013). doi:10.1007/s00498-013-0113-7
57. Hespanha, J., Ortega, A., Vasudevan, L.: Towards the control of linear systems with minimum bit rate. In: Proceedings of the International Symposium on the Mathematical Theory of Networks and Systems, University of Notre Dame, 2002
58. Hinrichsen, D., Pritchard, A.J.: Mathematical Systems Theory I. Modelling, State Space Analysis, Stability and Robustness. Texts in Applied Mathematics, vol. 48. Springer, Berlin (2005)
59. Hoock, A.-M.: Topological entropy and invariance entropy for infinite-dimensional linear systems. J. Dyn. Control Syst. (2013) (to appear)
60. Ito, F.: An estimate from above for the entropy and the topological entropy of a $\mathscr{C}^1$-diffeomorphism. Proc. Jpn. Acad. **46**, 226–230 (1970)
61. Katok, A., Hasselblatt, B.: Introduction to the Modern Theory of Dynamical Systems. Encyclopedia of Mathematics and its Applications, vol. 54. Cambridge University Press, Cambridge (1995)
62. Kawan, C.: Invariance entropy for control systems. Dissertation, University of Augsburg (2010)
63. Kawan, C.: Upper and lower estimates for invariance entropy. Discrete Contin. Dyn. Syst. **30**(1), 169–186 (2011)
64. Kawan, C.: Invariance entropy of control sets. SIAM J. Control Optim. **49**(2), 732–751 (2011)
65. Kawan, C.: Lower bounds for the strict invariance entropy. Nonlinearity **24**(7), 1910–1936 (2011)
66. Kawan, C., Stender, T.: Growth rates for semiflows on Hausdorff spaces. J. Dyn. Differ. Equ. **24**(2), 369–390 (2012)
67. Keynes, H.B., Robertson, J.B.: Generators for topological entropy and expansiveness. Math. Syst. Theory **3**, 51–59 (1969)
68. Kloeden, P.E., Rasmussen, M.: Nonautonomous Dynamical Systems. Mathematical Surveys and Monographs, vol. 176. American Mathematical Society, Providence (2011)

69. Kolmogorov, A.N.: A new metric invariant of transient dynamical systems and automorphisms in Lebesgue spaces (Russian). Dokl. Akad. Nauk SSSR (N.S.) **119**, 861–864 (1958)
70. Kolyada, S., Snoha, L.: Topological entropy of nonautonomous dynamical systems. Random Comput. Dyn. **4**(2–3), 205–233 (1996)
71. Kozlovski, O.S.: An integral formula for topological entropy of $\mathscr{C}^{\infty}$-maps. Ergod. Theory Dyn. Syst. **18**(2), 405–424 (1998)
72. Kushnirenko, A.G.: An upperbound for the entropy of a classical dynamical system (English. Russian original). Dokl. Akad. Nauk SSSR **161**, 37–38 (1965)
73. Lang, S.: Analysis II. Addison-Wesley, Reading (1969)
74. Ledrappier, F., Young, L.-S.: The metric entropy of diffeomorphisms. I: Characterization of measures satisfying Pesin's entropy formula. II: Relations between entropy, exponents and dimension. Ann. Math. (2) **122**, 509–539, 540–574 (1985)
75. Lee, J.M.: Introduction to Smooth Manifolds. Graduate Texts in Mathematics, vol. 218. Springer, New York (2003)
76. Liberzon, D., Hespanha, J.P.: Stabilization of nonlinear systems with limited information feedback. IEEE Trans. Autom. Control **50**(6), 910–915 (2005)
77. Liu, P.-D.: Pesin's entropy formula for endomorphisms. Nagoya Math. J. **150**, 197–209 (1998)
78. Liu, P.-D.: Random perturbations of axiom A basic sets. J. Stat. Phys. **90**(1–2), 467–490 (1998)
79. Matveev, A.S., Savkin, A.V.: Estimation and Control over Communication Networks. Control Engineering. Birkhäuser, Boston (2009)
80. Megginson, R.E.: An Introduction to Banach Space Theory. Graduate Texts in Mathematics, vol. 183. Springer, New York (1998)
81. Meyer, K.R., Sell, G.R.: Melnikov transforms, Bernoulli bundles, and almost periodic perturbations. Trans. Am. Math. Soc. **314**(1), 63–105 (1989)
82. Minero, P., Franceschetti, M., Dey, S., Nair, G.N.: Data rate theorem for stabilization over time-varying feedback channels. IEEE Trans. Autom. Control **54**(2), 243–255 (2009)
83. Nair, G.N., Evans, R.J.: Exponential stabilisability of finite-dimensional linear system with limited data rates. Automatica (J. Int. Fed. Autom. Control) **39**(4), 585–593 (2003)
84. Nair, G.N., Evans, R.J.: Stabilizability of stochastic linear systems with finite feedback data rates. SIAM J. Control Optim. **43**(2), 413–436 (2004)
85. Nair, G.N., Evans, R.J., Mareels, I.M.Y., Moran, W.: Topological feedback entropy and nonlinear stabilization. IEEE Trans. Autom. Control **49**(9), 1585–1597 (2004)
86. Nair, G.N., Fagnani, F., Zampieri, S., Evans, R.J.: Feedback control under data rate constraints: An overview. Proc. IEEE **95**, 108–137 (2007)
87. Noack, A.: Dimensions- und Entropieabschätzungen sowie Stabilitätsuntersuchungen für nichtlineare Systeme auf Mannigfaltigkeiten (German). Dissertation, Technical University of Dresden (1998)
88. Patrão, M., San Martin, L.A.B.: Semiflows on topological spaces: Chain transitivity and semigroups. J. Dyn. Differ. Equ. **19**(1), 155–180 (2007)
89. Pazy, A.: Semigroups of Linear Operators and Applications to Partial Differential Equations. Applied Mathematical Sciences, vol. 44. Springer, New York (1983)
90. Pesin, Y.B.: Characteristic Lyapunov exponents and smooth ergodic theory (Russian). Uspehi Mat. Nauk **32** (4) (196), 55–112, 287 (1977)
91. Pogromsky, A.Y., Matveev, A.S.: Estimation of topological entropy via the direct Lyapunov method. Nonlinearity **24**(7), 1937–1959 (2011)
92. Qian, M., Zhang, Z.-S.: Ergodic theory of Axiom A endomorphisms. Ergod. Theory Dyn. Syst. **1**, 161–174 (1995)
93. Robinson, C.: Dynamical Systems. Stability, Symbolic Dynamics, and Chaos, 2nd edn. Studies in Advanced Mathematics. CRC Press, Boca Raton (1999)
94. Ruelle, D.: An inequality for the entropy of differentiable maps. Bol. Soc. Brasil. Mat. **9**(1), 83–87 (1978)
95. San Martin, L.A.B., Tonelli, P.A.: Semigroup actions on homogeneous spaces. Semigroup Forum **50**, 59–88 (1995)

96. Savkin, A.V.: Analysis and synthesis of networked control systems: Topological entropy, observability, robustness and optimal control. Automatica J. IFAC **42**(1), 51–62 (2006)
97. Schreiber, S.J.: On growth rates of subadditive functions for semiflows. J. Differ. Equ. **148**(2), 334–350 (1998)
98. Siegmund, S.: Spektral-Theorie, glatte Faserungen und Normalformen für Differentialgleichungen vom Carathéodory-Typ (German). Dissertation. Augsburger mathematisch-naturwissenschaftliche Schriften, Wißner Verlag, Augsburg (1999)
99. Sinai, Ja.: On the concept of entropy for a dynamic system (Russian). Dokl. Akad. Nauk SSSR **124**, 768–771 (1959)
100. Sontag, E.D.: Finite-dimensional open-loop control generators for nonlinear systems. Int. J. Control **47**(2), 537–556 (1988)
101. Sontag, E.D.: Universal nonsingular controls. Syst. Control Lett. **19**(3), 221–224 (1992); Errata, Syst. Control Lett. **20**, 77 (1993)
102. Sontag, E.D.: Mathematical Control Theory. Deterministic Finite-Dimensional Systems, 2nd edn. Texts in Applied Mathematics, vol. 6. Springer, New York (1998)
103. Sontag, E.D., Wirth, F.R.: Remarks on universal nonsingular controls for discrete-time systems. Syst. Control Lett. **33**(2), 81–88 (1998)
104. Stoffer, D.: Transversal homoclinic points and hyperbolic sets for nonautonomous maps. I. Z. Angew. Math. Phys. **39**(4), 518–549 (1988)
105. Sussmann, H.J.: Single-input observability of continuous-time systems. Math. Syst. Theory **12**(4), 371–393 (1979)
106. Sussmann, H.J., Jurdjevic, V.: Controllability of nonlinear systems. J. Differ. Equ. **12**, 95–116 (1972)
107. Tatikonda, S., Mitter, S.: Control under communication constraints. IEEE Trans. Autom. Control **49**(7), 1056–1068 (2004)
108. Temam, R.: Infinite-Dimensional Dynamical Systems in Mechanics and Physics, 2nd edn. Applied Mathematical Sciences, vol. 68. Springer, New York (1997)
109. Vera, G.F.: Conjuntos de Control de Sistemas Lineales y Afines. Conjuntos Isócrones de Sistemas Invariantes (Spanish). Dissertation, Universidad Católica del Norte, Antofagasta, Chile (2009)
110. Wirth, F.R.: Robust Stability of Discrete-Time Systems under Time-Varying Perturbations. Dissertation, University of Bremen (1995)
111. Wirth, F.R.: Universal controls for homogeneous discrete-time systems. In: Proceedings of the 33rd IEEE CDC, New Orleans, 1995
112. Wirth, F.R.: Dynamics and controllability of nonlinear discrete-time control systems. In: 4th IFAC Nonlinear Control Systems Design Symposium (NOLCOS'98), Enschede, 1998
113. Wong, W.S., Brockett, R.W.: Systems with finite communication bandwidth constraints-II: Stabilization with limited information feedback. IEEE Trans. Autom. Control **44**(5), 1049–1053 (1999)
114. Xie, L.: Topological entropy and data rate for practical stability: A scalar case. Asian J. Control **11**(4), 376–385 (2009)
115. Young, L.-S.: Large deviations in dynamical systems. Trans. Am. Math. Soc. **318**, 525–543 (1990)

Index

absolute determinant, 221, 235
accessibility
    local, 13, 29, 124, 179, 197, 208
    strong, 138

Banach space, 17, 35, 124, 249
Bowen-ball, 151, 154, 158, 171, 193, 255
Bowen-metric, 154, 255

Carathéodory differential equation, 8, 236
chain control set, 31, 168, 178, 181, 208
chain recurrence, 20, 104, 208, 245, 251
chain transitivity, 15, 20, 170, 208, 245, 251
Christoffel symbols, 205, 233
cocycle
    additive, 103, 119, 174, 196, 209, 218, 261
    subadditive, 140, 217, 261
cocycle property, 5, 10, 12
coder-controller, 72, 84
control flow, 3, 15, 16, 104, 119, 140, 141, 145, 167, 174, 177, 193
control set, 16, 26, 63, 94, 122, 179, 181, 182, 204
    inner, 64, 136
cut-off function, 10, 23, 26, 28, 111, 119, 130, 242

derivation operator, 143, 224
dimension
    Hausdorff, 253
    lower capacitive, 115, 186, 254
    upper capacitive, 94, 107, 254, 260
divergence, 118, 235
domain of attraction, 29

ellipsoid, 98, 158, 222
entropy
    escape, 154, 165, 172, 189, 219
    metric, 52, 107, 121
    topological, 50, 52, 55, 60, 64, 90, 107, 115, 152, 255
    topological feedback, 68
    topological, of nonautonomous systems, 154
Euclidean
    ball, 163, 196
    metric, 110, 203
    norm, 60, 125
    space, 36, 98, 141, 158, 221, 247, 258
evolution operator, 97

Floquet theory, 128, 218
Fréchet derivative, 35, 37

geodesic, 113, 132, 156, 160, 187, 233

Hilbert space, 18, 94
Horn's inequality, 141, 222
hyperbolic matrix, 94, 136
hyperbolic set, 167, 209

invariant cover, 69, 78
inverse pendulum, 182
Ito's estimate, 107

Kalman rank condition, 16, 179, 180

lemma
    Bowen-Ruelle volume, 170
    fundamental, of Floquet theory, 128

LECTURE NOTES IN MATHEMATICS

Edited by J.-M. Morel, B. Teissier; P.K. Maini

Editorial Policy (for the publication of monographs)

1. Lecture Notes aim to report new developments in all areas of mathematics and their applications - quickly, informally and at a high level. Mathematical texts analysing new developments in modelling and numerical simulation are welcome.
   Monograph manuscripts should be reasonably self-contained and rounded off. Thus they may, and often will, present not only results of the author but also related work by other people. They may be based on specialised lecture courses. Furthermore, the manuscripts should provide sufficient motivation, examples and applications. This clearly distinguishes Lecture Notes from journal articles or technical reports which normally are very concise. Articles intended for a journal but too long to be accepted by most journals, usually do not have this "lecture notes" character. For similar reasons it is unusual for doctoral theses to be accepted for the Lecture Notes series, though habilitation theses may be appropriate.

2. Manuscripts should be submitted either online at www.editorialmanager.com/lnm to Springer's mathematics editorial in Heidelberg, or to one of the series editors. In general, manuscripts will be sent out to 2 external referees for evaluation. If a decision cannot yet be reached on the basis of the first 2 reports, further referees may be contacted: The author will be informed of this. A final decision to publish can be made only on the basis of the complete manuscript, however a refereeing process leading to a preliminary decision can be based on a pre-final or incomplete manuscript. The strict minimum amount of material that will be considered should include a detailed outline describing the planned contents of each chapter, a bibliography and several sample chapters.
   Authors should be aware that incomplete or insufficiently close to final manuscripts almost always result in longer refereeing times and nevertheless unclear referees' recommendations, making further refereeing of a final draft necessary.
   Authors should also be aware that parallel submission of their manuscript to another publisher while under consideration for LNM will in general lead to immediate rejection.

3. Manuscripts should in general be submitted in English. Final manuscripts should contain at least 100 pages of mathematical text and should always include

   - a table of contents;
   - an informative introduction, with adequate motivation and perhaps some historical remarks: it should be accessible to a reader not intimately familiar with the topic treated;
   - a subject index: as a rule this is genuinely helpful for the reader.

   For evaluation purposes, manuscripts may be submitted in print or electronic form (print form is still preferred by most referees), in the latter case preferably as pdf- or zipped ps-files. Lecture Notes volumes are, as a rule, printed digitally from the authors' files. To ensure best results, authors are asked to use the LaTeX2e style files available from Springer's web-server at:

   ftp://ftp.springer.de/pub/tex/latex/svmonot1/ (for monographs) and
   ftp://ftp.springer.de/pub/tex/latex/svmultt1/ (for summer schools/tutorials).

Additional technical instructions, if necessary, are available on request from lnm@springer.com.

4. Careful preparation of the manuscripts will help keep production time short besides ensuring satisfactory appearance of the finished book in print and online. After acceptance of the manuscript authors will be asked to prepare the final LaTeX source files and also the corresponding dvi-, pdf- or zipped ps-file. The LaTeX source files are essential for producing the full-text online version of the book (see http://www.springerlink.com/openurl.asp?genre=journal&issn=0075-8434 for the existing online volumes of LNM). The actual production of a Lecture Notes volume takes approximately 12 weeks.

5. Authors receive a total of 50 free copies of their volume, but no royalties. They are entitled to a discount of 33.3 % on the price of Springer books purchased for their personal use, if ordering directly from Springer.

6. Commitment to publish is made by letter of intent rather than by signing a formal contract. Springer-Verlag secures the copyright for each volume. Authors are free to reuse material contained in their LNM volumes in later publications: a brief written (or e-mail) request for formal permission is sufficient.

Addresses:
Professor J.-M. Morel, CMLA,
École Normale Supérieure de Cachan,
61 Avenue du Président Wilson, 94235 Cachan Cedex, France
E-mail: morel@cmla.ens-cachan.fr

Professor B. Teissier, Institut Mathématique de Jussieu,
UMR 7586 du CNRS, Équipe "Géométrie et Dynamique",
175 rue du Chevaleret
75013 Paris, France
E-mail: teissier@math.jussieu.fr

For the "Mathematical Biosciences Subseries" of LNM:

Professor P. K. Maini, Center for Mathematical Biology,
Mathematical Institute, 24-29 St Giles,
Oxford OX1 3LP, UK
E-mail: maini@maths.ox.ac.uk

Springer, Mathematics Editorial, Tiergartenstr. 17,
69121 Heidelberg, Germany,
Tel.: +49 (6221) 4876-8259

Fax: +49 (6221) 4876-8259
E-mail: lnm@springer.com